Nucleic Acids and Molecular Biology

Volume 35

AF811208

Series editor
Allen W. Nicholson
Philadelphia, PA 19122, USA

More information about this series at http://www.springer.com/series/881

César Menor-Salván
Editor

Prebiotic Chemistry and Chemical Evolution of Nucleic Acids

Springer

Editor
César Menor-Salván
NSF-NASA Center for Chemical Evolution
School of Chemistry and Biochemistry
Georgia Institute of Technology
Atlanta, GA, USA

Unidad de Bioquímica y Biología Molecular, Dept. Biologia de Sistemas
Universidad de Alcala
Madrid, Spain

ISSN 0933-1891 ISSN 1869-2486 (electronic)
Nucleic Acids and Molecular Biology
ISBN 978-3-030-06688-8 ISBN 978-3-319-93584-3 (eBook)
https://doi.org/10.1007/978-3-319-93584-3

© Springer International Publishing AG, part of Springer Nature 2018
Softcover re-print of the Hardcover 1st edition 2018
This work is subject to copyright. All rights are reserved by the Publisher, whether the whole or part of the material is concerned, specifically the rights of translation, reprinting, reuse of illustrations, recitation, broadcasting, reproduction on microfilms or in any other physical way, and transmission or information storage and retrieval, electronic adaptation, computer software, or by similar or dissimilar methodology now known or hereafter developed.
The use of general descriptive names, registered names, trademarks, service marks, etc. in this publication does not imply, even in the absence of a specific statement, that such names are exempt from the relevant protective laws and regulations and therefore free for general use.
The publisher, the authors, and the editors are safe to assume that the advice and information in this book are believed to be true and accurate at the date of publication. Neither the publisher nor the authors or the editors give a warranty, express or implied, with respect to the material contained herein or for any errors or omissions that may have been made. The publisher remains neutral with regard to jurisdictional claims in published maps and institutional affiliations.

Printed on acid-free paper

This Springer imprint is published by the registered company Springer Nature Switzerland AG.
The registered company address is: Gewerbestrasse 11, 6330 Cham, Switzerland

Preface

This book has been prepared with the aim to guide the reader into the current ideas upon chemical evolution of nucleic acids, i.e., all the chemical processes that occurred immediately before the emergence of Life, toward the formation, self-assembly, selection, and supramolecular organization of nucleic acids, and which, ultimately, allowed the emergence of a system capable of Darwinian evolution. Although the big question standing behind is *how life emerged in a geochemically plausible environment from chemically simple precursors,* this monograph does not intend to offer an answer to that question, which, perhaps, is one of the most persistent, elusive, and interesting questions of human thought. We focused it on the origin and early, abiotic evolution of nucleic acids. To discuss that, we recruited a group of influential scientists in topics covering from the geochemistry and mineralogy of phosphate and the prebiotic origin of small molecules to the chemical evolution of nucleic acid polymers and the emergence of complex systems. The book is written in such a way that it provides a perspective on the latest knowledge in this topic for general readers with a good background in General Chemistry, as well as for educated experts in all fields within Astrobiology and Origins of Life Studies.

The book begins with a remembrance of Stanley Miller, one of the pioneers of the experimental Prebiotic Chemistry. The following six chapters are devoted to the chemical origin of building blocks of nucleic acids (nucleobases and nucleosides) and its organization, the influence of the geochemistry and mineralogy in the prebiotic chemistry of nucleic acids, the origin of organic phosphate, and the fate of nucleic acids in the low-temperature hydrothermal systems; in the following two chapters, two relevant questions upon the early, abiotic chemical evolution of nucleic acid polymers are discussed; the final two chapters introduce the reader into some concepts upon the emergence of complexity in prebiotic evolution.

To prepare a monograph is a challenging but, in my opinion, a necessary task. After one century of scientific quest for the origins of life, we have accumulated enough data and ideas to create monographs and textbooks on the topic. The books pave the way from the data and observations accounted in the scientific journals to the culture and knowledge. Books present in a tangible form a process that we scientists perform inside our heads. A monograph is a good opportunity and a space of freedom to present the ideas, models, and theories of experts without the usual constraints of regular papers. That is exactly what we tried in this book: the reader will find a valuable, firsthand account of the research and ideas of the contributing authors, in a more discursive format than the corseted papers. I think it is important to note that there are no competing ideas or theories around the Chemical Evolution and Origins of Life. Competition is an illusion created by human egos, but in the study of chemical evolution, as Prof. Nicholas V. Hud once said, *our role as prebiotic chemists is to provide chemical and geologically plausible model scenarios for conditions and molecules that allow the robust formation of proto-biopolymers.* There are several possible model scenarios that could possibly have led to the emergence of life, and we have no way to know what or how many of these possible scenarios, reactions, and chemical pathways have been active during the origins of life epoch, around 4 billion years ago on Earth. Prebiotic Chemistry and Chemical Evolution is in constant feedback with fields as Geology and Space Sciences. Precisely, the advance in space science is testing and validating the robustness of our proposed models and, maybe in the near future, space exploration will find a real environment in which Chemical Evolution is active, giving us definite answers to our questions. Until that momentous event, all models and reactions within prebiotic chemistry that are plausible from the point of view of geochemistry and geology should be considered and accounted for future reference. Hence, the preparation of scientific books, in my opinion, is a responsibility with the future students and readers interested in this topic and an indicator of the maturation of the scientific field. To fulfill this responsibility is not an easy task, however, in the context of our system of science. Nowadays, top-notch scientists are immersed in a hyper-competitive system in which the priority is the publication of papers in high impact factor journals, nearly followed by the writing of grant proposals, the fulfilling of endless bureaucracy, and, in most cases, teaching responsibilities. Hence, in an unbalanced system that does not encourage the communication of science and the generation of knowledge (just the generation of papers), they have very little support and even less time available to write books or monograph chapters. Given this context, I hope the reader will appreciate, as I do, the generous and laudable effort invested by the authors in the preparation of this book.

When Prof. Allen W. Nicholson, from Temple University at Philadelphia, invited and encouraged me to edit a book devoted to the prebiotic chemistry of nucleic acids, I realized that it would not be easy to completely cover the models that should be represented in a book on that topic, in part for the reasons explained above. I apologize if the reader feels that some topics are underrepresented. I appreciate the support and patience until finalization of the book, proofreading, and formatting of Sabine Schwarz and Anand Venkatachalam, from Springer Nature, and the staff of SPi Global. My special thanks go to my colleagues, friends, and the staff of the Center for Chemical Evolution and the Georgia Institute of Technology, without whom this book would not have existed.

Atlanta, GA César Menor-Salván
July 2018

Outline

The origin of life is one of the big questions of human knowledge and a challenging topic for scientific research. Chemical evolution can be defined as all the chemical processes and interactions conducive to self-assembly and supramolecular organization, leading to an increase in complexity and the ultimate emergence of life. This volume addresses the formation of and initial events in the chemical evolution of the nucleic acids, including the geochemical roots of reaction components. The volume presents a review of the seminal work of Stanley Miller, and how the development of organic chemistry in the nineteenth century led to the development of the field of prebiotic chemistry, representing the frontier between organic chemistry, geochemistry, and biochemistry. Also covered are current topics on the organization of the nucleic acids: the origin of the nucleobases and nucleosides, phosphorylation and polymerization reactions, and the self-assembly and supramolecular organization in the context of life's origins.

Keywords Origins of life • Evolution • RNA-world • Nucleotide synthesis • Chemistry of purines and pyrimidines • Phosphorylation

Stanley L. Miller: A Personal Retrospective

I met Stanley Miller in 1965 when I was a first-year graduate student in the Department of Chemistry at the University of California at San Diego (UCSD). At that time, I was interested in the structure of supercooled water and was trying to use X-ray crystallography to determine if water had an increasing icelike structure when it was supercooled to lower and lower temperatures below its freezing point. I was using the X-ray equipment in the laboratory of a UCSD chemistry Professor Joseph Kraut who had allowed me the use of his instrumentation, between 10 pm and 8 am. Because it took a while to get X-ray images of one sample, I often wondered around looking at what some of the professors had posted on their doors and walls. I was surprised to find a chemistry professor who was also working at those late hours. The professor was Stanley Miller, and we eventually struck up many conversations (this was at times tough because Stanley was then a heavy chain-smoker).

Although I remembered reading something about Stanley Miller making amino acids in an experiment simulating the early Earth, I did not know exactly how and when he did this experiment. Stanley showed me the apparatus and how it worked and gave me some of his papers to read. Stanley was also interested in what I was doing and what I had found. After explaining my project and saying I was finding it very frustrating and challenging to carry out the measurements, he offered his advice. Although my project was interesting, I would probably not get my PhD until I was 40 years old. Being only 22 at the time, his comment was not encouraging to say the least.

As my frustrations with my project increased, I became more and more intrigued by the experiments Stanley had done and was continuing to do. I was especially captivated when he mentioned how little was actually known about the conditions on the abiotic Earth and that the field was ripe for some new ideas and novel experiments.

The more I thought about these comments the more interesting they became. I still remember my excitement fantasizing about what the Earth was like over 4 billion years ago. I went to the library and used Chemical Abstracts (this was well before the Internet and Google Scholar) and tried to read everything I could find about the early

Earth, which at that time was a subject more in the realm of geologists. I read Harold Urey's book *The Planets: Their Origin and Development* (1952) which impressed me about how important chemistry was in trying to understand the history of our planet. Urey was also a member of chemistry faculty at UCSD, and I also talked with him about the early Earth's chemistry. One thing I remember about our discussion was Urey's use of thermodynamics and equilibrium constants to constrain the mixture of gases that could have been present in the early atmosphere, especially if hydrogen was present. I mentioned this to Stanley, who in response said that there could be thermodynamic equilibrium constants for organic reactions that could be used to place further constraints on concentrations of reduced species in the atmosphere, especially ammonia. We discussed the reactions involved in amino acid formation in his experiments, and Stanley hinted that maybe I should look into this. I jumped at the chance.

So I went back to the library to read about any equilibrium reactions involving organic compounds, especially parts of the Strecker Reaction that had Stanley had proposed as the pathway by which amino and hydroxyl acids were produced in his experiments. The first step was the reaction of an aldehyde or ketone with hydrogen cyanide and ammonia to produce the corresponding amino and hydroxyl nitriles. This was a reversible reaction, which was constrained at least for amino acids by the concentrations of the three reaction components. Constraining all these components was complex and unlikely to yield anything useful because the number of combinations of the concentrations of the three constituents was impossibly huge.

I thought maybe finding out how the various amino acids used in biology were discovered might yield some interesting insights. I again went back to the library and to begin a search of the various reactions by which amino acids had been first synthesized. Searching Chemical Abstracts for reactions that might be relevant, I began compiling a list of possible reactions. One in particular caught my eye: the apparent equilibrium between aspartic acid, fumaric acid, and ammonia.

A report in 1926 (Dessaignes 1850) showed that aspartic acid decomposed by deamination to fumaric acid and ammonia in the presence of bacteria. The reaction was reversible, and equilibrium between all three species was attained. When I showed this to Stanley, he immediately suggested that maybe the equilibrium could be used to constrain the amount of ammonia on the early Earth. The problem was that it was not known how rapidly the deamination reaction took place at Earth's environmental temperatures, and whether equilibrium could be established without the presence of bacteria. Stanley and I discussed the several possible experimental scenarios to determine the rate of deamination at various temperatures. Finally, he said pick the one you think is best and test it out.

This was one of the hallmarks of my graduate student relationship with Stanley: discuss ideas and then I would be basically on my own with planning and setting up experiments as well as finding analytical methods to measure the extent of any reaction.

This was a valuable lesson for me. It gave me independence to do what I wanted and to think on my own. As I started setting up experiments and analytical methods I consulted with Stanley only occasionally. I planned experiments at various pH values and when I asked Stanley about the best buffers he simply took a book from his bookcase and told me to look it over. The book was *Determination of pH: Theory and Practice* by Roger G. Bates (1964).

The book had information on a large number of buffers and more importantly how their pH varied with temperature. I set out to manually graph (computers on campus consisted of large mainframe machines) the temperature vs. pH of buffers I thought I would use. When I showed these graphs to Stanley, he immediately asked where I had gotten them. When I said that I had made them, he looked at me with surprise. I would eventually found out that Stanley had a lot of trouble using graph paper and never learned to type or use a computer keyboard. Much later in our acquaintance he confided to me that the problem was his mother who insisted on him taking piano lessons, which he hated and found paralyzing.

I set out to make up solutions at various pH values of aspartic acid along with ones with fumaric acid and ammonium chloride. Stanley taught me to seal the solutions in glass tubes. I started heating the sealed tubes at various temperatures using refluxing solutions of organic solvents that boiled at various temperatures. I made sure to heed Stanley's warning not to use solvents with boiling points which were more two times the vapor pressure of water at that temperature, or the tubes could explode.

As my experiments continued and I started to get preliminary data, weeks went by without seeing much of Stanley. I usually came into the lab between 9 and 10 AM while Stanley came in late in the afternoon or early evening often after I had left. One day when we overlapped he asked me how my experiments were going. I showed him my carefully tabulated data and graphs of the results. Stanley looked this over and told me this is exactly what he had hoped for. He then said he thought we should immediately write a paper on how the aspartic acid/ammonia-fumarate equilibrium could be used to set limits on the concentration of ammonia in the primitive oceans and atmosphere if we assumed that the ratio of aspartic acid to fumaric acid was 1.0. He assigned me the task of writing the first draft, and when I showed this to him, he responded that a lot of work was required (this was one of the first scientific manuscript I had ever written for possible publication in a science journal). Stanley told me surprisingly that we should work on this together on finishing the manuscript as soon as possible.

I took about 2 weeks for the two of us to have what Stanley thought was an acceptable first draft. He said he would send this to some of his colleagues for comments. Very favorable comments came back. After we made the revisions and corrections suggested Stanley told me that we should submit the manuscript to *Science*. This almost knocked me over because I had no idea that Stanley would recommend such a high-ranking journal. I was assigned to get everything together that *Science* required, and we submitted the manuscript in the fall of 1967. Much to my relief the reviews came back very positive, and after some suggested changes, the paper was accepted and published in January of 1968 (Bada and Miller 1968).

Stanley then suggested I do a few more experiments and, after I completed these, to start writing my thesis in the form of manuscript drafts (I later did the same thing to each of my own graduate students). After he approved the thesis, my doctoral committee reviewed everything and the thesis was accepted. I got my PhD in the summer of 1968, a little over 3 years after I had started at UCSD.

When I sit back and think about this 50 years later, it was clear that this would have never happened on such a short timescale if Stanley had not been my thesis adviser. After I got my PhD, I went to Harvard and within a year I was offered an Assistant Professorship at the Scripps Institution of Oceanography (a part of UCSD). I accepted and headed back to La Jolla.

Stanley and I started to interact extensively, and we published several jointly authored papers together. One memorable paper was based on frozen dilute hydrogen cyanide solutions Stanley had kept for 25 years. When he first mentioned them to me over dinner at our favorite Chinese restaurant, I immediately asked didn't he think it was time to see if anything of interest had been made. He enthusiastically replied yes. We divided up the analyses with his laboratory looking for nucleobases while mine searched for amino acids. Together our results showed that both of these classes of prebiotic compounds had been synthesized in significant yields. These results were published in 2000 (Levy et al. 2000).

During these periods I also started to see a completely different person than the one I had known while a graduate student. Stanley had a sarcastic wit, had a passion for books about Winston Churchill's role in World War II (especially the parts about naval warfare), and was a traveler to exotic places: he once took the Siberia Express across the former USSR from Moscow to Vladivostok and took a trip from Pakistan to Afghanistan via the Khyber Pass. When I told him I was going to take research trips to an island off the coast of Alaska and later to work with Mary Leaky at Olduvai Gorge in Tanzania, he immediately asked if he could come along. These were incredible times and I especially remember the anxiety Stanley had when we found out that a spitting cobra had been chased out of a small cabin we were staying in before we got to Olduvai. During these various trips, Stanley took lots of photographs (actually slides). Upon his return to La Jolla, he would invite friends to his house and over pizza and beer were treated to the slides and stories of his travels.

As should be obvious, Stanley and I had formed a special bond. When he had his first stroke in the fall of 1999, right after the ISSOL meeting in La Jolla, it was a shock to everyone. He had given up smoking in the 1980s and became an avid cyclist.

He was in such good shape that he would ride his bike down to my office at Scripps and then head to the main campus up a very steep hill. One passion Stanley had was to take cycling trips in Europe where he would ride 70–80 miles a day with a tour group and stay in first-class hotels that had Michelin 3 star restaurants. Unfortunately, he continued to have a series of strokes and became totally disabled.

I eventually was asked by the Chemistry Department to clear out Stanley's laboratory and office. It was ironic that just before he died in May 2007 we found in the material that had been moved to my laboratory a cardboard box with Stanley's writing that said "Spark Discharge Samples." Inside we found small boxes containing vials; those labels indicated they were from his classic 1952–1953 experiments as well as some from experiments done 1958 that he had never analyzed. Also labeled on the boxes were the page numbers in his laboratory notebooks (these are part of his collection in the Special Collections of the *Geisel Library* at UCSD). I was stunned! He had never mentioned these to anyone, except in a passing remark to Antonio Lascano.

Unfortunately, Stanley was unable to understand when I told him about this discovery. We set out to analyze one set from an apparatus design that we nicknamed the "volcanic experiment" that Stanley had tested as part of his PhD thesis work at the University of Chicago. For unknown reasons, he never used this apparatus design again. We found many more amino acids than Stanley had reported in his original *Science* 1953 paper. Fittingly, we published these results in *Science* (Johnson et al. 2008). We also published several other papers on the other extracts that Stanley had so carefully stored away.

I cherish the times, discussions, and trips Stanley and I had together. Our relationship went well beyond that of a mentor to that of a trusted colleague and dear friend. I deeply miss his input to prebiotic chemistry, as I know many others do as well.

Scripps Institution of Oceanography,
University of California, San Diego,
La Jolla, CA, USA

Jeffrey L. Bada
jbada@ucsd.edu

References

Bada JL, Miller SL (1968) Ammonium ion concentration in the primitive ocean. Science 159:423–425

Dessaignes V (1850) Formation de l'Acide Aspartique avec le Bimalate d'Ammoniaque. Compt Rendus 30:324–325

Johnson AP, Cleaves HJ, Dworkin JP, Glavin DP, Lazcano A, Bada JL (2008) The Miller volcanic spark discharge experiment. Science 322:404

Levy M, Miller SL, Brinton K, Bada JL (2000) Prebiotic synthesis of adenine and amino acids under Europa-like conditions. Icarus 145:609–613

Contents

1 **Nucleobases on the Primitive Earth: Their Sources and Stabilities** 1
H. James Cleaves II

2 **Condensation and Decomposition of Nucleotides in Simulated Hydrothermal Fields** 21
Ryan Lorig-Roach and David Deamer

3 **Mineral-Organic Interactions in Prebiotic Synthesis** 31
Steven A. Benner, Hyo-Joong Kim, and Elisa Biondi

4 **From the Dawn of Organic Chemistry to Astrobiology: Urea as a Foundational Component in the Origin of Nucleobases and Nucleotides** 85
César Menor-Salván

5 **Searching for Possible Ancestors of RNA: The Self-Assembly Hypothesis for the Origin of Proto-RNA** 143
Brian J. Cafferty, David M. Fialho, and Nicholas V. Hud

6 **The Origin of the Ionized Linker: Geochemical Predestination for Phosphate?** 175
Matthew A. Pasek

7 **Template-Directed Replication of Nucleic Acids Mediated by Viscous Environments** 199
Isaac Gállego and Christine He

8 **Folding and Catalysis Near Life's Origin: Support for Fe^{2+} as a Dominant Divalent Cation** 227
C. Denise Okafor, Jessica C. Bowman, Nicholas V. Hud, Jennifer B. Glass, and Loren Dean Williams

9 Connections Between Mathematical Models of Prebiotic Evolution and Homochirality 245
Celia Blanco and Irene A. Chen

10 Network Theory in Prebiotic Evolution 263
Sara Imari Walker and Cole Mathis

Chapter 1
Nucleobases on the Primitive Earth: Their Sources and Stabilities

H. James Cleaves II

Abstract Nucleobases are nitrogen heterocycles that are key structural components of biological nucleic acids. Some theories for the origins of life suggest a role for environmentally supplied organic compounds, including nucleobases, as part of a primordial RNA or pre-RNA world. Over the last 65 years, many potentially prebiotic synthetic mechanisms have been experimentally demonstrated for nucleobases, and their presence in extraterrestrial materials has been extensively verified, suggesting some of these are valid explanations for how the environment produces them. However, the abundance of nucleobases in primitive environments would depend on the balance of the rates of their environmental synthesis and decomposition. The literature regarding chemical aspects of these questions is briefly reviewed here.

1.1 Introduction

Nucleobases are components of modern nucleic acids whose role in nucleic acid function is important because they help store the digital information of the genetic code. This role is made possible by the high proportion of their molecular composition that can engage in stacking and hydrogen-bonding interactions (Fig. 1.1).

The chemical properties of the nucleobases are essential for the biological functions of nucleic acids. There are likely a limited number of molecules of any size which have the optimal proportions of such properties and which are also stable under a wide range of conditions and to a wide range of environmental insults, for

H. J. Cleaves II (✉)
Earth-Life Science Institute, Tokyo Institute of Technology, Tokyo, Japan

The Institute for Advanced Study, Princeton, NJ, USA

Blue Marble Space Institute of Science, Washington, DC, USA

NSF-NASA Center for Chemical Evolution, Georgia Institute of Technology, Atlanta, GA, USA
e-mail: henderson.cleaves@gmail.com

© Springer International Publishing AG, part of Springer Nature 2018
C. Menor-Salván (ed.), *Prebiotic Chemistry and Chemical Evolution of Nucleic Acids*, Nucleic Acids and Molecular Biology 35,
https://doi.org/10.1007/978-3-319-93584-3_1

example, radiational excitation (Bertinchamps et al. 2011; Boldissar and de Vries 2018). This chapter will focus on the two topics of the abiotic sources and stabilities of the nucleobases themselves, rather than higher-order derivatives (e.g., nucleosides and nucleotides and their analogues) or alternative “xeno nucleic acids” (XNAs) (Pinheiro et al. 2013; Joyce et al. 1987), consideration of which might suggest other ways to think about some of the issues described here. This chapter is not intended to be an exhaustive review of the nitrogen heterocycle prebiotic synthesis literature; readers interested in this are encouraged to read further elsewhere (Ruiz-Mirazo et al. 2014; Ruiz-Bermejo et al. 2013; Cleaves 2012).

1.1.1 Nomenclature, Structure, and Physical Chemistry

The nucleobases of DNA comprise the purines adenine and guanine and the pyrimidines thymine and cytosine, while those of RNA include adenine, guanine, and cytosine, but the pyrimidine uracil substitutes for thymine (Fig. 1.1). The nucleobases had already been isolated, identified, and synthesized in the laboratory during the nineteenth century and first decade of the twentieth century, well before their role in nucleic acids was understood (Mason 1991). The term *nucleobase* is generally reserved for those compounds found in natural nucleic acids, variants on those structures with similar structures and properties or which allow them to play similar roles in biological systems are typically termed *nucleobase analogues*.

The nucleobases are substituted aromatic N-heterocycles. The pyrimidines are six-membered aromatic rings containing two nitrogen atoms at the 1 and 3 positions of the ring. The purines are fused pyrimidines, containing an additional imidazole ring connected across the C5 and C6 atoms of the pyrimidine ring and containing N atoms in the 1, 3, 7, and 9 positions (see Fig. 1.1). As they contain various exchangeable ring nitrogen-associated protons, these compounds are organic bases. Due to the electronic content of their ring systems, the nucleobases are

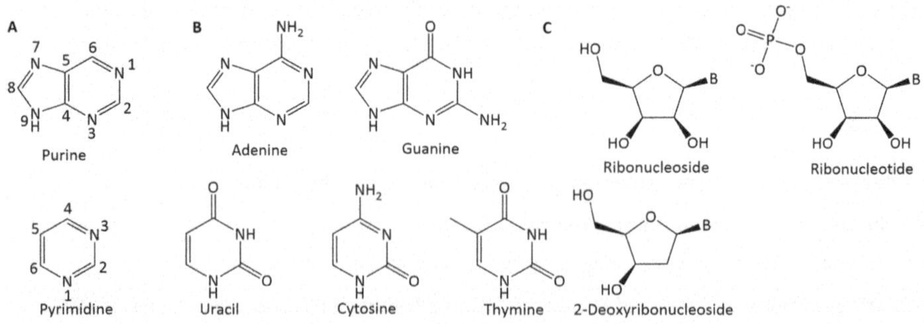

Fig. 1.1 The structures of the nucleobases and nucleic acids that they help constitute. A. The purine and pyrimidine ring systems and their atom numbering. B. The biological purines and pyrimidines. C. The structures of nucleosides and nucleotides containing the nucleobases

aromatic, and this leads to their ring structures being essentially planar. Their ring and exocyclic substitution patterns contain a great deal of structural information that enables them to present highly specific molecular recognition surfaces. The specific hydrogen bonding patterns nucleobases enable are also dependent on the stability of their tautomeric forms.

The nucleobases have attracted great attention in the study of biology since the revelation of the structure of DNA and the elucidation of the genetic code (Woese 1967; Orgel 1968; Crick 1968), which showed that they can be considered the "business end" of the molecules which make up the "alphabet" which specifies biological heredity. Via the spatially oriented presentation of their specific hydrogen bond donor and acceptor functionalities, they serve as a sort of digital information encoding system similar in some ways to that used to store and process information in modern computers. Nevertheless, though information storage and handling can be abstracted greatly, in biology the nucleobases are parts of metabolic chemical systems that operate in aqueous solution, and thus the nuances of their chemical properties, which are a result of their chemical structures, play important and subtle roles in their biological function.

Recent studies have shown that other molecules are able to play the molecular roles that nucleobases do in information storage and transmission (see, e.g., Switzer 1989; Malyshev et al. 2014). Many molecules, given water as a solvent, might also be able to engage in the interactions that enable nucleic acid function. Alternatively, few molecules in the idealized space of molecules may be aromatic, *and* predisposed to self-assembly because of their H-bond donor/acceptor patterns, or be able to do so in the dynamic context of nucleic acid structure (Cleaves and Bada 2012).

The nucleobases have relatively hydrophobic surfaces, though their molar solubility in water ranges over several orders of magnitude at room temperature and neutral pH, depending on their ring substitution. The degree to which the arrangement of H-bond donor and acceptor groups allows for self-pairing is to some degree responsible for their differences in solubility. Nucleobases which are capable of forming stable self-complementary patterns (such as guanine) tend to have very low solubility in water, though hydrophobic properties imparted by the ring systems also contribute, with the purines being generally less soluble than the pyrimidines.

The DNA guanine-cytosine and adenine-thymine pairs, in the context of nucleic acids, engage in H–bond-mediated base pairing, which is due to both the specific arrangement of H-bond donor and acceptor groups and the stacking arrangements of their π-bond systems. The guanine-cytosine and adenine-thymine base pairs engage in bonding involving three and two H-bonds, respectively. As hydrogen bonds require energy to disrupt, these serve as organizing elements for the formation of supramolecular structure. These energetic differences in affinity are exploited by biology at multiple levels: for example, thermophiles appear to preferentially use GC base pairs in their tRNA molecules (Wang et al. 2006), and there is some correlation between genomic DNA and structural RNA GC content and optimal growth temperature in prokaryotes (Musto et al. 2004).

1.1.2 The Relevance of Nucleobase to the Origins of Life

Concurrent with the elucidation of the genetic code and the crystallization of the concept of the central dogma of biology, a paradox was laid bare: if nucleic acids are required to make proteins and proteins to make nucleic acids, which logically would have come first? A cohesive conceptual resolution of this paradox was formulated almost simultaneously by Woese, Crick, and Orgel (Woese 1967; Crick 1968; Orgel 1968), which became known as the "RNA world hypothesis" (Gilbert 1986; Benner and Ellington 1991). Roughly speaking, the fundament of the hypothesis is that since it is possible that RNA could serve as both information storage molecule and catalyst, the modern system of information flow in biology may have grown outward from RNA in two directions: in one to make protein and in another to make DNA. Numerous models and experimental systems have now examined the chemical plausibility of this notion and provide various levels of support for it. Nevertheless, it quickly became clear [and the notion persists to this day (see, e.g., Pearce et al. 2017)] that the central issue in understanding the origins of life is the abiotic synthesis of RNA and thus the abiotic assembly of its components, including nucleobases, as precursors.

It should be made clear that there is no agreed-upon model for the origins of life, and there are competing models to the RNA world. Other models alternatively posit an origin of life based on simpler genetic polymers (Joyce et al. 1987), other types of inheritance (Segré et al. 2000), or chemical systems with only metabolic properties (Wächtershäuser 1988; Morowitz 1999). In addition, various authors have explored methods of building nucleosides directly from simpler components (Sanchez and Orgel 1970; Powner and Sutherland 2008; Powner et al. 2010); thus, the prebiotic synthesis or availability of nucleobases may not be directly relevant to the origin of the RNA world or the origins of life.

Finally, the biosynthesis of the nucleobases in most cases does not mirror the mechanisms of synthesis elucidated in studies of prebiotic chemistry (Zubay 1993). That biological nucleic acids include the nucleobases can be viewed as either predestined by abiotic chemistry or the serendipitous result of biological processes that allowed their discovery during evolution.

1.2 Prebiotic Synthesis of Nucleobases

1.2.1 General Background

Though there are examples of early work that sought to experimentally address the abiological synthesis of biochemicals from the ~100 years which preceded it, it is generally agreed that Miller's 1953 publication on the laboratory generation of amino acids from simulated reducing primitive Earth (Miller 1953) initiated the modern period of experimental origins of life research. This publication appeared within a month of Watson and Crick's proposed structure for the DNA double helix

(Watson and Crick 1953), and it was not long until the notion of the spontaneous "prebiotic" formation of organic compounds was applied to the nucleic acids. At the time of this writing, a simple Google Scholar search for the topic "prebiotic adenine synthesis" returned hundreds of citations. This extensive literature will not be exhaustively reviewed here, rather some of the key advances and open questions will be examined.

Miller's abiotic amino acid synthesis demonstration started an effort to discover methods of generating other components of modern biochemistry, preferably in the fewest steps and in the highest yield from the simplest reactants. This general schema intended to mimic primitive geochemical synthesis environments, where chemistry proceeded undirected. This type of chemistry is generally termed *prebiotic* or *abiotic* chemistry, with the former term stressing the potential role of a process to the origin of life and the latter only that a process occurs in the absence of biology. Both processes include all manner of unknown reaction and the latter all manner of explorable, but poorly understood, reaction.

What exactly constitutes a "plausibly prebiotic" reaction is the subject of some debate among chemists. Roughly speaking, such conditions could include any that produce organic compounds in the absence of biological agents. More restrictively, these are often limited to those that operate under conditions that allow for the existence of liquid water (which can include water above its atmospheric pressure boiling point) and should include only reactants that are themselves justifiable by some similarly environmentally plausible synthetic process. The synthesis should also provide some justification for the concentrations of the reactants and the various parameters that allow reactions to proceed (e.g., the pH, the temperature, the containment of the reactants for the required reaction period, the presence of required catalysts, etc.).

There is little consensus what the conditions on early Earth were at the time life began. It is documented that Earth has swung through several ice ages over the last 2.6 Ga. Earth's surface temperature is a complex function of the output of the Sun coupled with the atmospheric chemistry and dynamics of the Earth system. The Earth almost certainly accreted hot, and there was a period in Earth's earliest history (this could have been as little as 100 million years or as much as almost 1 billion years) during which its surface was too hot to host liquid water. During this period, most organic chemicals would have been rapidly degraded, potentially requiring organic synthesis or delivery to have occurred after that time. Nevertheless, at some point between ~4.4 Ga and 3.5 Ga, Earth cooled enough to allow the condensation of liquid water. There is also uncertainty in the pH of the early oceans once surface conditions allowed for liquid water (Halevy and Bachan 2017; Krissansen-Totton et al. 2018).

There is a great deal of uncertainty regarding the composition of Earth's primitive atmosphere. As will be discussed below in more detail, many prebiotic syntheses depend on reducing conditions for the gas phase (atmospheric) synthesis of precursors such as HCN (Schlesinger and Miller 1983b), and there is little consensus that conditions on the primitive Earth would have supported such reactions for extended periods of time (Rubey 1951; Tian et al. 2005). Extraterrestrially synthesized organic matter (Cronin 1989) delivered to the Earth's surface by infall during the tail end of planetary accretion (Chyba and Sagan 1992) offers one possible solution to this potential problem.

While early research regarding the synthesis of organic compounds on the primitive Earth focused on local planetary environmental conditions, the search for life beyond Earth has extended the range of conditions of interest to chemists. The solar system includes many heterogeneous reaction environments [e.g., ranging from bodies with highly reducing surface environments (e.g., Saturn's moon Titan) to those that have oxidizing surface conditions (e.g., Mars)]. It should be further borne in mind that bodies of various sizes such as asteroids, moons, and planets are themselves heterogeneous across scales and may harbor environmentally variable microenvironments, and both the larger bodies and microenvironments may also change over time.

1.2.2 Retrosynthetic Analysis

Organic chemists commonly employ the approach of "retrosynthetic analysis" when designing molecular syntheses (Corey 1988), which simply attempts to work backward to produce a road map of potential precursors and transformations from which a target molecule could be derived. There is often more than one plausible route, and such routes may be highly branched depending on the complexity of the target and the starting materials.

In prebiotic chemistry, it is generally assumed that the starting materials should be simple environmentally abundant compounds, consisting of a single heavy atom (in addition to hydrogen), for example, water, methane, NH_3, H_2S, etc. Cosmically abundant species containing more than one heavy atom such as N_2 (which makes up some 78% of Earth's present atmosphere and ~98% of Titan's) and CO_2 (which makes up roughly 95% of both Mars' and Venus' atmospheres) are also considered reasonable starting points due to their cosmic abundance.

There is a large body of research exploring how energetic input from various sources including ionizing radiation (Miyakawa et al. 2002b), heat (Yoshino et al. 1971), and electric discharges (Schlesinger and Miller 1983a) can result in the recombination of these simple starting molecules into more complex ones, e.g., those containing more heavy atoms, for example, HCN and HCHO. These are often removed from the site of energetic recombination to lower energy environments where they can undergo still further reactions. For example, Miller's simple experiment showed that NH_3, H_2, H_2O, and CH_4 could be recombined under the action of an electric discharge to give rise to HCN and HCHO (Miller 1955), which diffuse away from the discharge region and irreversibly dissolve into the aqueous phase of the reaction vessel.

1.2.3 Purine Nucleobase Synthesis from One-Carbon Compounds

1.2.3.1 Synthesis from HCN and Formamide

Not long after Miller's experiment, Oró and Kimball (Oró 1961) showed that adenine [which is formally a pentamer of HCN ($C_5H_5N_5$)] could be derived from alkaline aqueous HCN solutions at moderately elevated temperature and proposed a general mechanism, which was elaborated and extended by Orgel and co-workers (Ferris and Orgel 1965; Sanchez et al. 1967) (Fig. 1.2).

The proposed mechanism proceeds through HCN dimerization, trimerization, and tetramerization, giving rise to substituted imidazoles that can react with various simple carbon compounds to yield a suite of purines (including A and G, as well as xanthine, hypoxanthine, 2,6-diaminopurine, and isoguanine; see Fig. 1.2).

Subsequent studies revealed that the pathway to purine formation from HCN likely involves more than one mechanism and that the complex heterogeneous polymer that concentrated HCN produces in solution is itself a major source of adenine and guanine upon hydrolysis (Borquez et al. 2005). This suggests that there are thermally induced rearrangements of this material that produce purines by mechanisms unrelated to those explored by Orgel and Oró. Schwartz showed that adenine-8-carboxamide is also a product of polymerizing HCN, which also gives rise to A upon hydrolysis (Voet and Schwartz 1983) and that this may be simply one example of an 8-substituted precursor that can give rise to that ring system.

HCN is a ubiquitous cosmochemical compound and present in Titan's atmosphere (Gautier et al. 2011), comets (Mumma and Charnley 2011), and

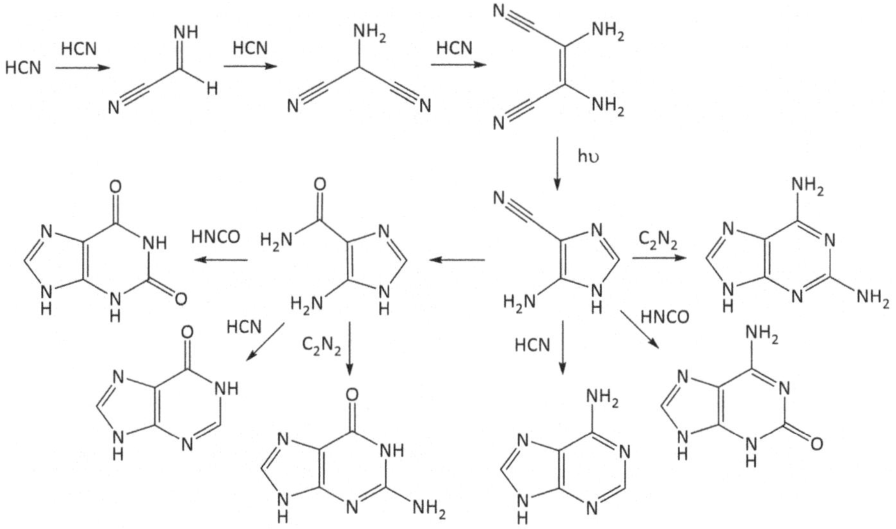

Fig. 1.2 Proposed mechanisms for the synthesis of purines from HCN

carbonaceous meteorites (Pizzarello 2012) and now observed in extrasolar and extragalactic settings as well (http://www.astro.uni-koeln.de/cdms/molecules/). It is reactive and hydrolyzes to give formamide and ultimately the salt ammonium formate (Miyakawa et al. 2002c):

$$HCN + H_2O \rightarrow HCONH_2$$
$$HCONH_2 + H_2O \rightarrow NH_4HCO_2$$

HCN is a gas at standard temperature and pressure. It has a pK_a of ~9.2 at 25 °C and thus readily dissolves in slightly alkaline water but does not concentrate well by evaporation from aqueous solutions below its pK_a. In contrast, formamide is completely miscible in water and has a boiling point above that of water, 210 °C, and thus in principle can be concentrated to purity as it is being produced from HCN hydrolysis by simple evaporation. The environmental and physical chemistry of HCN/formamide/water systems is likely complex enough to warrant more investigation.

Formamide, neat or in concentrated aqueous solution, can also serve as a precursor to various nucleobases, though higher concentrations (close to purity) and temperatures (typically above 110 °C) are required than for HCN condensation. It has been known for some time that $HCONH_2$ can be a starting material for both purine and pyrimidine synthesis (Bredereck et al. 1961), but this has more recently come into greater focus in a prebiotic context (Saladino et al. 2001; Barks et al. 2010; Adam et al. 2018). Formamide and hydrogen cyanide are linked by water-exchange reactions, and their products are similar.

1.2.3.2 One-Pot Purine Synthesis in Electric Discharges and Eutectics

While Oró and Orgel's purine synthesis mechanisms proceed from HCN, for some prebiotic chemists, this represents a purification of reagents too early in the synthetic process, and more heterogeneous syntheses deserve more emphasis. Additionally, few environments likely produce HCN exclusively, and the effects of congeners merit exploration.

The oligomerization of HCN in water is pH, temperature, and concentration dependent. It has been shown that HCN hydrolysis dominates below 0.1 M HCN (depending on pH and temperature), with oligomerization possible above that concentration (Sanchez et al. 1966a). The need for high concentrations of reactants led these authors to suggest that purine synthesis via HCN oligomerization could have more preferably taken place in low-temperature environments where eutectic concentration was possible.

Eutectic freezing of HCN solutions and the products of Miller-Urey type electric discharge reactions can give rise to various purines and pyrimidines (Miyakawa 2002a, b; Levy et al. 2000; Sanchez et al. 1966a; Schwartz et al. 1982). In these cases, it seems plausible that one or more of the previously suggested mechanisms may be at play, and the concentration effects that occur in eutectic brines enable them. Yields are typically not extremely high in such reactions, but they point to the ability of the direct production of these ring systems directly from simple precursors, and in heterogeneous combination, in a single environment.

1.2.4 Pyrimidine Synthesis

Pyrimidines have several possible retrosynthetic disconnects between carbon and nitrogen, though almost all reported prebiotic syntheses begin with a contiguous C3 compound. There are a few examples that start from only 1C compounds, but these depend on C–C bond formation reactions taking place in other contexts.

1.2.4.1 Pyrimidine Synthesis from C3 Precursors

Fox and Harada (1961) performed one of the first deliberate attempts at the prebiotic synthesis of uracil by heating malic acid and urea in polyphosphoric acid at elevated temperature. Oró (1963) later showed that uracil could be derived from the C3 compounds acrylonitrile, β-aminopropionitrile, and β-aminopropionamide reacted with urea in aqueous ammonia at 130 °C (Fig. 1.3). These C3 compounds and urea were all by then demonstrated products of Miller-Urey type syntheses. These reactions involve an oxidation across the nascent C5–C6 pyrimidine bond, which was achieved thermally in that case, though this has also been accomplished photochemically (Chittenden and Schwartz 1976).

Not long after, Orgel and colleagues showed how the biological pyrimidines could be produced from other simple compounds generated in electric discharges, namely, cyanoacetylene (CAA) and cyanate (Sanchez et al. 1966b; Ferris et al. 1968) (Fig. 1.3). This synthesis was later extended to give various pyrimidines including the biological ones (e.g., 2, 4-substituted pyrimidines including uracil, cytosine, isocytosine, 2,4-diaminopyrimidine, and various thio derivatives, depending on the reactants and conditions) when the hydrolysis product of CAA, cyanoacetaldehyde, was reacted with urea derivatives (urea, guanidine, thiourea, etc.) (Robertson and Miller 1995b; Robertson et al. 1996; Nelson et al. 2001).

Various other syntheses have added thymine to the roster of prebiotically accessible pyrimidines (Stephen-Sherwood et al. 1971; Choughuley et al. 1977), making use of the propensity for aldehydes to add to the C5 position of pyrimidines to give 5-hydroxymethyl substituted pyrimidines (Robertson and Miller 1995c), followed by reduction. Biological RNA molecules contain several modifications that allow them to carry out their cytosolic roles (see, e.g., http://mods.rna.albany.edu/), and there are now a number of plausibly prebiotic reactions which allow the introduction of some of these modifications (Robertson and Miller 1995a; Schneider et al. 2018).

1.2.4.2 Pyrimidine Synthesis from C1 Precursors

Voet and Schwartz (1982) showed that uracil could also be derived directly from HCN polymer. Thus, HCN offers a direct route to the nucleobase purines and at least

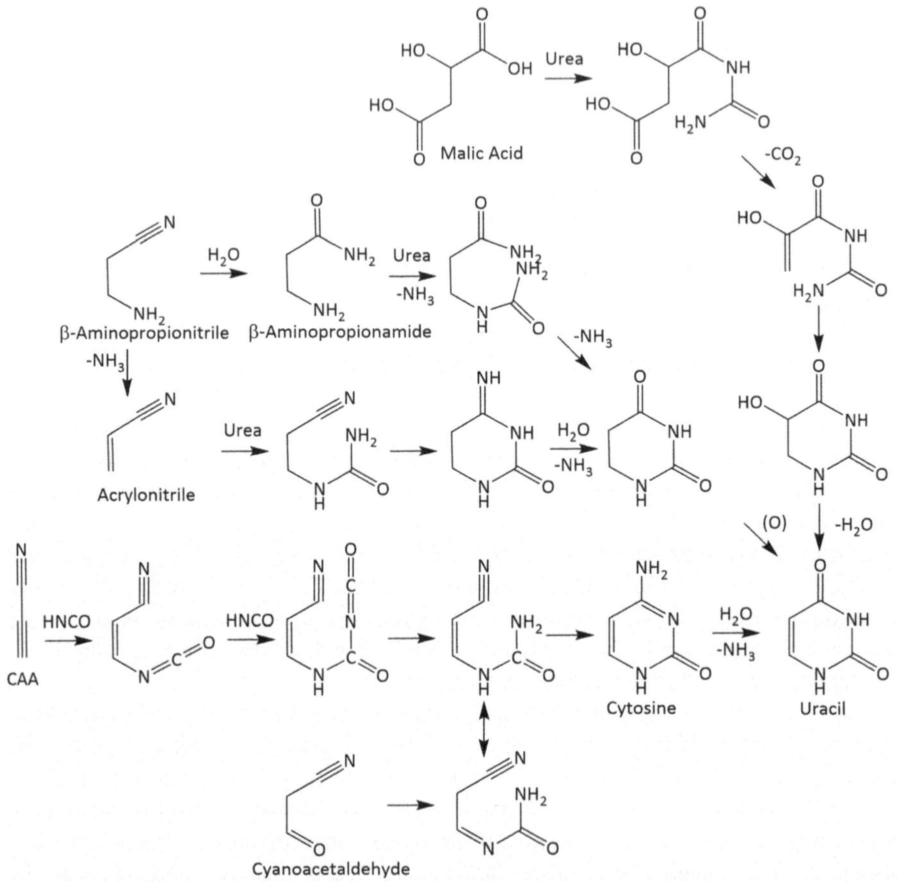

Fig. 1.3 Some proposed mechanisms for the synthesis of pyrimidine nucleobases discussed in the text

one of the pyrimidines. In addition to uracil, it had also been found that other non-biological pyrimidines, including 5-amino and 5-oxo-subsituted pyrimidines (Miyakawa et al. 2002a) as well as the pyrimidine biosynthetic precursor orotic acid, are derivable from HCN (Ferris et al. 1978). Searches for cytosine in such reactions have to date been unsuccessful (Miyakawa et al. 2002a).

1.3 Extraterrestrial Nucleobases

The solar system harbors many environments where organic synthesis evidently occurs. Perhaps the most tangible and compelling examples come from the carbonaceous chondrite meteorites, which are vestiges of the chemistry of the early solar system (Pizzarello and Shock 2010). These meteorites contain a variable though significant fraction (up to ~2% in weight) of organic material (Alexander et al. 2007) which includes a high-molecular weight intractable component (typically constituting 70+ % of the organic material) as well as a variable though often complex (Schmitt-Kopplin et al. 2010) suite of small molecules.

That much of this material is extraterrestrial in origin is now beyond dispute. It was not long after one of the most immediately collected pristine samples, the Murchison meteorite, fell to Earth in 1969 that it was shown fairly conclusively that such materials contain organic compounds of biological relevance, including amino acids (Kvenvolden et al. 1970, 1971). Soon after, it was noted that the types and abundances of amino acids closely corresponded to those produced in some Miller-Urey electric discharge experiments, suggesting common synthesis mechanisms between the two samples (Wolman et al. 1972), e.g., Strecker-like aqueous-phase synthesis from simple precursors including HCN, aldehydes, and ketones, and amines including ammonia. It has also been suggested that Fischer-Tropsch-type synthesis from simple gases such as CO and NH_3 over metal catalysts (as might have been present in the asteroidal parent bodies of carbonaceous chondrites) could be an alternative source of these compounds (Hayatsu et al. 1972), among other proposed mechanisms.

A search for nitrogen heterocycles in carbonaceous chondrites began soon after, with various reports suggesting the presence of nucleobases and other N-heterocycles (Hayatsu et al. 1975; Stoks and Schwartz 1981, 1982). The isotopic composition of nucleobases isolated from carbonaceous chondrites has now been measured, and the data support their being of extraterrestrial origin (Martins et al. 2009). Further, surveys across carbonaceous chondrite petrologic types not only show a wide distribution of these compounds but trends in speciation and abundance which appear to correlate with the degree of aqueous alteration of the meteorites (Callahan et al. 2011). This study also showed the presence of a few purine isomers not typically found in biology including 2,6-diaminopurine and purine itself in low abundance. Pyrimidines were not detected in the same abundance, though this could be attributable to spatial heterogeneity of the samples.

The types and distribution of purines detected were suggested to be concordant with those detected from HCN synthesis, and notably the abundance of purine, which is typically the major purine product of formamide-based syntheses (Bredereck et al. 1961), was very low. However, the 8-hydroxymethyl purine derivatives were not detected, which does not agree well with reports of the dominance of these as products when HCN is polymerized in the presence of HCHO (Schwartz and Bakker 1989). HCHO has now been suggested to be a major organic precursor to carbonaceous chondrite organics (Cody et al. 2011), so there are various data points that do not appear to support a single or simple origin for these compounds in carbonaceous chondrites. There may be still other as-yet-unexplored synthetic pathways that may help explain these discrepancies. Additionally, as there is some correlation of the degree of post-accretional aqueous alteration with the distribution of purines, it is possible that post-synthetic processes skew initial product distributions formed by one of the commonly suggested processes.

1.4 The Stability of the Nucleobases

To have been useful for the origins of life, nucleobases need not only have had a route available for their synthesis, but they must have been stable enough to accumulate for whatever subsequent process enabled the onset of replication. DNA, which is typically more stable than RNA, degrades rapidly under geological conditions, and the kinetics of its decomposition follow predictable Arrhenius kinetics (Allentoft et al. 2012). Even when shielded from biologically mediated degradation under relatively favorable preservational conditions (dry, −5 °C), a maximal survival time for amplifiable and sequenceable DNA fragments using present technologies of 0.4–1.4 Ma has been estimated (Willerslev et al. 2004). However, amplifiable DNA fragments pervade Earth's critical zone even in the most hostile environments (Schulze-Makuch et al. 2018), largely due to the rapidity with which life continuously synthesizes them: they are dynamically stable in the context of living systems. Once the regenerative process of biological reproduction ceases, nucleic acids become yet one more type of compound in the environment subject to degradation.

Nucleic acids tend to degrade by a few principle pathways, including depurination and depyrimidination and backbone strand-scission. The measured concentration of nucleobases liberated from degraded nucleic acids drops off quickly in sediment cores over decadal time scales (Van der Velden and Schwartz 1974). As nitrogen assimilation is an extremely bioenergetically costly process, after nucleobases are liberated from nucleic acids in the environment, they tend to be consumed or degraded by microorganisms. Thus, the rapid production and recycling of nucleobases in the biosphere gives little insight into their abiotic decomposition and preservation in the terrestrial environment.

1.4.1 Thermal Decomposition

Controlled laboratory studies offer a simple means to estimate the survival of nucleobases in abiological environments. Non-biological degradation mechanisms may be slower but are similarly unforgiving for nucleobase survival. Abiological non-catalyzed rates of hydrolysis of the nucleobases and several analogues were measured by Levy and Miller (1998), and decomposition in the dry state was measured by Minton and Rosenberg (1964). Levy and Miller's measurements were made in the context of the surging notions of the origin of life near subsea hydrothermal vents (Baross and Hoffman 1985; Miller and Bada 1988) and underscored the idea that lower temperatures are generally more beneficial to the long-term survival of nucleobases. Their half-lives in neutral water were all found to be less than 100 years at 100 °C (measured $t_{1/2}$ T ~ 56 years, U ~ 12 years, A ~ 1 year, G ~ 0.8 years, and C ~ 19 days), with rates generally increasing under more acidic and more alkaline conditions. At 0 °C, the measured $t_{1/2}$ values were T ~ 2×10^9 years, U ~ 3.8×10^8 years, G ~ 1.3×10^6 years, A ~ 6×10^5 years, and C ~1.7×10^4 years It was pointed out that these rates are comparable to the present rate of seawater circulation through submarine hydrothermal vents (~10^7 years), and thus marine hydrothermal circulation may present an upper bound to the accumulation of nucleobases in the primitive oceans. Importantly, these authors did not study the influence of potential soluble or insoluble catalysts on degradation but noted that some cation-substituted clays markedly increase the rate of deamination of adenine (Strašák and Šeršeň 1991).

In contrast, nucleobases are considerably more stable in the dry state. Minton and Rosenberg (1964), for instance, extrapolated the half-lives of A and C as ~ 10^6 years at 25°, those of G and U as between 10^4 and 10^5 years, and that of T as less than 10^3 years, in contrast with the half-lives measured by Levy and Miller for A and G of ~10^4 years and that of C as 3.4×10^2 years at 25 °C in neutral solution.

The hydrolytic deamination of cytosine presents an interesting example of the importance of considerations of nucleobase stability in astrobiology. The relatively extreme instability of C has been cited variously as a problem for RNA world models (Shapiro 1999), a driver for early evolution (Lewis et al. 2016), and an explanation for the lack of detection of C in carbonaceous meteorites (Levy and Miller 1998).

1.4.2 Decomposition by Ionizing Radiation

In addition to thermally mediated decomposition, radiation presents yet another environmental limitation to nucleobase accumulation. The primitive solar system was undoubtedly a higher-radiation environment, as many primordial radionuclides were still abundant (Draganić et al. 1990). It also seems likely that the early Sun had a greater output in short wavelength regions of the electromagnetic spectrum, and on Earth in particular, UV radiation may have penetrated the atmosphere more efficiently before the advent of a significant ozone layer (Cleaves and Miller 1998).

In most cases, high-energy radiation is significantly more destructive of nucleobases in solution than in the dry state. This is mainly due to the generation of highly reactive species including solvated electrons, hydrogen atoms, hydroxyl radicals, and peroxides (Doane 2017). All of these have a high reactivity with the π electron systems of the nucleobases (Bertinchamps et al. 2011). Interestingly, relative to some other nitrogen heterocycles, the nucleobases appear to be able to especially efficiently eliminate UV-induced electronic excitation by internal conversion (Boldissar and de Vries 2018).

The source of the radiation and relative location of target nucleobases is also important. While UV radiation, galactic cosmic radiation, and solar energetic particles have a low penetrance through various materials including water and minerals, many types of pervasive high-energy particles and waves have significant penetrance or are generable within rock matrices or pore water spaces via the decay of various nuclides. Kminek and Bada (2006) conducted a quantitative study of the stability of amino acids to ionizing radiation under Mars-like conditions. They showed that radiolytic survival is inversely proportional to the molecular weight of the target, likely simply due to larger molecular cross-sectional area, and suggested that it is highly unlikely that there would be much small organic molecule survival in the top 1.5 m of regolith after exposure during the last 3 Ga, despite the prevailing low temperatures. Laboratory studies of unprotected nucleobases irradiated in the gas phase suggest survival times on the order of a few hours in space in Earth's vicinity and at most a few million years in dense molecular clouds (Peeters et al. 2003), the major variable being the energetic flux intensity and degree of shielding.

In contrast, it is evident that carbonaceous chondrites do contain indigenous nucleobases (Stoks and Schwartz 1982; Callahan et al. 2011); thus, it is possible to surmise that these bodies first either inherited these compounds during accretion or harbored conditions conducive to their synthesis and subsequently preserved them over the last 4.5 Ga of solar system history under extremely cold, dry, and radiation-sheltered conditions.

1.5 Conclusions and Future Directions

The nucleobases, and related N-heterocycles, undoubtedly are the products of abiotic synthesis in primitive solar system environments, as evidenced by their presence in carbonaceous meteorites. The molecular mechanism by which they were produced in them is open questions but may include one or more of the routes which have been demonstrated in various laboratories: synthesis from low molecular weight, high-energy compounds such as HCN, formamide, and CAA, either in stepwise fashion or from the degradation of complex heterogeneous polymeric materials derived from similar compounds. The synthesis of these materials can be induced by various energy sources (UV light, electric discharges, heat), provided there is a sufficient abundance of carbon and nitrogen, and the reactive compounds can be concentrated by some environmental process (e.g., eutectic freezing or evaporative drying).

However, the nucleobases also degrade in various geochemical settings by thermal and radiation-induced processes, and their potential roles in the origins of life would depend on the synthetic, depositional, and preservational conditions under which they arose. High carbon-flux, reducing environments appear to be the most amenable to synthesis and cold, dry environments the most conducive to long-term preservation. However, it is possible that the incorporation of nucleobases into higher-order structures and processes that may have led to living systems requires dynamic environments that are not conducive to long-term preservation. Consequently, the origins of life, if dependent on these types of compounds, may be relatively fast in a geological context (Pearce et al. 2017).

Acknowledgments This work was partially supported by a JSPS KAKENHI Grant-in-Aid for Scientific Research on Innovative Areas "Hadean Bioscience," grant number JP26106003. This project was also supported by the ELSI Origins Network (EON), which is supported by a grant from the John Templeton Foundation. The opinions expressed in this publication are those of the authors and do not necessarily reflect the views of the John Templeton Foundation.

References

Adam ZR, Hongo Y, Cleaves HJ, Yi R, Fahrenbach AC, Yoda I, Aono M (2018) Estimating the capacity for production of formamide by radioactive minerals on the prebiotic Earth. Sci Rep 8 (1):265

Alexander CMOD, Fogel M, Yabuta H, Cody GD (2007) The origin and evolution of chondrites recorded in the elemental and isotopic compositions of their macromolecular organic matter. Geochim Cosmochim Acta 71:4380–4403

Allentoft ME, Collins M, Harker D, Haile J, Oskam CL, Hale ML, Campos PF, Samaniego JA, Gilbert MTP, Willerslev E, Zhang G, Scofield RP, Holdaway RN, Bunce M (2012) The half-life of DNA in bone: measuring decay kinetics in 158 dated fossils. Proc R Soc B 279 (1748):4724–4733. https://doi.org/10.1098/rspb.2012.1745

Barks HL, Buckley R, Grieves GA, Di Mauro E, Hud NV, Orlando TM (2010) Guanine, adenine, and hypoxanthine production in UV-irradiated formamide solutions: relaxation of the requirements for prebiotic purine nucleobase formation. Chembiochem 11:1240–1243

Baross J, Hoffman S (1985) Submarine hydrothermal vents and associated gradient environments as sites for the origin and evolution of life. Orig Life Evol Biosph 15:327–345

Benner SA, Ellington AD (1991) RNA world. Science 252:1232

Bertinchamps AJ, Ahnström G, Hüttermann J, Cadet J, Köhnlein W, Coquerelle T, Teoule R, Cramp WA, Ehrenberg A, Elliott JP (2011) Effects of ionizing radiation on DNA: physical, chemical and biological aspects. Springer, Berlin

Boldissar S, de Vries MS (2018) How nature covers its bases. Phys Chem Chem Phys 20:9701–9716

Borquez E, Cleaves HJ, Lazcano A, Miller SL (2005) An investigation of prebiotic purine synthesis from the hydrolysis of HCN polymers. Orig Life Evol Biosph 35:79–90

Bredereck H, Effenberger F, Rainer G (1961) Eine neue einfache Purin-Synthese. Angew Chem 73:63–63

Callahan MP, Smith KE, Cleaves HJ, Ruzicka J, Stern JC, Glavin DP, House CH, Dworkin JP (2011) Carbonaceous meteorites contain a wide range of extraterrestrial nucleobases. Proc Natl Acad Sci 108:13995–13998

Chittenden GJ, Schwartz AW (1976) Possible pathway for prebiotic uracil synthesis by photodehydrogenation. Nature 263:350–351

Choughuley AS, Subbaraman AS, Kazi ZA, Chadha MS (1977) A possible prebiotic synthesis of thymine: uracil-formaldehyde-formic acid reaction. Biosystems 9:73–80

Chyba C, Sagan C (1992) Endogenous production, exogenous delivery and impact-shock synthesis of organic molecules: an inventory for the origins of life. Nature 355:125–132

Cleaves HJ (2012) Prebiotic chemistry: What we know, what we don't. Evol Educ Outreach 5:342–360

Cleaves HJ, Bada J (2012) The prebiotic chemistry of alternative nucleic acids. In: Seckbach J (ed) Genesis – in the beginning. Springer, Dordrecht, pp 3–33

Cleaves HJ, Miller SL (1998) Oceanic protection of prebiotic organic compounds from UV radiation. Proc Natl Acad Sci U S A 95:7260–7263

Cody GD, Heying E, Alexander CMO, Nittler LR, Kilcoyne ALD, Sandford SA, Stroud RM (2011) Establishing a molecular relationship between chondritic and cometary organic solids. Proc Natl Acad Sci

Corey EJ (1988) Retrosynthetic thinking – essentials and examples. Chem Soc Rev 17:111–133. https://doi.org/10.1039/CS9881700111

Crick FHC (1968) The origin of the genetic code. J Mol Biol 38:367–379

Cronin JR (1989) Origin of organic compounds in carbonaceous chondrites. Adv Space Res 9:59–64

Doane TA (2017) A survey of photogeochemistry. Geochem Trans 18:1

Draganić IG, Draganić ZD, Adloff JP (1990) Radiation and radioactivity on Earth and beyond. CRC Press, Boca Raton, FL

Ferris JP, Orgel LE (1965) Aminomalononitrile and 4-amino-5-cyanoimidazole in hydrogen cyanide polymerization and adenine synthesis. J Am Chem Soc 87:4976–4977
Ferris JP, Sanchez RA, Orgel LE (1968) Studies in prebiotic synthesis. 3. Synthesis of pyrimidines from cyanoacetylene and cyanate. J Mol Biol 33:693–704
Ferris JP, Joshi PC, Edelson EH, Lawless JG (1978) HCN: a plausible source of purines, pyrimidines and amino acids on the primitive earth. J Mol Evol 11:293–311
Fox SW, Harada K (1961) Synthesis of uracil under conditions of a thermal model of prebiological chemistry. Science 133:1923–1924
Gautier T, Carrasco N, Buch A, Szopa C, Sciamma-O'Brien E, Cernogora G (2011) Nitrile gas chemistry in Titan's atmosphere. Icarus 213:625–635
Gilbert W (1986) Origin of life: The RNA world. Nature 319
Halevy I, Bachan A (2017) The geologic history of seawater pH. Science 355(6329):1069–1071
Hayatsu R, Studier MH, Matsuoka S, Anders E (1972) Origin of organic-matter in early solar system. 6. Catalytic synthesis of nitriles, nitrogen bases and porphyrin-like pigments. Geochim Cosmochim Acta 36:555–571
Hayatsu R, Studier MH, Moore LP, Anders E (1975) Purines and triazines in murchison meteorite. Geochim Cosmochim Acta 39:471–488
Joyce GF, Schwartz AW, Miller SL, Orgel LE (1987) The case for an ancestral genetic system involving simple analogues of the nucleotides. Proc Natl Acad Sci USA 84:4398–4402
Kminek G, Bada JL (2006) The effect of ionizing radiation on the preservation of amino acids on Mars. Earth Planet Sci Lett 245:1–5
Krissansen-Totton J, Arney GN, Catling DC (2018) Constraining the climate and ocean pH of the early Earth with a geological carbon cycle model. Proc Natl Acad Sci 115(16):201721296
Kvenvolden K, Lawless J, Pering K, Peterson E, Flores J, Ponnamperuma C, Kaplan IR, Moore C (1970) Evidence for extraterrestrial amino acids and hydrocarbons in Murchison meteorite. Nature 228:923–926
Kvenvolden K, Lawless JG, Ponnamperuma C (1971) Nonprotein amino acids in Murchison meteorite. Proc Natl Acad Sci U S A 68:486–490
Levy M, Miller SL (1998) The stability of the RNA bases: implications for the origin of life. Proc Natl Acad Sci U S A 95:7933–7938
Levy M, Miller SL, Brinton K, Bada JL (2000) Prebiotic synthesis of adenine and amino acids under Europa-like conditions. Icarus 145:609–613
Lewis CA, Crayle J, Zhou S, Swanstrom R, Wolfenden R (2016) Cytosine deamination and the precipitous decline of spontaneous mutation during Earth's history. Proc Natl Acad Sci 113 (29):8194–8199
Malyshev DA, Dhami K, Lavergne T, Chen T, Dai N, Foster JM, Correa IR, Romesberg FE (2014) A semi-synthetic organism with an expanded genetic alphabet. Nature 509:385–388
Martins Z, Botta O, Fogel ML, Sephton MA, Glavin DP, Watson JS, Dworkin JP, Schwartz AW, Ehrenfreund P (2009) Extraterrestrial nucleobases in the Murchison meteorite. Orig Life Evol Biosph 39:214–214
Mason SF (1991) Chemical evolution: origin of the elements, molecules, and living systems. Clarendon Press, Oxford
Miller SL (1953) A production of amino acids under possible primitive Earth conditions. Science 117:528–529
Miller SL (1955) Production of some organic compounds under possible primitive Earth conditions. J Am Chem Soc 77:2351–2361
Miller SL, Bada JL (1988) Submarine hot springs and the origin of life. Nature 334:609–611
Minton A, Rosenberg E (1964) The effect of temperature on the preservation of purine and pyrimidine bases. Geochim Cosmochim Acta 28(12):1953–1959
Miyakawa S, Cleaves HJ, Miller SL (2002a) The cold origin of life: B. Implications based on pyrimidines and purines produced from frozen ammonium cyanide solutions. Orig Life Evol Biosph 32:209–218

Miyakawa S, Yamanashi H, Kobayashi K, Cleaves HJ, Miller SL (2002b) Prebiotic synthesis from CO atmospheres: implications for the origins of life. Proc Natl Acad Sci U S A 99:14628–14631

Miyakawa S, Cleaves HJ, Miller SL (2002c) The cold origin of life: A. Implications based on the hydrolytic stabilities of hydrogen cyanide and formamide. Orig Life Evol Biosph 32:195–208

Morowitz HJ (1999) A theory of biochemical organization, metabolic pathways, and evolution. Complexity 4:39–53

Mumma MJ, Charnley SB (2011) The chemical composition of comets—Emerging taxonomies and natal heritage. Astron Astrophys 49:471–524

Musto H, Naya H, Zavala A, Romero H, Alvarez-Valın F, Bernardi G (2004) Correlations between genomic GC levels and optimal growth temperatures in prokaryotes. FEBS Lett 573(1–3):73–77

Nelson KE, Robertson M, Levy M, Miller SL (2001) Concentration by evaporation and the prebiotic synthesis of cytosine. Orig Life Evol Biosph 31:221–229

Orgel LE (1968) Evolution of the genetic apparatus. J Mol Biol 38:381–393

Oró J (1961) Mechanism of synthesis of adenine from hydrogen cyanide under possible primitive earth conditions. Nature 191:1193–1194

Oró J (1963) Studies in experimental organic cosmochemistry. Ann NY Acad Sci 108(2):464–481

Pearce BK, Pudritz RE, Semenov DA, Henning TK (2017) Origin of the RNA world: the fate of nucleobases in warm little ponds. Proc Natl Acad Sci, p 201710339

Peeters Z, Botta O, Charnley SB, Ruiterkamp R, Ehrenfreund P (2003) The astrobiology of nucleobases. Astrophys J Lett 593(2):L129

Pinheiro VB, Loakes D, Holliger P (2013) Synthetic polymers and their potential as genetic materials. BioEssays 35:113–122

Pizzarello S (2012) Hydrogen cyanide in the Murchison meteorite. Astrophys J Lett 754:L27

Pizzarello S, Shock E (2010) The organic composition of carbonaceous meteorites: the evolutionary story ahead of biochemistry. Cold Spring Harb Perspect Biol 2:a002105

Powner MW, Sutherland JD (2008) Potentially prebiotic synthesis of pyrimidine beta-D-ribonucleotides by photoanomerization/hydrolysis of alpha-D-cytidine-2′-phosphate. Chembiochem 9:2386–2387

Powner MW, Sutherland JD, Szostak JW (2010) Chemoselective multicomponent one-pot assembly of purine precursors in water. J Am Chem Soc 132:16677–16688

Robertson MP, Miller SL (1995a) Prebiotic synthesis of 5-substituted uracils – a bridge between the RNA world and the DNA-protein world. Science 268:702–705

Robertson MP, Miller SL (1995b) An efficient prebiotic synthesis of cytosine and uracil. Nature 375:772–774

Robertson MP, Miller SL (1995c) Prebiotic synthesis of 5-substituted uracils: a bridge between the RNA world and the DNA-protein world. Science 268:702–705

Robertson MP, Levy M, Miller SL (1996) Prebiotic synthesis of diaminopyrimidine and thiocytosine. J Mol Evol 43:543–550

Rubey WW (1951) Geologic history of sea water. Geol Soc Am Bull 62:1111–1148

Ruiz-Bermejo M, Zorzano M-P, Osuna-Esteban S (2013) Simple organics and biomonomers identified in HCN polymers: An overview. Life 3:421–448

Ruiz-Mirazo K, Briones C, de la Escosura A (2014) Prebiotic systems chemistry: new perspectives for the origins of life. Chem Rev 114(1):285–366

Saladino R, Crestini C, Costanzo G, Negri R, Di Mauro E (2001) A possible prebiotic synthesis of purine, adenine, cytosine, and 4(3H)-pyrimidinone from formamide: implications for the origin of life. Bioorg Med Chem 9:1249–1253

Sanchez RA, Orgel LE (1970) Studies in prebiotic synthesis. V. Synthesis and photoanomerization of pyrimidine nucleosides. J Mol Biol 47:531–543

Sanchez R, Ferris J, Orgel LE (1966a) Conditions for purine synthesis: did prebiotic synthesis occur at low temperatures? Science 153:72–73

Sanchez RA, Ferris JP, Orgel LE (1966b) Cyanoacetylene in prebiotic synthesis. Science 154:784–785

Sanchez RA, Ferris JP, Orgel LE (1967) Studies in prebiotic synthesis. II. Synthesis of purine precursors and amino acids from aqueous hydrogen cyanide. J Mol Biol 30:223–253
Schlesinger G, Miller SL (1983a) Prebiotic synthesis in atmospheres containing CH_4, CO, and CO_2. I. Amino acids. J Mol Evol 19:376–382
Schlesinger G, Miller SL (1983b) Prebiotic synthesis in atmospheres containing CH_4, CO, and CO_2. II. Hydrogen cyanide, formaldehyde and ammonia. J Mol Evol 19:383–390
Schmitt-Kopplin P, Gabelica Z, Gougeon RD, Fekete A, Kanawati B, Harir M, Gebefuegi I, Eckel G, Hertkorn N (2010) High molecular diversity of extraterrestrial organic matter in Murchison meteorite revealed 40 years after its fall. Proc Natl Acad Sci 107:2763–2768
Schneider C, Becker S, Okamura H, Crisp A, Amatov T, Stadlmeier M, Carell T (2018) Prebiotic methylations and carbamoylations generate non-canonical RNA nucleosides as molecular fossils of an early Earth. Angew Chem
Schulze-Makuch D, Wagner D, Kounaves SP, Mangelsdorf K, Devine KG, de Vera JP, Schmitt-Kopplin P, Grossart HP, Parro V, Kaupenjohann M, Galy A (2018) Transitory microbial habitat in the hyperarid Atacama Desert. Proc Natl Acad Sci 115(11):2670–2675
Schwartz AW, Bakker CG (1989) Was adenine the first purine? Science 245:1102–1104
Schwartz AW, Joosten H, Voet AB (1982) Prebiotic adenine synthesis via HCN oligomerization in ice. Biosystems 15(3):191
Segré D, Ben-Eli D, Lancet D (2000) Compositional genomes: prebiotic information transfer in mutually catalytic noncovalent assemblies. Proc Natl Acad Sci U S A 97:4112–4117
Shapiro R (1999) Prebiotic cytosine synthesis: a critical analysis and implications for the origin of life. Proc Natl Acad Sci U S A 96:4396–4401
Stephen-Sherwood E, Oro J, Kimball AP (1971) Thymine: a possible prebiotic synthesis. Science 173:446–447
Stoks PG, Schwartz AW (1981) Nitrogen-heterocyclic compounds in Meteorites – significance and mechanisms of formation. Geochim Cosmochim Acta 45:563–569
Stoks PG, Schwartz AW (1982) Basic nitrogen-heterocyclic compounds in the Murchison meteorite. Geochim Cosmochim Acta 46:309–315
Strašák M, Šeršeň F (1991) An unusual reaction of adenine and adenosine on montmorillonite. Naturwissenschaften 78(3):121–122
Switzer CMSBS (1989) Enzymatic incorporation of a new base pair into DNA and RNA. J Am Chem Soc 111:8322–8323
Tian F, Toon OB, Pavlov AA, De Sterck H (2005) A hydrogen-rich early Earth atmosphere. Science 308:1014–1017
Van der Velden W, Schwartz AW (1974) Purines and pyrimidines in sediments from Lake Erie. Science 185(4152):691–693
Voet AB, Schwartz AW (1982) Uracil synthesis via HCN oligomerization. Orig Life 12:45–49
Voet AB, Schwartz AW (1983) Prebiotic adenine synthesis from HCN—Evidence for a newly discovered major pathway. Bioorg Chem 12:8–17
Wächtershäuser G (1988) Before enzymes and templates: theory of surface metabolism. Microbiol Rev 52:452–484
Wang HC, Susko E, Roger AJ (2006) On the correlation between genomic G+C content and optimal growth temperature in prokaryotes: data quality and confounding factors. Biochem Biophys Res Commun 342(3):681–684
Watson JD, Crick FHC (1953) Molecular structure of nucleic acids: a structure for deoxyribose nucleic acid. Nature 171:737–738
Willerslev E, Hansen AJ, Rønn R, Brand TB, Barnes I, Wiuf C, Gilichinsky D, Mitchell D, Cooper A (2004) Long-term persistence of bacterial DNA. Curr Biol 14(1):R9–R10
Woese CR (1967) The genetic code: the molecular basis for genetic expression. Harper & Row, New York

Wolman Y, Haverland WJ, Miller SL (1972) Nonprotein amino acids from spark discharges and their comparison with the Murchison meteorite amino acids. Proc Natl Acad Sci U S A 69:809–811

Yoshino D, Hayatsu R, Anders E (1971) Origin of organic matter in early solar system. 3. Amino acids – catalytic synthesis. Geochim Cosmochim Acta 35:927–938

Zubay G (1993) To what extent do biochemical pathways mimic prebiotic pathways? Chemtracts Biochem Mol Biol 4:317–323

Chapter 2
Condensation and Decomposition of Nucleotides in Simulated Hydrothermal Fields

Ryan Lorig-Roach and David Deamer

Abstract This chapter describes recent studies in which nucleic acid oligomers are synthesized in simulated hydrothermal field conditions using cycles of dehydration and rehydration to promote ester bond synthesis. Such conditions involve elevated temperatures and acidic pH ranges that are also conducive to depurination of nucleotides. For this reason it is important to establish the extent to which depurination occurs and whether this limits yields of oligomers. Here we review condensation reactions that occur in mixtures of AMP and UMP undergoing multiple dehydration cycles in acidic conditions, and report new results related to depurination under the same conditions. Although depurination could be detected, the reaction was inhibited by the presence of a phospholipid. Furthermore, a fraction of the original AMP remains in subsequent cycles, suggesting that depurination does not proceed to completion. We conclude that even though decomposition of mononucleotides occurs in hydrothermal cycling, purine nucleotides will continue to be available to participate in polymerization.

2.1 Introduction

All life today exists in a steady state in which uphill synthetic condensation reactions are balanced against downhill decomposition, usually spontaneous or by catalyzed hydrolysis. A similar balance was likely to have occurred on the prebiotic Earth, suggesting that the origin of life involved a steady state in which a thermodynamically uphill synthesis of polymers balanced spontaneous polymer hydrolysis. Therefore, one of our goals has been to discover conditions in which biologically relevant polymers are synthesized but also to determine the steady state in which their synthesis is balanced by hydrolysis.

R. Lorig-Roach · D. Deamer (✉)
Department of Biomolecular Engineering, University of California, Santa Cruz, CA, USA
e-mail: deamer@soe.ucsc.edu

© Springer International Publishing AG, part of Springer Nature 2018
C. Menor-Salván (ed.), *Prebiotic Chemistry and Chemical Evolution of Nucleic Acids*, Nucleic Acids and Molecular Biology 35,
https://doi.org/10.1007/978-3-319-93584-3_2

Our approach is guided by considering geochemical conditions on the early Earth. To this end, we are testing whether polymerization of mononucleotides can be driven by hydration-dehydration (HD) cycles that occur in small pools produced by hydrothermal conditions associated with volcanic activity. Field studies in analogue environments in Kamchatka, Iceland, Hawaii, and Northern California have shown that such pools range in temperature from 80 °C to near boiling (Deamer et al. 2006). They are mildly acidic, typically between pH 2 and 3. HD cycles occur around the edges of the pools in time frames of minutes to hours, and the pools themselves evaporate and refill over weeks to months. The question to be addressed here concerns how the chemical composition of mononucleotides changes when they are exposed to such conditions.

We have established a laboratory simulation of HD cycles which we consider to be essential drivers of polymerization and increasing complexity. Previous studies reported two conditions that go beyond simple wet-dry cycles. We found that polymerization can occur in the presence of lipids that assemble into membranous compartments (Rajamani et al. 2008; De Guzman et al. 2014) but is also enhanced if the cycles include monovalent salts (Da Silva et al. 2015). We will first describe condensation reactions of mononucleotides, then consider two kinds of spontaneous decomposition reactions that also must occur: hydrolysis of ester bonds and depurination.

2.1.1 A Laboratory Simulation of HD Cycles

We constructed a chamber that simulated the cycling conditions associated with geothermal hot springs (Fig. 2.1). Glass vials containing reaction mixtures to be tested are placed in 24 wells in an aluminum disk, and the chamber is filled with an inert gas such as carbon dioxide. The disk is heated to a desired temperature, and rotation of the disk is controlled by a programmed stepper motor. As the disk rotates in steps of 7.5 degrees, the samples are dehydrated by a flow of dry carbon dioxide through four ports on either side. Each vial is dried for 60 min, then the rotation brings the vial under a port through which water is injected at a constant rate of 0.2 mL per step by a programmable syringe pump. The CO_2 maintains anaerobic conditions and carries away water molecules produced during condensation reactions, thereby inhibiting back reactions of hydrolysis. In a typical run, every sample undergoes four cycles of wetting and drying.

Some studies do not require a full complement of 24 vials with individual experiments, so smaller scale experiments use glass slides with two wells on each slide that hold 0.1 mL of the reaction mixture. Four slides are arranged on a laboratory hot plate set at the desired temperature range, and a plastic flow box with an inlet at the rear allows carbon dioxide to flow into the reaction mixture at ~10 cubic centimeters per second. Eight small tubes (1 mm diameter) penetrate the top of the flow box and are placed so that each tube is directly over a well. Polyethylene tubing leads the tubes to eight one milliliter syringes controlled by a programmable

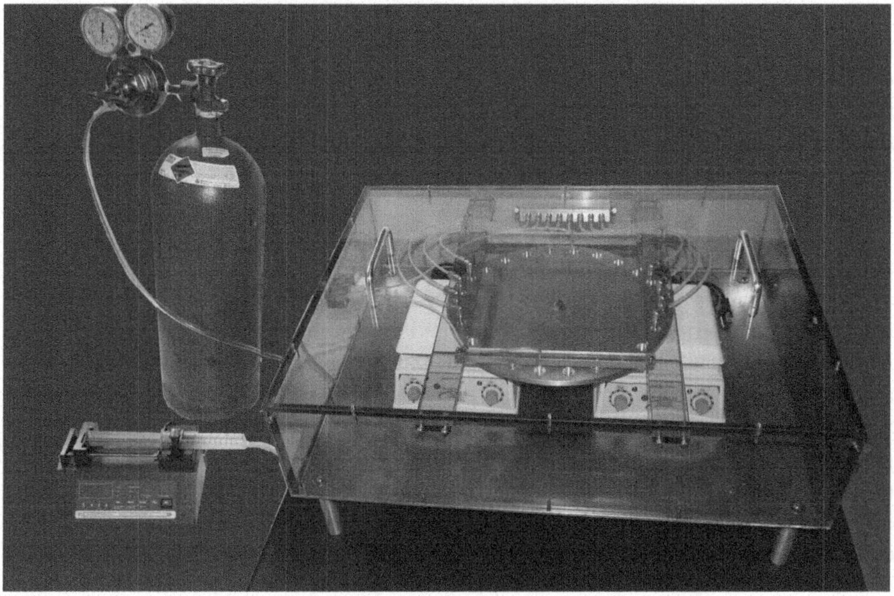

Fig. 2.1 Simulation chamber. The chamber is filled with carbon dioxide to exclude oxygen, and water is delivered to samples by a programmable syringe pump

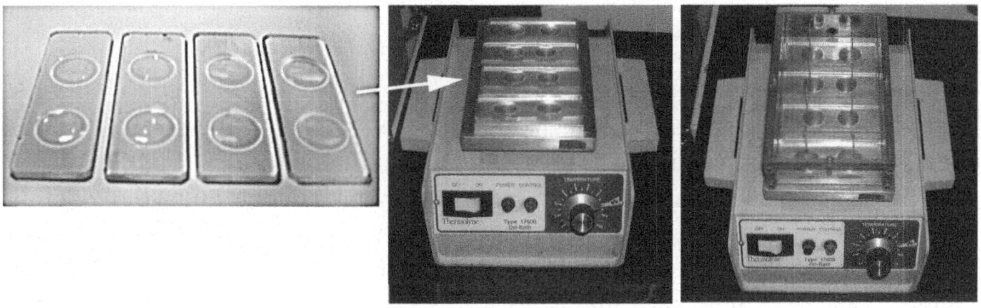

Fig. 2.2 Small simulation apparatus (See text for description)

perfusion pump. Each syringe injects 0.1 mL of water into its well at defined intervals, typically every 30 min (Fig. 2.2).

A typical reaction mixture was 0.1 mL of 10 mM mononucleotides (AMP and UMP mixed in 1:1 mole ratios) with and without addition of a phospholipid or a monovalent salt. The AMP and UMP were purchased as free acids, not sodium salts, so the initial pH was 2.5. We used lysophosphatidylcholine as the phospholipid because it is present as micelles in solution but forms a multilamellar liquid crystalline matrix when dried. Da Silva et al. (2015) reported that monovalent salts promote polymerization of mononucleotides. Ammonium chloride produced somewhat greater yields, so it was chosen for the studies reported here.

The polymer products were isolated by standard precipitation in ethanol or purification with spin tubes designed for oligonucleotide purification. Similar amounts of products were obtained with either procedure, consistent with the presence of polymers that behaved like RNA. Polymer yields of the reaction were determined by UV spectrophotometry, and product composition was monitored by gel electrophoresis in 4% agarose gels. All of the gels are shown as inverted images to increase contrast for visibility.

Hydrolysis of RNA was studied with a 1:1 mixture of polyadenylic acid (polyA) and polyuridylic acid (polyU) that form a double helix at room temperature but melt into single strands at the elevated temperatures required to activate condensation reactions. The starting amount was 20 μg, and the amount remaining after each cycle was recovered by ethanol precipitation. The values were normalized as percent of the amount recovered in the absence of cycling (0 cycles in the figures). Depurination was investigated with ribo- and deoxyribonucleotides under the same conditions as polymerization, and the effect of a phospholipid or a monovalent salt was determined by adding 10 mM lysophosphatidylcholine (LPC) or 0.1 M NH_4Cl to the initial reaction mixture. In a typical experiment, the reactants were exposed to four cycles of wetting and drying. At the end of the cycle series, the samples were dissolved in 0.1 mL of water, and depurination products were analyzed by HPLC.

2.2 Results and Discussion

2.2.1 Stability of RNA Under Simulation Conditions

We first investigated the rate at which a known RNA sample undergoes hydrolysis during HD cycles at acidic pH ranges and elevated temperatures. Figure 2.3, left panel, shows a gel and nanodrop analysis of the poly(AU) duplex going through four HD cycles. We found that most of the poly(AU) survived three cycles, but about half was hydrolyzed after the fourth cycle. If LPC, AMP, and UMP were present during cycling, there was little measurable hydrolysis after four cycles (Fig. 2.3, right panel).

2.2.2 Condensation Reactions

If the HD cycles are run with monovalent salts present, yields of polymer are greater than those obtained with lipid enhancement. Da Silva et al. (2015) reported that 0.1 M sodium, potassium, and ammonium chloride all promote polymerization, with NH4Cl having the greatest effect. Figure 2.4 compares the relative yields of polymer products from condensation reactions in the presence of ammonium chloride or lysophosphatidylcholine (LPC). Polymers do accumulate during cycling, and the results confirm that a monovalent salt increases the yields.

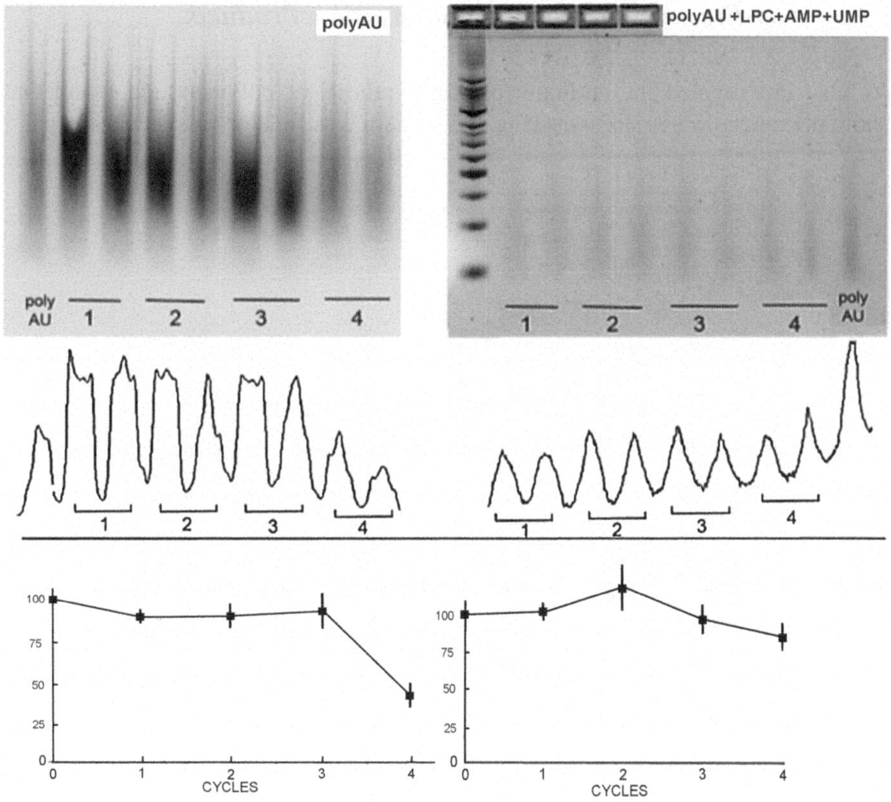

Fig. 2.3 Hydrolysis of poly(A)-poly(U) duplex RNA. The gels and scans are duplicate samples exposed to four HD cycles. The Y axis shows the normalized amount of polymer remaining after each cycle, with 100 representing the yield that was not exposed to cycling. Wells labeled polyAU are standard amounts (2 micrograms) of the original poly(A)-poly(U) duplex RNA. The ladder shows 50 nt increments

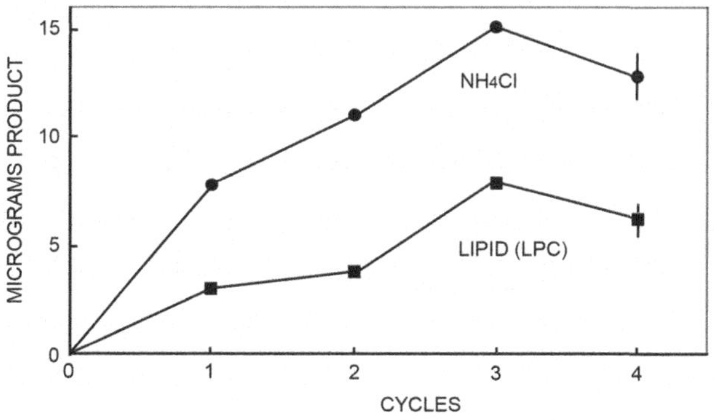

Fig. 2.4 Total polymer yields combined from four samples increased with the number of cycles in two separate experiments with 1:1 AMP and UMP as monomers. Ammonium chloride was also effective in promoting polymerization, sometimes producing a greater yield than that of lipid-enhanced polymerization

2.2.3 Gel Electrophoresis of Condensation Products

We also investigated the optimum pH for synthesis of polymers and found that acidic pH ranges were necessary (Fig. 2.5). There was little or no synthesis at pH 5

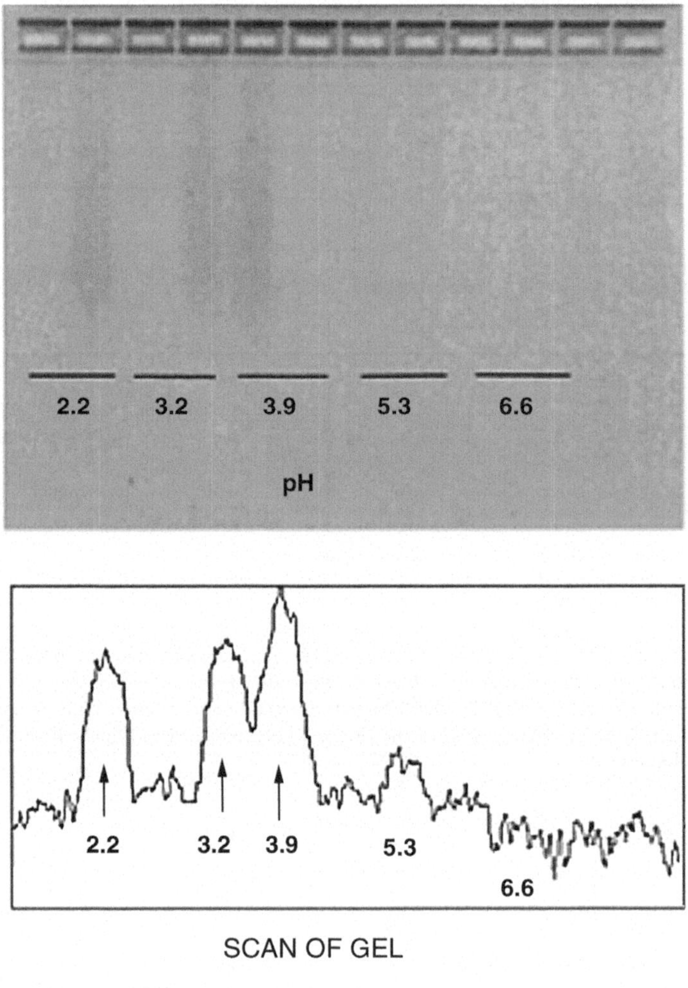

Fig. 2.5 Gel electrophoresis of the products of four HD cycles run with AMP and UMP as monomers and the effect of varying pH. The initial acidic pH of the free acid nucleotides was adjusted upward by addition of sodium hydroxide

and above, which is consistent with the condensation mechanism being an acid-catalyzed synthesis.

2.2.4 *Effect of Temperature, pH, and Ionic Solutes on Depurination*

As described above, the synthesis of RNA-like polymers under simulated hydrothermal conditions is promoted by the presence of either lipids or ammonium chloride salts and is most efficient in acidic pH between 2 and 3, at elevated temperatures greater than 80 °C. However, the hydrolytic cleavage of the N-glycosyl bond of purine nucleotides also occurs at a certain rate in these conditions. To determine the extent of depurination, we monitored release of adenine during cycling of a 10 mM mixture of AMP and UMP. For comparison, we also monitored depurination of deoxyribonucleotides (dAMP and TMP). The extent of depurination was calculated after separation and quantification of products by HPLC.

Figure 2.6 shows elution profiles obtained after three HD cycles for ribo- and deoxyribonucleotides samples, respectively. Comparing 6A and 6B makes it immediately apparent that dAMP is much more labile to depurination than AMP. Figure 2.7 shows the kinetics of depurination of AMP and dAMP occurring in the simulated hydrothermal conditions. We observed ~10% depurination of AMP after two 30 min cycles (Fig. 2.7a) and a steady state of roughly 20% after eight cycles. In the presence of NH4Cl, ribonucleotides were more sensitive to depurination than the controls and reached 70% after eight cycles.

Figure 2.7b shows depurination of deoxyribonucleotides over time. Deoxyribonucleotides were significantly less stable than ribonucleotides, approaching 80% depurination after a single cycle. Because depurination was so extensive, another

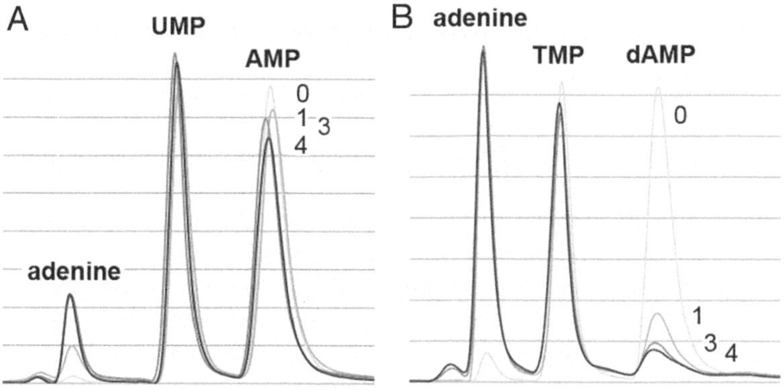

Fig. 2.6 Elution profiles of samples with (**a**) AMP and UMP or (**b**) dAMP and TMP as starting monomers after three HD cycles at 85 °C, pH 2 followed by rehydration with water every 30 min

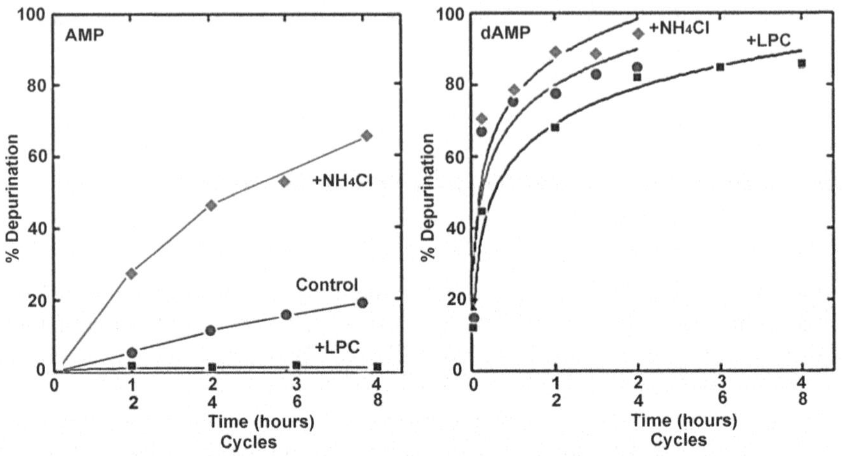

Fig. 2.7 Kinetics of depurination experiments using (**a**) AMP and UMP or (**b**) dAMP and TMP as starting monomers. Percentage of depurination was calculated as adenine released and quantified by HPLC over the initial amount of AMP present at the beginning of the experiments. The reaction mixtures were cycled at 85 °C, pH 2, followed by rehydration with water every 30 min in the presence of either monomers by themselves (red circle) or LPC 10 mM (blue squares) or NH_4Cl 0.1 M (green diamonds)

experiment was performed in which the products of a single cycle were determined 2 and 7 min after complete drying. Depurination was 20% and 70% after 2 and 7 min, respectively, approaching an apparent steady state of ~90% after four cycles.

Depurination of AMP was markedly reduced when a lipid was present. Even after 4 h of 30 min HD cycles, less than 5% had undergone depurination. The presence of lipids had less of an effect on depurination of dAMP, which approached a steady state of ~80% after eight cycles.

2.2.5 *Hydrolysis and Depurination Are Sufficiently Slow for Polymers to Accumulate*

In earlier studies, we reported the synthesis of polymers from nonactivated monomeric ribonucleotides in conditions that simulate prebiotic hydrothermal ponds undergoing natural fluctuations of hydration and dehydration. However, the conditions of this simulated environment (acidic pH and elevated temperature) also favor the decomposition of monomers and polymers. We therefore investigated a variety of conditions that might be expected to affect stability of monomers and polymers.

We began by addressing rates of hydrolysis of ester bonds. A model RNA polymer—poly(A)-poly(U)—exposed to HD cycles at acidic pH does ultimately break into shorter fragments that cannot be precipitated by ethanol, but hydrolysis is inhibited under conditions that promote synthesis, in which AMP, UMP, and a phospholipid are present in the mixture (Fig. 2.3). The protective effect could be explained if condensation of the monomers during each cycle tended to balance hydrolysis rates. The results of a separate series of experiments were consistent with this possibility, because polymeric products were found to accumulate during multiple HD cycles in mixtures of AMP and UMP in the presence of a phospholipid or monovalent salts (De Guzman et al. 2014; Da Silva et al. 2015). We conclude that the hydrolysis reaction expected from thermodynamic considerations can be reversed by HD cycles in acidic pH ranges, in which polymer synthesis rates exceed rates of hydrolysis.

We also addressed depurination, the hydrolytic cleavage of the N-glycosyl bonds that releases purine bases from nucleic acids (Lindahl 1993). In a recent study, Mungi and Rajamani (2015) demonstrated that acidic pH and elevated temperatures can cause significant depurination of AMP in conditions that promote condensation reactions. This confirmed previous studies reporting that cleavage occurs at elevated temperatures (Stockbridge et al. 2010) and that the N-glycosyl bonds of ribonucleotides are significantly more stable than those of deoxyribonucleotides (Rios et al. 2015).

If nonenzymatic synthesis of RNA-like polymers was necessary to initiate the chemical evolution leading to a putative RNA world, it is clear that decomposition reactions such as depurination must be sufficiently slow so that monomers remain available to undergo polymerization. In the present study, we observed that depurination of ribonucleotides occurs in the first HD cycles, but instead of continuing to completion, it reaches a steady state of approximately 60% for ribonucleotides and 90% for deoxyribonucleotides. The presence of a phospholipid significantly protected ribonucleotides and limited depurination to ~2% after four HD cycles.

We conclude that although hydrolysis and depurination do occur in conditions simulating those of a prebiotic hydrothermal pool, the rates of condensation reactions are sufficiently fast for polymers to accumulate in a kinetic trap. Furthermore, conditions such as the presence of lipids can slow hydrolysis and depurination even further. Because membranous compartments are essential for the origin of cellular life (Szostak et al. 2001; Damer and Deamer 2015), the protective effects of lipids are likely to be a fruitful focus of future research.

References

Da Silva L, Maurel M-C, Deamer D (2015) Salt-promoted synthesis of RNA-like molecules in simulated hydrothermal conditions. J Mol Evol 80:86–97

Damer B, Deamer D (2015) Coupled phases and combinatorial selection in fluctuating hydrothermal pools: a scenario to guide experimental approaches to the origin of cellular life. Life (Basel) 5:872–887. https://doi.org/10.3390/life5010872

Deamer D, Singaram S, Rajamani S, Kompanichenko V, Guggenheim S (2006) Self-assembly processes in the prebiotic environment. Philos Trans R Soc Lond B 361:1809–1818

DeGuzman V, Vercoutere W, Shenasa H, Deamer D (2014) Generation of oligonucleotides under hydrothermal conditions by non-enzymatic polymerization. J Mol Evol 78:251–262
Lindahl T (1993) Instability and decay of the primary structure of DNA. Nature 362:709–715
Mungi CV, Rajamani S (2015) Characterization of RNA-like oligomers from lipid-assisted nonenzymatic synthesis: implications for origin of informational molecules on early earth. Life (Basel) 5:65–84
Rajamani S, Vlassov A, Benner S, Coombs A, Olasagasti F, Deamer D (2008) Lipid-assisted synthesis of RNA-like polymers from mononucleotides. Orig Life Evol Biosph 38:57–74
Rios AC, Yua HT, Tor Y (2015) Hydrolytic fitness of N-glycosyl bonds: comparing the deglycosylation kinetics of modified, alternative, and native nucleosides. J Phys Org Chem 28 (3):173–180
Stockbridge RB, Schroeder GK, Wolfenden R (2010) The rate of spontaneous cleavage of the glycosidic bond of adenosine. Bioorg Chem 38:224–228
Szostak JW, Bartel DP, Luisi PL (2001) Synthesizing life. Nature 409:387–390

Chapter 3
Mineral-Organic Interactions in Prebiotic Synthesis

The Discontinuous Synthesis Model for the Formation of RNA in Naturally Complex Geological Environments

Steven A. Benner, Hyo-Joong Kim, and Elisa Biondi

Abstract A common criticism of "prebiotic chemistry research" is that it is done with starting materials that are too pure, in experiments that are too directed, to get results that are too scripted, under conditions that could never have existed on Earth. Planetary scientists in particular remark that these experiments often arise simply because a chemist has a "cool idea" and then pursues it without considering external factors, especially geological and planetary context. A growing literature addresses this criticism and is reviewed here. We assume a model where RNA emerged spontaneously from a prebiotic environment on early Earth, giving the planet its first access to Darwinism. This "RNA First Hypothesis" is not driven by the intrinsic prebiotic accessibility; quite the contrary, RNA is a "prebiotic chemist's nightmare." However, by assuming models for the accretion of the Earth, the formation of the Moon, and the acquisition of Earth's "late veneer," a reasonable geological model can be envisioned to deliver the organic precursors needed to form the nucleobases and ribose of RNA. A geological model having an environment with dry arid land under a carbon dioxide atmosphere receiving effluent from serpentinizing igneous rocks allows their conversion to nucleosides and nucleoside phosphates. Mineral elements including boron and molybdenum prevent organic material from devolving to form "tars" along the way. And dehydration and activation allows the formation of oligomeric RNA that can be stabilized by adsorption on available minerals.

S. A. Benner · H.-J. Kim · E. Biondi (✉)
The Westheimer Institute for Science and Technology, Gainesville, FL, USA

Firebird Biomolecular Sciences LLC, Alachua, FL, USA

The Foundation for Applied Molecular Evolution, Alachua, FL, USA
e-mail: ebiondi@ffame.org

© Springer International Publishing AG, part of Springer Nature 2018

C. Menor-Salván (ed.), *Prebiotic Chemistry and Chemical Evolution of Nucleic Acids*, Nucleic Acids and Molecular Biology 35,
https://doi.org/10.1007/978-3-319-93584-3_3

3.1 Introduction

The chemical reactions that led to the first Darwinian systems on Earth did not occur in Pyrex glassware. Rather, they occurred in a geological context, in rock assemblages, basins, or other environments constructed from and bounded by mineral species. The precursors for those reactions also formed in geological contexts, in rock formations or, for simpler molecules, in an atmosphere above early Earth or in space. Even in space, however, organics may have been made on mineral grains that later came to Earth via meteorites.

A common criticism made of classical prebiotic chemistry by planetary scientists is that this geological context is too severely ignored (Sahai et al. 2016). Even non-geologists, like Robert Shapiro, have been harsh. For example, Shapiro, criticizing classical prebiotic chemistry performed by organic chemists, likened it to a golfer who himself having played 18 holes, believes that the ball could have propelled itself around the same course on its own without human intervention (Shapiro 2007). In this view, classical prebiotic chemistry examined compounds that were too pure, in experiments that were too contrived, to get results that were too scripted from environments that could never have actually existed on a planet.

Some of this criticism is unfair. Chemists have good reason to work with pure organic substances and to regularly intervene in the ongoing chemistry to isolate and purify reaction products before putting them into the next Pyrex flask for the next step. The products of every reaction must be analyzed for anything to be learned. That analysis cannot be done if a reaction generates dozens (or hundreds) of products, each in small amounts. Last, to directly use an unanalyzed product mixture as the starting point for another reaction sequence is, by hard experience, a recipe for disaster.

Further, chemists who do seek guidance from the planetary science community quickly discover that that guidance is generally sparse, often contradictory, and usually supported by few data (Schoonen and Smirnov 2016). While progress is being made in ways that we will describe below in some detail, the Hadean planetary environment remains too unconstrained to dictate specific environments where prebiotic chemistry must operate. Prebiotic chemists must choose among various models for the early Earth, constrain their experiments by that choice, and change those constraints if/when the preferred model changes.

For these reasons, most chemists feel justified when we follow John Sutherland who, after suggesting that the complicating reactivity of aminooxazole might be mitigated by having it sublime into an atmosphere and then "rain into" a separate environment that fortuitously contained a substantial amount of glyceraldehyde, wrote that "although the issue of temporally separated supplies of glycolaldehyde and glyceraldehyde remains a problem, a number of situations could have arisen that would result in the conditions of heating and progressive dehydration followed by cooling, rehydration and ultraviolet irradiation. Comparative assessment of these models is beyond the scope of this work" (Powner et al. 2009).

A reverse criticism is also possible. Few geologists are conversant with the reactivity of complex organic molecules. Thus, much prebiotic chemistry that emerges from geology-focused laboratories does little more than reduce carbon dioxide, create methane or ammonia, or create low molecular weight carboxylic acids such as acetic acid and straight chain fatty acids (Schoonen et al. 2004). Unfortunately, even if we stipulate that all of these reactions occurred in the Hadean, problems remain that make the origin of life seem (nearly) impossible. These problems arise from problematic reactivities of more complex organic molecules, not in the production of acetate or ammonia.

For example, Robert Hazen, a particularly daring geologist in this respect, went so far as to incubate pyruvic acid ($CH_3CO\text{-}COOH$) with minerals at high pressures of carbon dioxide, expecting to see carbon fixation (Hazen 2005). He expressed surprise to see that he had created only a complex mixture of organic products ("tar") that defied analysis. This surprise reflected a lack of experience with the reactivity of pyruvic acid (an "alpha-keto acid"), which is especially prone to form tars. The same lack of experience applies to proposals for a mineral-supported "reverse citric acid cycle," which feature prominently in the "wish list" of many early Earth scientists (Russell and Martin 2004). This process also has molecules that share a propensity to form tars well before they fix CO_2.

Such is the situation with multidisciplinary research in general (Benner et al. 2013). This review seeks to describe the current relation of the Hadean geological environment to prebiotic chemistry. To direct this discussion, we have chosen just one model for the origin of Darwinism on Earth: the "RNA First Model." This model holds that Darwinism emerged on Earth through the creation of RNA molecules that carried information that could be replicated, with errors, with the errors themselves being replicable. This is the simplest combination of features needed for a system to use random variation followed by natural selection to increase its information to capture increasingly large fractions of the surrounding resources in competition with other Darwinian systems, as well as to compete with spontaneous degradation and devolution.

3.1.1 Decisions Must Be Made

Any model for the origin of life requires that we make decisions about what geology was available at the time that life emerged. These decisions cannot possibly rely on any statement of "proof." Leaving aside the fact that proof is unknown in science for any interesting proposition, it is especially unknown for historical events ("What happened?") that occurred over 4 billion years ago (Ga). A decade ago, too few constraints were available for either the geology or the time of life emerging to have much of a productive discussion.

The combinatorial explosion of parameter space does not even allow us to say that we are "agnostic" with respect to key decisions. Avoiding decisions leaves us with too little guidance among too many possibilities. Nor are we allowed to declare that we disavow "belief" in our scientific thought processes and therefore are "objective" in our consideration of "all options." In fact, humans cannot be ideally objective or absent of belief; to claim otherwise simply offers an example of the Feynman observation that people are easy to fool, and the easiest person to fool is yourself (Feynman 1974).

Rather, productive science requires that the believing and nonobjective scientist first understands what those beliefs and biases are. Science then becomes the exercise of managing those. Self-management is preferred, but if that fails, science has a mechanism for the community to do the needed management. This allows science to succeed even though it is done by scientists.

3.1.2 The Logic Of Plausibility

We outline here what decisions guide our assembly of a prebiological geo-origins model. Here, we must manage the word "plausible," which appears in phrases such as "this organic compound is prebiotically plausible." "Plausible" in the field of origins has come to have a cultural definition, which in turn does little more than expressing the desire of a scientist to have the compound available in a prebiotic environment.

A survey of the contexts where "plausible" is used is beyond the scope of this review, but some examples are useful. For example, if a molecule can be seen by microwave spectroscopy of a nebula in our galaxy where stellar formation is active, then it is often said to be "prebiotically plausible." This is the case for cyanoacetylene and glycolaldehyde, two relatively complex molecules in addition to primitives such as water, methane, hydrogen cyanide, and carbon dioxide.

Assuming that molecules found in interstellar gas clouds were available on a prebiotic Earth is not entirely defensible. As discussed below, during both the initial accretion of the Earth and in the subsequent Moon-forming event, essentially none of these molecules would have survived in any form more complex than their constituent atoms.

More defensible would be the view that organic materials are delivered to Earth by meteorites, including the carbonaceous chondrites that continue to fall on Earth. These organics have been the focus of much analysis. Further, they arrived continuously following the cataclysmic events in the natural history of Earth, did not experience extreme heating upon arrival, and could indeed have offered a reservoir of organic material for prebiotic synthesis. They may be used to define what was "plausibly" available by way of organic species on early Earth.

Concerned about the amounts and distribution of these, we take a narrower view. *Here, we chose to consider an organic species as "plausible" only if it can be generated by Earth-based processes.*

3.1.3 The Correspondence Principle

The Correspondence Principle in origins is analogous to the Correspondence Principle in physics (Van Vleck 1928), which holds that as one moves from the quantum to the macroscopic, the predictions of quantum mechanics must converge on the predictions of classical mechanics. In the field of bio-origins, the analogous Correspondence Principle starts with modern life as the end point. It then extrapolates modern biological chemistry back in time to ancient biological chemistry (Benner et al. 1989). That extrapolation, supported by paleogenetic experiments (Jermann et al. 1995; Benner et al. 2002; Benner 2017b), suggests that the last common ancestor of all life that we know had these features:

1. It used RNA in key catalytic reactions (such as in the biosynthesis of proteins).
2. It exploited RNA cofactors in a metabolically diverse metabolism that included carbon-carbon bond forming reactions, oxidation reactions, reduction reactions, methyl transfer reactions, and phosphoryl transfer reactions.
3. It used triphosphates as activated species to drive reactions that are thermodynamically unfavorable in water.

The Correspondence Principle here would argue that the origin of life coming from unknown chemistry produced RNA, RNA cofactors, and triphosphates at the outset. This would allow the uninterrupted emergence of the molecules that we observe in modern life from the unknown model that we are building for its origins. Further, this Correspondence Principle holds that the form of RNA that originated is the same as the RNA present in the most distant ancestor whose RNA we can infer the structure of. According to the Correspondence Principle, the RNA that emerged from the origins contained bridging oxygens (not nitrogens) in the phosphate backbone, which was built from four nucleotides (not six or more) (Benner et al. 2016) that included thymine (not 2-thiothymine), and the backbone was linked by way of the 3′- and 5′-oxygens, inter alia.

For example, many consider replacing ribose by a four-carbon sugar such as threose, perhaps because it is easier to make prebiotically than ribose (Orgel 2000b; Herdewijn 2001; Ichida et al. 2005). Hud and others consider alternative hydrophobic organic molecules as "molecular midwives" that, because of their intrinsic hydrophobic nature, help assemble a stack of nucleotides (Hud et al. 2007). Some time ago, Orgel all but abandoned triphosphates as high-energy building blocks for RNA because they did not polymerize well, replacing the pyrophosphate leaving group with 2-methylimidazole (Inoue and Orgel 1982).

In many cases, changing the target structure away from the RNA suggested by the Correspondence Principle creates something of a "mission creep." For example, the Szostak laboratory, seeking to finally solve the half-century old problem of template directed "uncatalyzed" synthesis of RNA, moved from the methylimidazole of Orgel to an aminoimidazole leaving group (Li et al. 2017).

They also replaced the 3′-oxygen of modern ribose by a 3′-nitrogen (Izgu et al. 2016) and got better performance from 2-thiothymidine instead of standard thymidine (Zhang et al. 2013). Some spectacular examples of success have emerged. However, much of the work seeking prebiotic routes to the modern RNA building blocks must be set aside, as 3′-amino-2-thiothymine ribonucleoside requires a different prebiotic synthesis "concept" than the ones that we have been working so hard to create.

Thus, targeting the exact structure of modern RNA is useful simply because it is a clear target. For example, considering some of the most productive collaborations in this field, the Sutherland laboratory, supported by Simons philanthropy, attempts to find ways to make nucleosides such as cytidine and thymidine, not thiothymidine, with oxygens on the 2′-, 3′-, and 5′- positions, not nitrogens, and with ribose as the sugar, not threose. Analogous statements can be made by the recent work of Carell, which generated adenosine attached to ribose by N-9, exactly as found in the RNA extrapolated for the most ancient life on Earth, not some other purine attached in some other way (Becker et al. 2016). Both laboratories consider their chemistry to be successful when it delivers these exact molecules.

In this review, we embrace the Correspondence Principle, in part because alternatives are being heavily explored by other laboratories, in part because we appreciate the clarity of the goal. Our model must deliver RNA building blocks exactly as they are found in modern life, put phosphates onto the correct positions, and condense to give the natural RNA, preferably initially, but certainly as an end product.

3.1.4 The "RNA First Hypothesis"

So, why is RNA our target? In part, it is because of the results cited above that arise if we extrapolate modern molecular biology back in time to infer the molecular biology of the last common ancestor of all life on Earth. However, the RNA First Hypothesis (RFH) was introduced by Alex Rich (1962) for other reasons. Rich saw this as a way to manage two problems in particular:

(a) The "chicken or egg" problem, from terran molecular biology: Proteins are needed to make nucleic acids, while nucleic acids are needed to make proteins.
(b) A lack of molecular mechanisms by which proteins might transfer their information to descendant proteins.

Rich proposed that if RNA had catalytic power, an RNA system could pass its information to descendants by RNA catalysis. This is the strong version of the "RNA World Hypothesis," which holds only that an earlier episode of Darwinism on Earth (not necessarily the first) used RNA as its only genetically encoded component of biological catalysis. This hypothesis is supported by the ribosome and RNA cofactors, inter alia. In 1989, the Benner lab built a model for a metabolism of the last riboorganism based on an analysis of modern biochemistry (Benner et al. 1989).

The discovery by Cech, Altman, and others that RNA could catalyze the splicing of introns and the cleavage of pre-tRNA molecules strengthened this model (Cech and Bass 1986; Guerrier-Takada et al. 1983). RNA can catalyze reactions. Further, the structure of the ribosome showed what had been suspected from "wet" biochemistry years earlier: The machine that makes proteins across the known biosphere uses RNA as the catalyst (Cech 2000).

We have remarked on the challenges of obtaining a single biopolymer that can do both genetics and catalysis (Benner 2017a). A catalytic molecule *must* fold, so as to surround its substrates and its transition state. A genetic molecule should *not* fold, so that it can template the formation of its complement. A catalytic molecule *should* change its behavior rapidly with small changes in sequence so that it can effectively search sequence space. A genetic molecule should *not* change its behavior with changes in sequence, so as to not disrupt the biophysics that allow faithful transmission of genetic information. A catalytic molecule should have *many* building blocks, so that they can catalyze many different kinds of reactions. A genetic biopolymer should have *few* building blocks, as this allows for the highest fidelity of information transfer.

RNA is a biopolymer that appears to make an effective compromise between these competing molecular attributes. Proteins, by way of contrast, do not. Accordingly, RNA is a reasonable target biopolymer to first support Darwinism. This, in turn, allows us to reserve for later the prebiotic synthesis of amino acids and peptides. This reservation has strategic value, as amino acids and peptides are both more stable than RNA and RNA components, are easier to form abiotically, and are therefore far less challenging (and less interesting) targets.

3.1.5 What Makes a Model Persuasive

Accordingly, our goal is to construct a model for a process that converts compounds that were likely to have been available on early Earth, in environments that are likely to have been available on early Earth, and to support this by experimental work. We next must ask what we should consider as a "solution to the problem. When will these experiments end (Galison 1987)? What form must a proposed model have to be convincing to the communities that the problem has been "solved"?

This is, of course, a matter of culture, not logic. Here, however, we must also make a choice about what metrics we will use for progress. *In this work, we have chosen the metric to be the number of steps that require human intervention as the process is performed in the laboratory, which translates into the continuity of the process.* Of course, for the analogous process to work as a solution in the natural environment, no intervention at all is allowed.

3.2 The Planetary History

3.2.1 Choices Made Concerning Planetary History

Any model for the origin of life in a geological context requires that we make choices about what geology was available at the time that life emerged. A decade ago, too few constraints were available for either the geology of early Earth or the time when life emerged to have a productive discussion. However, today, the situation is changed. We rely on the preponderance of evidence to make these choices. We do not, because we cannot, rely on any concept of "proof." Rather, we simply recognize that our view of the availability of prebiotic resources is quite different depending on those choices. It is preferable to make them explicitly than implicitly, writing out the evidence that we rely upon to make those choices.

3.2.1.1 Deciding What Cosmogenic Factors to Build into the Model

Several cosmogenic events are of special importance when we consider the possible geological context on early Earth where RNA might have emerged. We consider these individually.

Initial Accretion

It is generally accepted that the Earth was formed stepwise, with material from the pre-solar cloud of gas gravitationally collapsing to form "embryonic" planets having approximately the size of the Moon or, later, Mars, which then gravitationally accreted. While the specific material that formed the Earth can no longer be analyzed as such, meteorites that (until recently) escaped accretion are generally used as proxies for this material.

Interestingly, the Earth and its Moon both have isotopic compositions that are indistinguishable from enstatite meteorites (Javoy et al. 2010; Qin et al. 2010; Warren 2011; Zhang et al. 2012; Dauphas et al. 2015; Dauphas and Schauble 2016; Young et al. 2016). Enstatites are very reduced, with their iron being present as a metal or as a sulfide; essentially no iron oxide is present. Enstatites are named for their dominant mineral enstatite ($MgSiO_3$), which is the end member of a series of pyroxene silicate minerals that grade to ferrosilite ($FeSiO_3$). In a ratio of 9:1, this is a dominant mineral of the Earth's mantle, including its peridotite, which features prominently in discussions of the origins of life. It is also observed with forsterite (Mg_2SiO_4) in spectra obtained from outside of the solar system (Molster et al. 2001).

These isotopic similarities between enstatites and Earth include ^{17}O, which is the same in both the Earth and the Moon. However, these similarities also extend to isotopes that record different stages of the accretion of the Earth, including the isotopes of lithophilic ("rock loving") elements such as oxygen, calcium, titanium,

Fig. 3.1 Asteroid "21 Lutetia" appears to be an E-type chondrite of the type that dominated Earth's early accretion. This image was obtained by ESA's Rosetta mission. Source: ESA 2010 MPS for OSIRIS Team MPS/UPD/LAM/IAA/RSSD/INTA/UPM/DASP/IDA—https://www.flickr.com/photos/europeanspaceagency/4781143008/

and neodymium; the moderately siderophilic ("iron loving") elements chromium, nickel, and molybdenum; and strongly siderophilic elements such as ruthenium. This isotopic similarity suggests that the material that accreted to create the Earth was always dominated by "E-type" enstatite chondrites. These contributed perhaps half of the accreting material in the first 60% of the accretion, with the remaining accretion being essentially all enstatites.

This model has the advantage of allowing the impactor that formed the Moon to have essentially the same isotopic ratio as the proto-Earth that it impacted. This, in turn, relaxes the constraints of models for lunar formation (Fig. 3.1).

Formation of the Moon

Many lines of evidence suggest that the Moon was formed by the impact on a growing Earth of a Mars-sized body that led to substantial vaporization of the rocks. The Moon congealed from this vapor. Because the vaporization was planetwide, little doubt exists that the origin of life must postdate the formation of the Moon. Analysis of the surface of the Moon, which has been preserved in composition (ignoring the time during which it was reworked by continuing impacts), is likely to give us as well an idea of the surface composition of the Earth immediately after the Moon was formed. Especially important are the potassium-rare earth element-phosphate (KREEP) species on the Moon.

The redox state (in geologist terms, the oxygen "fugacity" of the Hadean mantle) has been constrained by the analysis of zircon ($ZrSiO_4$) crystals that have survived from the Hadean. Trail and his coworkers measured the ratio of Ce^{3+}/Ce^{4+} in these zircons (Trail et al. 2011). They found that the mantle in which they were formed had an oxidation state similar to the oxidation state of the modern Earth. This corresponds to a fugacity near that of the fayalite-quartz-magnetite (FQM) redox state. This represents essentially complete loss of Fe^0 to the core by this time. Other evidence suggests that surface water was also available deep in the Hadean.

The "Late Veneer" (LV)

When the metallic iron sank to form the core of the Earth, it took the siderophilic elements with it, leaving behind a crust and mantle that were depleted in these elements relative to chondrites (Becker 2006). Thus, the mantle and the crust are far more oxygenated than the core, are dominated by silicate, and have a higher redox potential, as measured by the fugacity of oxygen.

However, a puzzle arises because the abundances of siderophilic metals in the mantle and crust are much greater than predicted by experimental data that measure the partitioning of siderophilic elements between liquid iron and silicate (Mann et al. 2012; Rubie et al. 2015a, b). Indeed, the relative amounts of these siderophilic elements in the terrestrial and Martian mantels are nearly the same as those in chondrites (Brasser and Mojzsis 2017).

One solution to this puzzle is the assumption that after core formation was complete [~4.45 Ga (Allègre et al. 2008)], small-scale accretion [an additional 1% (weight)] delivered more siderophilic elements to the crust under conditions that did not give them the opportunity to descend into the iron core. This "late veneer" period remains controversial (Willbold et al. 2011, 2015; Touboul et al. 2012, 2014; Puchtel et al. 2014; cf. Righter et al. 2015) but is supported by the preponderance of evidence.

The amount of the siderophilic elements that made it to the Moon appears to be smaller, perhaps only 0.02 to 0.035 weight percent (Day and Walker 2015; Kruijer et al. 2015). This means that the Earth, relative to the Moon, got 1950 $\pm$ 650 times more of these siderophilic elements than would be expected based simply on their relative gravitational cross sections.

To explain this, various authors have suggested that the late veneer came from a very few, mostly large, later impacts, ranging in size from the asteroid 4 Vesta (535 km diameter) to our Moon. An impactor of this size would itself have had an iron core, with its siderophilic elements separated from its surface silicate shell.

Thus, as long as the iron core of the impactor did not hit the Earth "square on" (where its core would simply have joined Earth's), the impact would have fragmented the impactor into smaller particles, shearing the core of the impactor into fragments of molten iron that would descent "in a hail" to the surface of the Earth (Brasser et al. 2016; Genda et al. 2017), together with its siderophilic elements. This accounts for the amounts of siderophilic elements "polluting" Earth's mantle/crust and the relative paucity of these on the Moon. A glancing blow at 45 °C is, of course, at the center of the probability distribution of impact angles.

Why is this important to prebiotic chemistry? The delivery of molten iron to the Earth's atmosphere after the core had formed, the initially accreted Fe^0 had been lost to the core, the mantle had achieved a FQM redox state, and the oceans had formed, would give a pulse of reducing power in the atmosphere of the early Earth. The reaction between iron and water produces dihydrogen, and Genda et al. (2017) estimate that 90 bars of H_2 would be produced for an impactor having the size of the Moon. Ultraviolet radiation from a young Sun would erode that H_2 over a period of ~200 million years (Genda et al. 2017; Hamano et al. 2013).

This model appears to be supported by the preponderance of evidence, and we can constrain our prebiotic chemistry models using it. It has the prebiotically important consequence of creating an atmosphere that, at least for a time, is more reducing than the mantle and crust, and more reducing than the volcanic gasses that might emerge from the mantle and crust at a FQM redox state. This planetary thermodynamic disequilibrium can be exploited. An atmosphere with substantial amounts of H_2 (and, therefore, some methane and ammonia) is a productive source of hydrogen cyanide (HCN), cyanamide (HNCNH), cyanoacetylene (HCCCN), and other species that we will incorporate into our model for the prebiotic generation of RNA. At the same time, it allows minerals to have higher oxidation states, including silicates, borates, phosphates, and molybdates.

This productive opportunity, however, is short-lived (if 200 million years is seen as a short period of time). For prebiotic chemistry to exploit it, it must occur within less than 200 million years after the impact that created the late veneer.

The Late Heavy Bombardment (LHB)

There is little doubt that as time went forward, the number of impacts on the Earth decreased as its local region of the solar system was "clean swept" by the Earth itself. What remains unclear is whether this first-order exponential decreasing of impacts was interrupted by a spike in impacts 3.9 to 4.1 Ga. Originally proposed in 1974 based on an analysis of lunar rocks collected by the Apollo mission (Tera et al. 1974), this "late heavy bombardment" (LHB) has been revisited often in the past four decades (Kring and Cohen 2002; Morbidelli et al. 2012; Marchi et al. 2013).

A LHB is troubling for prebiotic chemists, as it implies that the prebiotic sphere had less time to generate life than the full lifespan of the Earth. If the late bombardment of the Earth was heavy enough, it could have sterilized the planet (Maher and Stevenson 1988). Failing that, a late heavy bombardment would have forced the proto-biosphere into refugia that constrain the nature and diversity of its metabolism (Abramov and Mojzsis 2009).

This problem becomes only worse as carbon isotope ratios indicative of life are found in increasingly more ancient strata. Indeed, if the most ancient life-indicating $^{12}C/^{13}C$ ratios are accepted at face value at the dates as early as 4.1 Ga (Mojzsis et al. 1996; Rosing 1999; Bell et al. 2015), life would have effectively had to arise on Earth instantly after any late heavy bombardment, and possibly because of it.

However, a critical review of the literature finds that the evidence for the late heavy bombardment is equivocal, at best. The originally proposed "terminal lunar cataclysm" was based on 18 samples collected by the Apollo mission from a lunar highland where the U-Pb fractionation and rubidium-strontium isochrones dated them at ca. 3.9 Ga. These samples were interpreted as having come from "widely separated areas," therefore implying multiple large impacts well after the formation of the Moon.

This view was immediately contradicted (Hartmann 1975). The contradiction noted that the few samples were not from widely separated areas. Further, it was noted that older lunar rocks would be hard to find based on the apparent history of cratering on the Moon. More recently, Chapman et al. reviewed this and other evidence, pointing out that the samples were likely to be strongly biased against older examples (Chapman et al. 2007).

Nevertheless, the late heavy bombardment remains a staple in the thinking of prebiotic chemists whose focus lies outside of geology. For example, the Sutherland group references it as contributing to several features of their prebiotic organic chemical models (Sutherland 2016).

Here, we choose a model that holds that the number and intensity of impacts declined smoothly over time, without any noticeable paucity of those impacts before 4.1 Ga, and no spike in those in the 4.1–3.9 Ga. In part, this is because of an absence of data to contradict the more reasonable model of monotonic decline. In particular, we find the analysis of Boehnken and Harrison (2016) sufficiently persuasive for us to discount the $^{40}Ar/^{39}Ar$ "plateau" dates; we see it likely that diffusion of this noble gas under repeated reworking of the lunar surface would create the "illusion" of a late bombardment. These $^{40}Ar/^{39}Ar$ data stand behind a number of current arguments for the LHB. Here, Moon-derived meteorites, which are presumed to come from all around the Moon, were exploited (Cohen et al. 2000; Kring and Cohen 2002; Marchi et al. 2013).

3.2.2 The Geological Consequence of the Model Derived from These Choices

We then turn to some interrelated questions that are addressed by this model for the early Earth that have relevance to prebiotic processes for the formation of RNA:

(A) What was the global inventory of free water on Earth relative to the global topography? Specifically, did they permit dry land?
(B) Related to this, what was the degree and nature mixing of materials within and between the crust and mantle, perhaps different from modern plate tectonics?
(C) Related to this, what was the redox potential of the global crust and the global mantle, and the consequent composition and redox state of volcanic gasses and magma?
(D) Related to this, what was the redox potential of the global atmosphere?
(E) Related to this, what mineral species were available globally?

All of these questions have time-dependent answers; the states of the atmosphere, crust, mantle, topography, and other parameters were all likely different 4.5 Ga than 3.5 Ga, when life presumably had become a historical fact. However, because they ask about global averages, they are not complex questions. They are difficult to answer only because theoretical models for planetary accretion that might allow us to work "forward in time" are poorly constrained, physical records from the Hadean are largely lost, and "backwards in time" approaches from the biosphere become ambiguous well before the first living organism can be modeled.

Several classical views provide answers to these questions. The first argues that the inventory of water on Earth was high enough to threaten a flooded planet, especially if plate tectonics and other global processes had not created a substantial topography. Here, the classical answer held that plate tectonics took hundreds of millions of years to begin and may not have begun in the modern sense of the term until quite late, perhaps even requiring billions of years to get started (Condie 2018). This encourages the possibility that the early Earth was a "waterworld." This, in turn, threatens the loss by dilution into a global ocean of any progress making RNA under prebiotic conditions.

Here again, we rely on data from crystals of Hadean zircon. Zircon is an exceptionally durable mineral that retains its primary chemistry with respect to most of its elemental constituents and isotopes from the time of its crystallization (Cherniak et al. 1997). This includes uranium and thorium, which are easily incorporated into the zircon crystal, while their radioactive daughter product lead atoms are not. This allows relatively precise dating of zircon crystals.

These zircons retain also the oxygen isotopes from the reservoir in which they are formed, and inclusions. Together, they suggest that the Earth did contain at least some dry land. *Our model uses this dry land as a key locale to form RNA.*

3.2.3 What Organic Species Are Available from the Atmosphere?

Dating back again to the time of Stanley Miller, many scientists have identified experimental conditions in which prebiotically interesting small organic molecules might be obtained by spark discharge, silent electrical discharge (Löb 1913), ultraviolet irradiation, or other energy sources. These are summarized in Table 3.1.

These small organics are believed to be useful to create RNA precursors. For example, purine and pyrimidine nucleobases may be formed from hydrogen cyanide (HCN) (Sekimoto and Takayama 2012) and NH_4CN (Miyakawa et al. 2002), cyanoacetylene (Sanchez et al. 1966; Ferris et al. 1968), and cyanamide ($NCNH_2$) (Sanchez and Orgel 1970; Levy et al. 1999) [but see critical evaluation in Shapiro (1995, 1999)]. These, in turn, are formed by lightning or UV in atmospheres containing CO_2, N_2, and H_2O (Yuasa et al. 1984; Bada et al. 2016). Indeed, a Georgia Tech group developed an undergraduate laboratory teaching recipe for

Table 3.1 Prebiotic atmosphere required to generate the following compounds

Atmosphere-generated	Downstream prebiotic	Role in RNA formation	Other roles in prebiotic chemistry	Atmosphere components	References
HCN	Formamide	Nucleobases in RNA	Formamide to solve "water problem"	N_2, CH_4	Sekimoto and Takayama (2012)
HCHO	Carbohydrate	Ribose in RNA		H_2O, CO_2	Cleaves (2008)
HNCNH	Urea	Nucleobases in RNA	Solvent to solve the "water problem"	N_2, CH_4	Miyakawa et al. (2002)
HCCCN	Cyanoacetylene	Nucleobases in RNA		N_2, CH_4	Sanchez et al. (1966), Ferris et al. (1968)
O=C=S	H_2S	None known	Peptide condensation		Leman et al. (2006)

making prebiotic adenine from these (using diaminomaleonitrile instead of HCN) (Anumukonda et al. 2011).

Many modern models for the prebiotic origins of RNA components rely on these molecules. For example, Sutherland's model for the prebiotic synthesis of cytidine requires cyanoacetylene (Powner et al. 2009). The Saladino model for the prebiotic formation of nucleobases depends heavily on the production of substantial amounts of formamide, which in turn requires large amounts of HCN to be formed (Saladino et al. 2003, 2011). Our model relies on formamide and/or urea as a solvent/reaction environment to solve the "water problem," which reflects the instability of many bonds in RNA with respect to hydrolysis in water (Neveu et al. 2013).

The amounts of small organic molecules produced depend very much on the extent to which the atmosphere is reduced. For example, the production of HCN drops rapidly as the overall oxidation potential of the gas mixture increases (Pearce et al. 2017; Pearce and Pudritz 2015). Thus, the amounts of HCN required for various of these models are simply not available if the atmosphere has a redox potential that corresponds to the FQM redox potential of a mantle. *We chose to manage this problem by relying on the model outlined above for the formation of the late veneer, which allows that atmosphere to be reducing enough to enable these RNA precursors to be formed in it.*

3.2.4 Local Variation in Mineral Species Available from the Mantle and Crust

As discussed below, valuable minerals for our models for the prebiotic synthesis of RNA are obtained from a mantle having an FQM fugacity. In some cases, as with silicate and borate, the oxidation state comes naturally; elemental silicon and elemental boron are not likely to be found in any mineralogical environment other than iron metal. In some cases, the useful oxidation state seems accessible even in a more reducing environment. This includes ferrous iron (Fe^{2+}) and phosphate, although a lower oxidation state of phosphorus is key to its mobilization by the mechanism proposed by Matthew Pasek (see below).

Global statements about planetary geology are not sufficient, however. More complex and, unfortunately, likely more relevant to the origins of life are questions that ask about the potential for early Earth to have *diverse* local environments. Even the simplest models for the emergence of RNA involve multiple models that have differing access to water, different redox potentials, and locally distinctive minerals. For example, the discontinuous model for the synthesis of RNA (Neveu et al. 2013) proposes:

1. To create simple organic molecules (formaldehyde, hydrogen cyanide) in an atmosphere containing at least some methane, substantial carbon dioxide, and water.
2. Then, have them rain into another environment, where they are transformed by percolation through a different environment, alkaline basalts containing olivine, igneous tourmaline, and igneous apatites (Piccoli and Candela 2002), followed by runoff water carrying their product to a different environment.
3. Evaporate basins under an atmosphere rich in carbon dioxide and volcanic gasses, where borate, phosphate, and sulfate minerals are present.

Sleep pointed out that this change does not create a Shapiro-objectionable requirement for human intervention (Sleep et al. 2011). Except for the atmosphere, environments (2) and (3) are known on modern Earth near each other geographically.

Further, material easily moves from one environment to the other by gravity, even on modern Earth. For example, water percolating through rocks that contain olivine and tourmaline in the Sierra Nevada creates serpentinization, producing as much as 9.1 μM/kg per hour (200 °C, 300 bar) or 0.3 mmol/kg total inorganic carbon (Jones et al. 2010). The water flows out of these rocks into Death Valley, where borate from eroding tourmalines gives abundant borate minerals in the evaporite. Such a complex environment also appears to have been present on Mars, where remnants of evaporite minerals include gypsum facie containing borate minerals (http://www.manyworlds.space/index.php/(2016/12/14/with-the-discovery-of-boron-on-mars-the-package-of-chemicals-needed-for-life-is-complete/). If we can establish that such diversity in environments could have been present during the Hadean, the prebiotic chemist can go to work.

The possibility of proximal mineralogical diversity is key to prebiotic chemistry. Molecular systems do not see a planet. Indeed, nine orders of magnitude separate the nanometer dimensions of a molecule from the meter dimensions of a typical rock formation. However, no matter how difficult it is to model "global" properties of the Hadean, modeling local environments is worse. That modeling must begin with a picture of the Hadean Earth as a whole, suffer all of its ambiguities, and then tackle spatial variation, a large task indeed.

3.2.5 *When Is a Mineral Considered Impossible*

As interesting as a discussion about what minerals might be present is a discussion that guides us to expect that certain minerals might be *excluded* from the early Earth environment. For example, Hazen et al. (2008) have proposed various natural histories of mineral evolution. The idea is driven in part by empirical "mineral counting" in facie of different ages with various attempts at mitigating "preservation bias" (the fact that some minerals are harder, less soluble in water, or more likely to survive eons of geological time), sampling bias, or other factors that make it appear as if fewer minerals were present long ago than today. It is also driven by theoretical models that identify a dozen preplanetary refractory mineral species ("ur-minerals") that might have been present on early Earth.

These authors have divided Earth's history into three eras and ten stages of "mineral evolution." Each has been proposed to have seen significant changes in the planet's near-surface mineralogy, including increases in the number of mineral species; shifts in the distribution of those species; systemic changes in major, minor, and trace element and isotopic compositions of minerals; and the appearance of new mineral grain sizes, textures, and morphologies. These changes include changes that have been affected by the biosphere, including its production of atmospheric dioxygen.

We consider the likelihood for an early environment to contain specific minerals in the discussion below, each time that mineral is proposed to be part of a process that mitigates problems in the prebiotic chemistry of RNA. Interestingly, at the 2017 ISSOL meeting (the XVIII International Conference on the Origins of Life), Hazen introduced the "Benner rule" to mitigate the severity by which specific mineral species might be rejected through their incompatibility with the broader picture of mineral evolution. In this rule, Hazen noted that if Benner, or some other prebiotic chemist, found exceptional value in the reactivity of a mineral that was excluded by a global view of mineral evolution, he would consider revising that view. This implies that, given sufficient reasons to look for it, the mineral will actually be found.

3.3 Classical Literature on Minerals

Given these choices about the natural history of Earth, its atmospheric history, and the history of its crust and mantle, what can minerals do to help the prebiotic chemistry research based on the Correspondence Principle and the RNA First Model for the origins of Darwinism on Earth?

In 1947, two distinguished authors mentioned that clays might have been involved in early prebiotic chemistry on Earth as concentrators and catalysts. Unfortunately, these were both "false starts." Thus, Bernal's mention of clay was casual (Bernal 1951) and he never developed the suggestion. Goldschmidt, the founder of geochemistry, prepared a more elaborate analysis independently, published in 1952 after being edited by N. W. Pirie (Goldschmidt 1952). The article is prescient, mentioning the geological ubiquity of carbon dioxide, the chirality of quartz, and the possible prebiotic relevance of micas, carbonates, and clays. Unfortunately, Goldschmidt died before he could develop his insights.

Indeed, experiments from the classical period of prebiotic chemistry rarely began with a mineralogical context. Instead, these experiments focused on the organic molecules that might have emerged from the atmosphere. Löb (1913) examined an atmosphere model whose composition contained CO_2. Miller (1953) was more influenced by the composition of the atmosphere of Jupiter and therefore examined more reducing environments. Oró provided an extensive review of these classical experiments (Oró 1965), all without any detailed discussion of minerals.

3.3.1 Classical Prebiotic Chemistry Involving Minerals: Clays

The first modern effort to bring geology into prebiotic chemistry also began with clays, but in roles more exotic than as concentrators and catalysts. Cairns-Smith mused on the challenges of obtaining biopolymers out of prebiotic soups, and commented favorably on the self-organizing properties of minerals. He then suggested that clays might themselves support Darwinism, replicating themselves. In this replication, the defects in their structures would hold information that is transmitted to descendant clays, with errors, where those errors could be further transmitted. This, he suggested, would allow a mineral species to evolve without a genetic biopolymer (Cairns-Smith 1977, 1982, 2005).

The Cairns-Smith approach proved, again unfortunately, to be a false start. No working system has yet emerged where minerals in two or three dimensions replace linear biopolymers to support Darwinism. However, it is not clear that much effort has actually been devoted to explore this possibility. Nevertheless, Cairns-Smith's introduction of the term "genetic takeover" has had considerable impact, for example, on heroes in the field like Albert Eschenmoser (1997). The view, of course, requires renunciation of the Correspondence Principle.

James Ferris returned to explore the less exotic possible roles of clays in prebiotic chemistry in the decades to follow. His focus was montmorillonite $[(Na,Ca)_{0.33}(Al,Mg)_2(Si_4O_{10})(OH)_2{\cdot}nH_2O]$. This clay arises from the aqueous erosion of igneous rocks containing feldspars and therefore might have occurred on early Earth. Perhaps the most interesting outcome of his work was the observation that montmorillonite catalyzes the assembly of RNA building blocks having a 5′-phosphate activated by being pre-conjugated to an imidazole. Again, this imidazole was not the triphosphate that the Correspondence Principle would lead us to prefer. However, it is among the best examples for the formation of long RNA molecules without direction from a template (Ferris and Ertem 1992, 1993; Ertem and Ferris 1996, 1997; Ferris 2005).

Clay has allowed the coupling of RNA building blocks in other experiments. For example, Burcar et al. (2015) observed clay-based polymerization of RNA building blocks in the presence of a water-soluble carbodiimide reagent (1-ethyl-3-(3-dimethylaminopropyl)carbodiimide) that creates the imidazolide in water.

Montmorillonite has been explored by the prebiotic community as an adsorbent. For example, montmorillonite can bind to and adsorb macromolecular RNA, including catalytic RNA, protecting it from degrading agents and in some cases even enhancing its catalytic activity (Biondi et al. 2007a, b; Stephenson et al. 2016). The interactions between clays and liposomes have been studied (Hanczyc et al. 2003). Adsorption onto clays may protect other biomolecules (Ertem et al. 2017), including peptides (Lahav et al. 1978).

3.3.2 Classical Prebiotic Chemistry Involving Minerals: Pyrites and Other Sulfides

A separate line of research involving minerals in classical prebiotic chemistry focused on various transition metal sulfide minerals, whose redox potential allows them to transform organic molecules. This was popularized in the 1990s by Günter Wächtershäuser with a focus on pyrite. Here, ferrous ion (in the +2 oxidation state, e.g., FeS) reacts with hydrogen sulfide to form pyrite (FeS_2, with iron in its +4 oxidation state) as it delivers electrons that can reduce organic species (Wächtershäuser 1988a, b, 1990a, b, 1993; Blöchl et al. 1992).

Huber carried this concept further by considering how this process on iron-nickel sulfide surfaces might activate amino acids to form peptides (Huber and Wächtershäuser 1997, 1998). Russell and his associates took this concept further, proposing that sulfide mineral assemblages near ocean floor hydrothermal vents might be venues for the origin of Darwinism (Russell et al. 1994; Russell and Hall 1997).

For the most part, experimental work with pyrites and other sulfides has delivered only simple compounds, such as acetate, methane, and ammonia (Brandes et al. 1998). However, many of these models, theoretically elaborated, lead to a "metabolism first" view for the origins of life. Here, rather than seeking to generate a genetic biopolymer by prebiotic chemistry, they propose to get a compositional form of Darwinism working through the reactions of assemblies of molecules (Markovitch and Lancet 2014). This work continues in prebiotic chemical research in the current millennium (Huber et al. 2003, 2012; Cody et al. 2000, 2001, 2004; Cody 2004, 2005).

At one level, we have no quarrel with the "metabolism first" concept. Even models that require a biopolymer to give biology its first access to standard Darwinism require a preceding set of connected organic transformations to generate the precursors for that biopolymer. The borate-molybdate formose process (see below) may be one example (Kim et al. 2011). However, as discussed below, such systems often "leak," losing material nonproductively to give (again) tar. Further, it is difficult to conceive real chemical systems that are, by "compositional inheritance," able to evolve to capture more resources to the exclusion of other systems (Anet 2004).

We have chosen not to pursue a "metabolism first" model in the work reviewed here. Nevertheless, we continue to be interested in "metabolism first" models that suggest actionable chemistry with specific organic molecules that might create "compositional inheritance." We have frequently offered to synthesize these molecules and deliver them to anyone wishing to test the actionable idea. We renew this offer here.

3.3.3 *Classical Prebiotic Chemistry Involving Minerals: Silicates, Oxides, and Others*

In a third independent line of work, classical prebiotic chemistry proposed roles in the origin of life for silica-containing minerals. Quartz (SiO_2), the basic silica mineral, was especially interesting because it forms chiral crystals. Adsorption of prebiotic molecules onto these was suggested as possibly being the origin of chirality in biomolecular chemistry, a suggestion mentioned by Goldschmidt (Bonner et al. 1974, 1975; Evgenii and Wolfram 2000).

The surfaces of more complex silica minerals have been suggested to be useful as containers, concentrators, activators, and stabilizers. This includes feldspars (Parsons et al. 1998).

Zeolites, with their large water content and large pores, have also been discussed (Smith 1998; Smith et al. 1999).

In still another line of research, the serpentinization of iron magnesium silicates is central to many models for early Earth chemistry, both on the land and beneath the ocean. Analogous to the Wächtershäuser pyrite formation, this serpentinization provides reducing power in many models for the reduction of CO_2 and the formation of ammonia from a variety of oxidized nitrogen species (Berndt et al. 1996; McCollom and Seewald 2001; Summers and Chang 1993). Mineral oxides and hydroxides have also been discussed, again as concentrators and catalysts (Holm et al. 1993; Weber 1995).

3.4 How to Tackle the RNA First Problem Using Mineralogy

Several excellent reviews capture the state of affairs at the start of the current millennium (Lahav 1994; Orgel 2000a; Schoonen et al. 2004). Minerals do well at providing small organic molecules, but few that are needed to manage the problems of the RNA First Hypothesis for the origins of Darwinism.

Indeed, at a Gordon Conference on the origin of life in 2002, Stanley Miller remarked that, in his view, very little progress had been made in this direction. This was, of course, just a few years after he had published a study measuring the instability of carbohydrates, an instability that had driven him to conclude that "stability considerations preclude the use of ribose and other sugars as prebiotic reagents . . .. It follows that ribose and other sugars were not components of the first genetic material. . ." (Larralde et al. 1995). This study dampened enormously the enthusiasm of the community for the "RNA first" hypothesis for the origin of life on Earth.

This contrasts with the enthusiasm that dominated the field just 25 years earlier. For example, in 1969, Burton et al., citing Calvin, declared that "the successful 'chance' synthesis of nearly all the biologically important monomeric substances (amino acids, sugars, fatty acids, even purines and pyrimidines) *has been accomplished*" [italics added] (Calvin and Calvin 1964; Burton et al. 1969).

Instead, the situation seemed grim at the start of the new millennium. Nowhere is this stated more directly than by Bregestovski (2015). After noting that the model is the "most accepted and widespread contemporary scenario of prebiotic evolution that led to the emergence of the first cells on our planet," Bregestovski remarked that the hypothesis "suffers from a number of *insurmountable* [italics added] problems of chemical and informational nature". Among these, he included the

(a) unreliability of the synthesis of starting components;
(b) catastrophic increasing instability of the polynucleotide molecules as they elongate;
(c) exceedingly low probability of meaningful sequences;
(d) lack of the mechanism that would generate membrane-bound vesicles able to divide regularly and permeable to the nitrogenous bases and other RNA components;
(e) absence of driving forces for the transition from the 'RNA world' to the much more complex 'DNA-RNA world.'

The conclusion was captured in the title of the paper: "RNA World, a highly improbable scenario of the origin and early evolution of life on Earth." Or, as Joyce and Orgel remarked in 1999, RNA is "the prebiotic chemist's nightmare" (Joyce and Orgel 1999).

3.4.1 Paradoxes

One way to direct research in a field that approaches "big questions" (Benner 2009, 2013; Malaterre 2013) does not ask a historical question ("How did life arise in the solar system?"), but rather asks a "contrary to fact" question: "What generally accepted facts and theories suggest that it is *impossible* for life to have arisen in our solar system?" This framework focuses attention on places where existing geologic, chemical, and biological theory seems to *rule out* the possibility of life emerging, despite the fact that it evidently did. In doing so, a focus on apparent "paradoxes" redirects activity away from the approach where a chemist has what he/she thinks is a "cool idea" to make an RNA precursor, runs some experiments to show that it works in Pyrex, and then constructs an argument to explain why it might be "prebiotically plausible."

The "RNA First" Model (Rich 1962; Benner et al. 1989) has many such apparent "paradoxes." We list some of these below, ranking these in an "even if" order.

3.4.1.1 The Tar Paradox

As an observation well-established by chemists and non-chemists alike, organic molecules devolve to give complex mixtures if given energy and left to themselves. These products are "tars" or "asphalts," better suited for paving roads than supporting Darwinism. Any scenario to create building blocks for RNA requires a way to allow organic material to escape this devolution. Here, we ask whether minerals can help.

3.4.1.2 The Phosphate Paradox

Even if precursors of RNA can escape devolution, one component of RNA seems geologically scarce according to generally accepted facts: phosphate. Free phosphate is likely scarce in alkaline serpentinizing prebiotic environments that contain calcium (Bach et al. 2006; Sleep et al. 2004, Andreani et al. 2013), as calcium and phosphate precipitate as insoluble minerals (e.g., $Ca_5(PO_4)_3(OH)$, hydroxyapatite). Yet some models for the prebiotic formation of RNA precursors require phosphate concentrations as high as 96 g/L (1 M) (Powner et al. 2009), since these models use phosphate as a buffer as well as a precursor to manage undesired reactivity of organic species that, absent large amounts of dissolved phosphate, would devolve to give tar. Can minerals help?

3.4.1.3 The Water Paradox

Even if organic precursors for RNA escape devolution, and *even if* phosphate is available to prevent this, many bonds in RNA are thermodynamically unstable in water (Fig. 3.2). This is also paradoxical, as the repeating phosphate backbone charge in RNA makes the molecule likely to perform *only* in water. The water-sensitive bonds might be made using high-energy dehydrating reagents, like carbodiimides. However, these themselves are unstable in water. Thus, any scenario for origins based on the RNA First Hypothesis must manage the apparent contradiction where life needs a solvent (water) that is inherently corrosive to the Darwinian biopolymer that is also needed. Can minerals help?

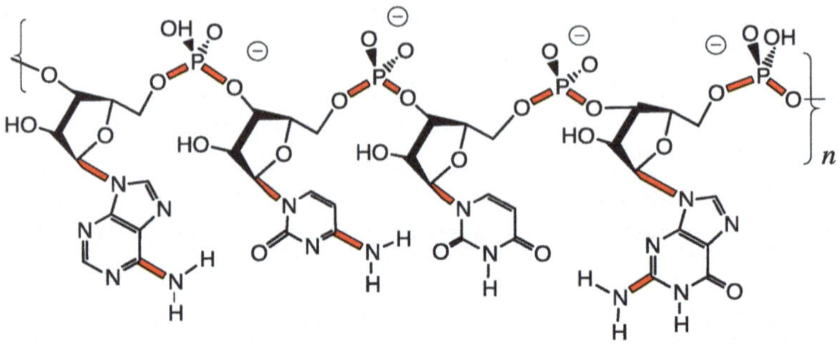

Fig. 3.2 Red bonds in this generic RNA structure are thermodynamically unstable in water. They can be formed in water from "activated" species (e.g., the phosphates may be activated as triphosphates or imidazolides). However, these activated species must also be unstable in water

3.4.1.4 The Chirality Problem

Theory suggests that to be evolvable, the building blocks of a biopolymer must be homochiral, to allow the semicontinuous changes in the behavior of a biopolymer as it moves across a sequence landscape required for Darwinism (Benner 2017b). Illustrating with proteins, changing an amino acid from tryptophan to arginine (for example) already influences the physical properties of the biopolymer (hence, the "semi"). However, semicontinuous change is impossible if, in addition to changing the biophysical properties of a side chain, the change also switches the orientation of the side chain. Unfortunately, *even if* a prebiotic chemistry makes building blocks, not tar, with phosphates, it rarely generates them with common chirality. Can minerals help?

3.4.1.5 The Information-Need Paradox

The amount of information required for a chemical system to gain access to replication with imperfections that are themselves replicable is difficult to estimate. However, by any current theory, biopolymers that might support Darwinism seem to be too long to have arisen spontaneously from the amounts of building blocks that might have escaped devolution in water. If a biopolymer is assumed to be needed for Darwinism, some way is needed to obtain high enough concentrations of building blocks to allow thermodynamically tolerable assembly of those biopolymers with adequate amounts of information. Can minerals help?

3.4.1.6 The Biopolymer Stability Problem

Even if organic precursors avoid the tar problem, *even if* they gain access to phosphate to form the linking bonds, *even if* the phosphate linkers in RNA can be made by dehydration, *even if* the nucleoside groups linked all have the same chirality, and *even if* enough can be gathered to give an RNA system that supports Darwinism (Mutschler et al. 2015), the RNA must survive long enough to initiate Darwinism. Unfortunately, RNA is not very stable, especially at high Mg^{++} where experimental studies of catalytic RNA suggest that its power is the strongest. Can minerals help?

Clearly, if minerals are to help, they have their job set out for them.

3.4.2 Minerals as a Source of Prebiotic Components

We begin with a very simple problem: How might the phosphate "mineral" components of RNA have been extracted from a geological environment? Available phosphate is required as a linking group in RNA. Phosphate is also found in many RNA cofactors. And at least one model for the formation of an RNA precursor, from Sutherland, requires large amounts of dissolved phosphate to manage the "tar" paradox (Powner et al. 2009).

An abundant source of geological phosphate is apatite, a calcium phosphate that has a fluoride, chloride, or hydroxide group filling a spot in the structure to give chlorapatite, fluorapatite, and hydroxylapatite. Apatite has the general formula $A_5(XO_4)_3Z$, where Z = Cl, Fl, or OH (and, less commonly, Br and I), and A represents Ca^{2+} or various cations that can substitute for Ca^{2+} in the structure (e.g., Sr^{2+}, Pb^{2+}, Ba^{2+}, Mg^{2+}, Mn^{2+}, Fe^{2+}, rare earth elements $(REE)^{3+}$, Eu^{2+}, Cd^{2+}, and Na^{+}). Hydroxyapatite is, of course, widely used in contemporary biology; it is a principal component of vertebrate bone.

Apatite is found in many environments. Igneous apatite is seen in mafic igneous rocks (i.e., basalts), but also in felsic rocks (i.e., andesite and rhyolite) and the ultramafic rocks of the Earth's crust and mantle. It is also found on the Moon with whitlockite ($Ca_9(MgFe)(PO_4)_6PO_3OH$), in meteorites (Fuchs 1969), and on Mars (Santos et al. 2013).

The presence of the calcium phosphates on the Moon makes them likely to have been available to prebiotic chemistry early in the Hadean. This is also consistent with the behavior of apatite in igneous environments. Although apatite has low abundance overall reflecting the low abundance of phosphate relative to silicate and aluminate, it is enriched in igneous melts due to its low solubility and the limited capacity of common rock-forming minerals to accept phosphate into their structures (Piccoli and Candela 2002 *op. cit.*).

Thus, the "phosphate problem" exists in prebiotic chemistry not because phosphate rocks are likely absent from the Hadean, but rather because phosphate is generally seen to be difficult to extract from phosphate minerals, such as apatite. However, many phosphate minerals have low solubility. Even magnesium phosphate, a common source of phosphate in biochemical laboratories, has low solubility (2.27×10^{-3} g/L at 20 °C). Likewise, most transition metal phosphates have low solubility, as exemplified by ferrous phosphate (the mineral vivianite), copper phosphate (the mineral libenthite), and zinc phosphate (the mineral hopeite). Thus, calcium alone is not the only challenge to models that use high concentrations of dissolved phosphate to manage other paradoxes in the RNA First Model; strontium, barium, and all transition metals, including ferrous iron, create problems.

Multiple attempts were made in the prebiotic chemistry of the last millennium to mobilize phosphate from phosphate minerals, usually apatite. These generally failed. For this reason, in much of the classical literature, discussed below, apatite is seen primarily as useful for adsorbing and concentrating prebiotically interesting molecules, including prebiotic nucleoside phosphates. However, several solutions have been proposed to "the phosphate problem" in the new millennium.

3.4.2.1 Solutions to the Phosphate Problem: Acidification

While many phosphate salts are quite insoluble in water, hydrogen phosphates tend to be more soluble. This is, of course, well-known to those who experience dental disease, where bacteria secreting acid slowly dissolve tooth enamel calcium phosphate. Thus, calcium dihydrogen phosphate $Ca(H_2PO_4)_2$ dissolves in water to ~ 18 g/L at 20 °C, contrasting with the estimated <0.02 g/L for apatite itself. Lowering the pH escapes the calcium precipitation "trap," although it needs not rescue models that require one molar phosphate at pH 7.0 [the conditions used in Powner et al. (2009)].

The principal challenge in exploiting acidity as a way of mobilizing phosphate from insoluble phosphate minerals is to find a source of acid. An atmosphere of 100 bar CO_2, considered possible for early Earth, would create a pH of ~4 in pure water oceans, absent buffering. This is sufficiently acidic to convert phosphate into mobile dihydrogen phosphate. However, the challenge comes in the next step. For phosphate ester bond formation to be thermodynamically accessible, the water must be evaporated. If the acidity comes from carbonic acid, evaporation would cause the carbonic acid to revert to carbon dioxide and water, effectively removing a proton in the process of raising the pH.

Volcanoes offer alternative sources of acidity, depending on the redox state of the crust and mantle. If the oxygen fugacity is sufficiently high (*general geological question C*), SO_2 would be prominent among volcanic gasses. Sulfur dioxide reacts with water to produce sulfurous acid, which is acidic enough to also dissolve phosphate. Further, sulfurous acid would more easily remain behind upon evaporation than carbonic acid.

3.4.2.2 Solutions to the Phosphate Problem: Change the Redox State

In a very influential insight, Pasek pointed out some time ago that the low oxidation state of phosphorus in the meteoritic mineral schreibersite ($(Fe,Ni)_3P$) could, upon corrosion, deliver phosphite and other low oxidation forms of phosphorus to an aquifer (Pasek 2016). Pasek estimated that 1–10% of the accessible phosphorus in the Hadean would have been in the form of schreibersite contributed by impact (*general geological question E*) (Gull et al. 2015). Phosphite minerals are generally more soluble than the corresponding phosphates (compare barium hydrogen phosphate and barium hydrogen phosphite at 0.1 and 6.9 g/L, respectively, at 20 °C). Further, phosphorus in the +3 oxidation state easily exchanges ligands to transiently form esters that could be stabilized by the subsequent oxidation of phosphorus +3 to phosphorus +5.

The Pasek model requires multiple geological microenvironments in proximity, of course. The first involves aqueous corrosion of low oxidation state phosphorus to give phosphite. The last requires oxidation. Minerals might provide the source of oxidation. For example, the manganese mineral pyrolusite (MnO_2) has manganese at its +4 oxidation state, which might serve as an oxidant. If the volcanic gasses have the correct oxygen fugacity, elemental sulfur (S_8) may also serve as an oxidant, giving the thiophosphate analog of RNA. Indeed, in solid-phase commercial DNA synthesis commonly used in the biotechnology industry, phosphorus in its 3+ oxidation state is converted to thiophosphate using elemental sulfur.

The challenge here is that in classical DNA synthesis, a phosphite triester is the intermediate. A phosphate diester, in contrast, does not have the same tautomeric form. Rather, it exists as RO-P(=O)(H)-OR. This does not behave well, making it a problematic intermediate in the assembly of oligonucleotides.

3.4.2.3 Solutions to the Phosphate Problem: Change the Solvent

The insolubility of phosphate minerals is, of course, discussed with reference to water as the solvent. Accordingly, one solution is to change the solvent. Some years back, we explored the ability of apatite to dissolve in formamide, without much success. However, more recently, the Hud group explored mixtures of semi-aqueous solvents in the hope of finding some that would mobilize the phosphate from apatite (Burcar et al. 2016).

Particularly successful were mixtures that combined urea with ammonium formate and water in a molar ratio of 1:2:4. Ammonium formate was proposed to be the end product of hydrolysis of HCN by way of formamide (Bada et al. 2016). Urea is produced in reactions (McCollom 2013), when ammonium cyanide (NH_4CN) is exposed to sunlight (Lohrmann 1972), or via hydrolysis of cyanamide (HNCNH), which may be formed by electrical discharge in somewhat reducing N_2-containing atmospheres (Fiore and Strazewski 2016). A small amount of formamide is apparently made upon heating the mixtures containing ammonium formate. Hud suggested that the composition of the mixture was stabilized around a eutectic point by evaporation of water and ammonium formate and/or loss of urea by precipitation.

Burcar et al. then tested the ability of both soluble phosphates (such as sodium phosphate) and insoluble phosphates (such as apatite) to donate phosphate to nucleosides in these mixtures. Perhaps as expected, the soluble phosphates gave mixtures of phosphorylated nucleosides at 65 °C over 19 days. No phosphorylation was seen with hydroxyapatite as a donor. However, when the temperature was raised, apatite also generated small amounts of phosphorylated products. The analogous phosphorylation was seen with glycerol as the phosphate acceptor. Further, addition of magnesium sulfate and free phosphate, at concentrations suggested to be comparable to those that dissolved apatite generated when treated with acidic heated surface waters, improved phosphorylation yields (Grosjean et al. 1995; Feely et al. 1998; Hedenquist et al. 2000).

While the reaction conditions here remain to be explored in greater depth, it appears as if the substantial part of the success is not due to changing the solvent per se, but rather due to the lowering of the pH in these mixtures, the mechanism outlined above. Thus, Burcar et al. report the formation of both struvite ($MgNH_4PO_4$• $6H_2O$) and newberyite ($Mg(HPO_4)\cdot 3H_2O$), minerals reflecting a lower pH.

3.4.2.4 Solutions to the Phosphate Problem: Use a Realistically Complex Mineralogical Background

The mechanism of these reactions notwithstanding, Hud's results show that mineral complexity in a geological environment has an impact on the outcome of organic reactions. Although today's oxygen-containing atmosphere makes the surface and near-surface environments quite different in many respects for what could have been possible during the Hadean, Earth has plenty of environments to explore natural mineral complexity in natural planetary environments. Unfortunately, the literature is poorly indexed. Accordingly, a degree of random reading is necessary to come across, serendipitously, mineral assemblages that are interesting enough to lead to experimental work.

For example, random reading came across a report of terran evaporite formations that contained together in one locale gypsum (calcium sulfate, $CaSO_4$•$2H_2O$), boracite ($Mg_3(B_7O_{13})Cl$), and lüneburgite ($Mg_3B_2(PO_4)_2(OH)_6$•$8H_2O$) (Müller and Fabricius 1978). The occurrence in natural evaporites of a mineral containing calcium, *but combined with sulfate* (gypsum) *rather than phosphate*, and a mineral containing phosphate, *but combined with magnesium and boron* (lüneburgite) *rather than calcium*, suggested that phosphate in nature need not be irrevocably tied up with calcium, even in mineral environments rich in calcium, *if borate is present*. We therefore asked whether borophosphate evaporite mineral assemblages could address the "phosphate problem" by segregating calcium and phosphate. da Silva and Holm have addressed the larger possible prebiotic relevance of borophosphate minerals in general (da Silva and Holm 2014).

As with the Hud laboratory, we began with experiments that showed that hydroxyapatite does not dissolve to any measurable extent in formamide, despite formamide being an excellent solvent for many organic solutes. Parallel experiments showed that synthetic apatite does not phosphorylate nucleosides in urea under standard conditions.

We then asked whether lüneburgite itself could act as a phosphorylating agent in nonaqueous conditions, needed for phosphate esters to be formed thermodynamically. Synthetic lüneburgite was made by mixing and evaporating boric acid, phosphoric acid, sulfuric acid, calcium oxide, and magnesium oxide in the needed ratios. The resulting material was analyzed by powder X-ray diffraction. Absent calcium, lüneburgite and (with excess magnesium) periclase (MgO) were formed. However, if calcium was added, powder X-ray diffraction found that lüneburgite and gypsum segregated as separate minerals. This spontaneous removal of phosphate from calcium in the *presence* of borate, but not in the *absence* of borate, reproduced in the laboratory what was observed in natural evaporites. This segregation may be understood as a consequence of the relative solubilities of the mineral species involved.

But is phosphate more accessible from lüneburgite than from hydroxyapatite when nucleosides are available to be phosphorylated? Here, the affinity of borate for the diol units of ribose and ribonucleosides suggested that ribonucleoside phosphorylation targets might *themselves* extract the borate from lüneburgite, with the accompanying release of phosphate that then can be used. Ribonucleosides have 2′,3′-diol units that can bind borate.

Simple NMR experiments showed this to be true. While the amount of inorganic phosphate released to pure water from lüneburgite alone was very small, uridine nucleoside released phosphate in increasing amounts. By extracting the borate from the mineral, uridine made the lüneburgite phosphate soluble, even in the presence of calcium-containing gypsum.

However, release of phosphate to water does not address the thermodynamic instability of phosphate esters there. Thus, we asked whether lüneburgite itself could phosphorylate nucleosides in nonaqueous media. This proved to be the case in urea, when the water was evaporated at 85 °C for 16 h. This simulated desert was then flooded to give uridine and uridine-5′-phosphate (85:15) with few side products. The same was seen for the other three ribonucleosides G, A, and C.

Interestingly, borate extracted from lüneburgite by ribonucleosides to be phosphorylated also mitigated a part of the "tar problem" in the reaction. *Without* borate from lüneburgite (instead using sodium dihydrogen phosphate), complex phosphorylation product mixtures were found. With lüneburgite, by complexing the 2′- and 3′- oxygens, borate directs the phosphorylation reaction to the 5′-hydroxyl group, preventing complex product formation (Kim et al. 2016).

3.4.2.5 Borate Minerals to Mitigate the "Tar Problem"

Whether any of these offered solutions to the phosphate problem are acceptable is a cultural statement. However, the phosphate problem is "easy," in that the phosphate unit is stable; it will not decompose. Nucleobases are similar, since they are aromatic ring systems whose kernels are stable under many conditions, although their amino functional groups can be lost by hydrolysis (Levy and Miller 1998).

However, the third entity, the ribose carbohydrate, is an archetypal source of tar. Indeed, as noted above, Stanley Miller concluded from the instability of carbohydrates that neither ribose nor any other sugar could be components of the first genetic material (Larralde et al. 1995).

Ribose (a 5-carbon sugar, or pentose $C_5H_{10}O_5$) is not a problem because it cannot be formed. On the contrary, it has been known for a half century that ribose can be formed from formaldehyde under alkaline conditions (Appayee and Breslow 2014). Ribose and other carbohydrates (with general formulae $C_nH_{2n}O_n$) are "polymers" of formaldehyde (HCHO, $C_1H_2O_1$), which almost certainly form in prebiotic atmospheres containing CO_2, H_2O, and electrical discharge and/or ultraviolet radiation (Cleaves 2008). Thus, HCHO can be the sole starting material to form many carbohydrates in a process known as the "formose reaction" (Butlerov 1861). This requires $Ca(OH)_2$ (pH of ~12.5) at elevated temperatures (65–80 °C).

Further, glycolaldehyde ($C_2H_4O_2$) is formed by electrical discharge in moist atmospheres, albeit in smaller amounts than HCHO (Löb 1913; Harman et al. 2013). This "C2" species reacts with HCHO (a "C1 species") to give glyceraldehyde (a "C3" species, $C_3H_6O_3$). This C3 species can react with C2 glycolaldehyde to give C5 species, including ribose (Ricardo et al. 2004). For formose cycling (Fig. 3.3), glycolaldehyde is a "catalyst," as one glycolaldehyde can fix many HCHO molecules. These reactions proceed on the hour time scale at pH ~10.5, somewhat milder conditions than for the formose reaction. Thus, the classical literature viewed the formose process as a solution to the problem of prebiotic carbohydrate formation (Dyson 1985).

Rather, ribose becomes a problem because it then goes on to react further under the conditions where it was formed to give complex mixtures of products containing only small amounts of ribose (Decker et al. 1982). As shown in Fig. 3.3, the reactions that can occur with carbohydrates are many and represent the carbohydrate version of the tar paradox. Thus, *even if* the formose process is manually supervised, it gives tar. And if it is unsupervised, anything productive reacts further to give saccharinic acids, furans, and other "tars", especially after HCHO is depleted.

How does naturally complex mineralogy mitigate this problem? Here, as with the phosphorylation with lüneburgite described above, the ability of mineral borate to coordinate to the adjacent hydroxyl groups of organic species offers a solution. In a series of experiments, including those where intermediates in Fig. 3.3 were prepared with ^{13}C substitution, Kim et al. dissected the formose processes (Kim et al. 2011). This dissection showed that minerals containing borate (e.g., kernite, ulexite, and colemanite) mitigate the intrinsic instability of ribose in several ways. First, for the reaction between glycolaldehyde and glyceraldehyde (i.e., C3 + C2 → C5), borate stabilizes the ribose formed as the preferred species under the conditions where it was formed. The same can be happening if glyceraldehyde is made in situ from glycolaldehyde and formaldehyde (i.e., C1 + C2 → C3). This is because ribose, in its cyclic hemiacetal form, presents two adjacent hydroxyl groups in a pre-organized conformation, allowing the resulting borate complex to be stabilized by an internal hydrogen bond. Indeed, the complex between ribose and borate is as stable as for any pentose.

This represents a partial solution to the carbohydrate "tar" problem with ≥25 mM borate. This work prompted Stephenson et al. to seek such concentrations of borate in clay of a Mars meteorite; these were found (Stephenson et al. 2013). These results, in turn, prompted the Mars Science Laboratory to seek borate in Gale Crater on

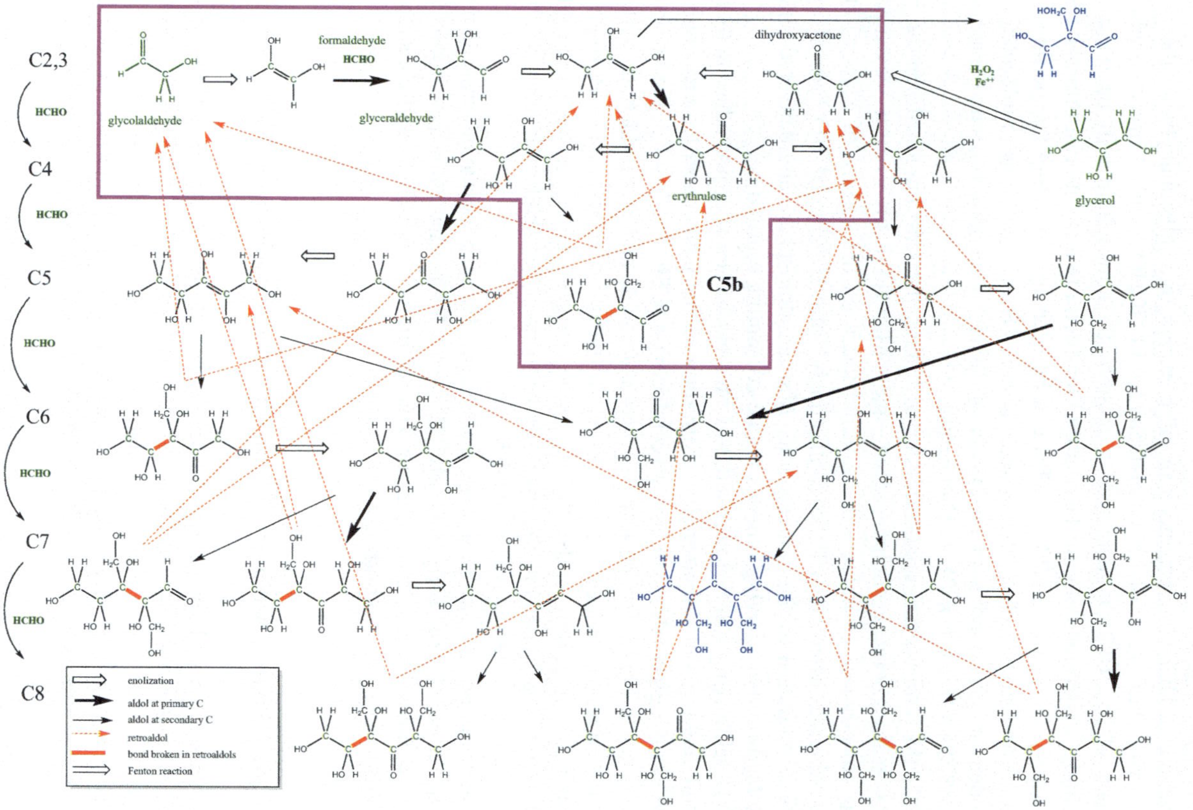

Fig. 3.3 Formose-like processes make useful carbohydrate. However, under conditions where they are formed, these react further, making the unmoderated process useless as a prebiotic source for ribose, the "R" in "RNA." The Benner lab has synthesized many of these intermediates, including with ^{13}C isotopic

Mars, where it was found associated with gypsum {http://www.manyworlds.space/index.php/2016/12/14/with-the-discovery-of-boron-on-mars-the-package-of-chemicals-needed-for-life-is-complete/}. This association is, of course, exactly the same as discussed above with lüneburgite.

However, as HCHO is produced in large excess over glycolaldehyde, the net stoichiometry of this reaction (2 glycolaldehydes +1 HCHO = only 1 ribose) implies that much of the HCHO generated in a prebiotic atmosphere will be wasted. Interestingly, experiments in our laboratory then found that borate can improve the stoichiometry (Neveu et al. 2013). This can be seen by extracting the portion of Fig. 3.3 contained within the magenta box. Here, just one molecule of glycolaldehyde presented with an excess of HCHO generates branched pentose **C5b** in a C2 + C1 + C1 + C1 → C5 process.

At 50 mM (and higher) concentrations of borate, the **C5b** branched pentose is quite stable as its borate complex; indeed, unless Mars holds life to eat it, we expect that a pentose-borate complex will be found today on the surface of Mars as an organic mineral, where it will be continuously formed from HCHO formed by ultraviolet irradiation of the contemporary Martian atmosphere and delivered onto eroding basaltic rocks having igneous borate-containing tourmaline.

This stoichiometry can be improved, as **C5b** lacks an acidic proton adjacent to its C=O unit, making it unable to enolize. Kim showed that **C5b** can suffer a retroaldol reaction to give glycolaldehyde and glyceraldehyde (**C5b** → C2 + C3, the vertical dotted arrows in Fig. 3.3). The resulting glycolaldehyde and glyceraldehyde molecules can then repeat the cycle, adding a total of five more HCHO molecules to give two **C5b** molecules. At the end of the next cycle, one glycolaldehyde has fixed eight HCHO molecules. Repeating the cycle, as many as 17 HCHO molecules can be fixed by a single molecule of glycolaldehyde.

But were borate minerals present on early Earth? For sure, borate minerals do not appear in Hazen's list of "early minerals." However, borates have now been found in 3.8 Ga Isua formations, and work in the Trail laboratory with Hadean zircons may yield information on this shortly.

3.4.2.6 Molybdate Chemistry and the Influence of Redox States

Unfortunately, although the pentose has the same collection of atoms as ribose, it is not ribose. Further, the retroaldol reaction fragments of **C5b** to give glycolaldehyde and glyceraldehyde place these two compounds "at risk" of further reaction into an unproductive manifold. Thus, the glycolaldehyde and glyceraldehyde molecules need not cooperatively react with formaldehyde to give ribose. Instead, two

---

Fig. 3.3 (continued) labels to trace reactions in this process (Benner and Kim 2015). *Sedimentary minerals containing borate* (e.g., kernite, ulexite, colemanite) constrain the complexity to the box by binding adjacent -OH groups in carbohydrate intermediates, giving the branched pentose **C5b** as a metastable product as its borate complex

glycolaldehyde molecules might react with each other to create a 4-carbon sugar; two glyceraldehyde molecules might react with each other to create a 6-carbon sugar. Further, to allow the retroaldol reaction to occur requires that the concentration of borate (relative to **C5b**) drops to release **C5b**. This risks the loss of carbon to "tar" at high pH by leakage from the cycle.

One solution is to find mechanisms that would rearrange the atoms in **C5b** to give ribose without fragmenting the molecule. This would be done by carbon and hydride shifts that occur while the entire molecule is kept together and does not allow pieces to move apart. Can mineral species help do this?

Again, the answer is "yes." Specifically, molybdenum will in its +6 oxidation state (molybdate, Mo^{6+}) catalyze the rearrangement of **C5b** at pH 6–7 to give linear pentoses, including ribose, stereospecifically (**C5b** is either erythro or threo), without formation of "tar" (Fig. 3.4) (Petrus et al. 2001). Again, this requires movement of material from one environment to another. Here, once **C5b** is formed at high pH (~10) in aqueous media, it must move in the aquifer into an arid basin beneath an atmosphere rich in CO_2. Here, the pH necessarily drops. The stability of the complex between **C5b** and borate drops as well as the pH drops. At pH 6–7, the rates of the tar-forming reactions of carbohydrates are quite slow, making stabilization by borate no longer necessary.

But is molybdate accessible on early Earth? Molybdate minerals do not appear in Hazen's list of "early minerals." However, recent experiments suggest that molybdate minerals were easily present, given our knowledge of the fugacity in the mantle at times deep in the Hadean (*general geological question C*). Again, silicate melts at different oxygen fugacities have been studied experimentally in the laboratory, and the amount of Mo in either 4+ or 6+ oxidation states measured. Molybdenum is shown in quartz melts to transition from Mo^{+4} to Mo^{+6} one log unit just *below* the iron-wustite buffer (Fig. 3.5) (Borisov 2016). Trail's work with zircons containing cerium has the terran mantle near the FMQ redox state, where $Mo^{+6} >> Mo^{+4}$. Even Mars, more reducing overall than the Earth, may have had Mo^{+6}. The fO_2 of ALH 84001 [4.1 Ga by Lu-Hf dating (Lapen et al. 2010)] has been estimated to be at a higher oxidation state (~ FMQ −2.7) (Righter et al. 2008), where $Mo(^{6+})$ dominates.

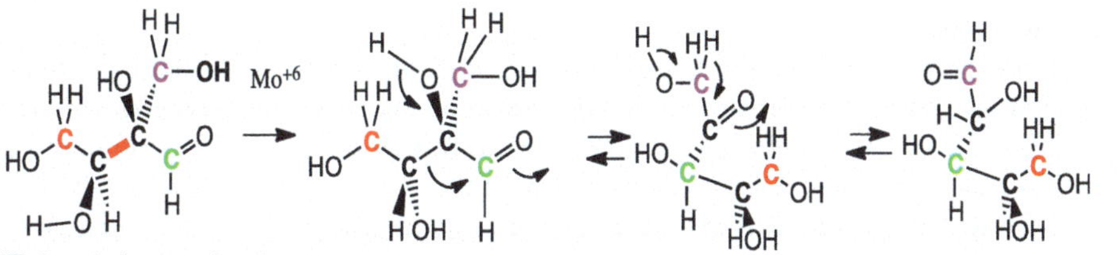

C5b, branched pentose from borate-moderated formose process

ribulose

ribose

Fig. 3.4 $Mo(^{6+})$ catalyzes the stereospecific interconversion of pentoses (the Bilik reaction) at neutral pH, without risk of tar formation, giving linear carbohydrates. At this lower pH, only the most stable borate complexes form

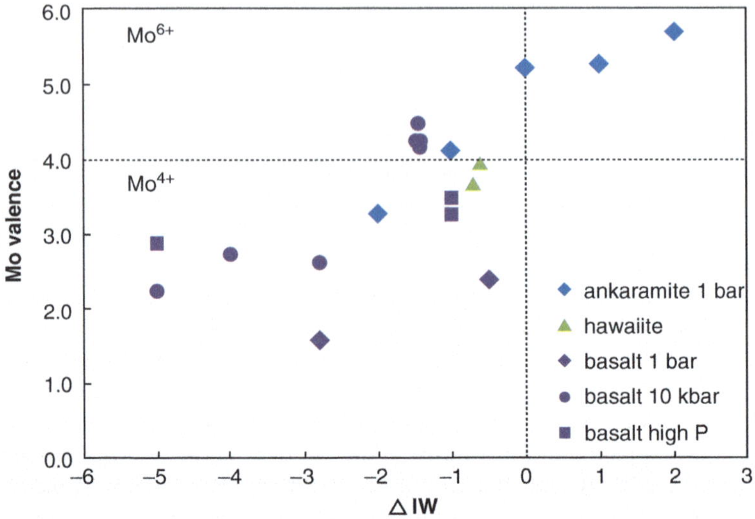

Fig. 3.5 Molybdenum average valence as a function of oxygen fugacity relative to the iron-wüstite buffer. Experiments were done at 16 and 24 GPa (Danielson et al. 2011)

Thus, Mo^{6+} seems to be likely to have been present to prebiotic chemistry in analogous arid regions on Hadean Earth. Here, with borate minerals, it would have been available to mitigate the tar problem associated with carbohydrates.

3.4.3 Minerals as Reservoirs of Organic Materials

We have mentioned that borate-ribose and other species might be organic minerals that serve as reservoirs of unstable compounds, stabilizing them and concentrating them for later use. Organic minerals are scarce today on the surface of modern Earth, as they are easily food for organisms in the biosphere. Nevertheless, several are known where they survive rapid consumption by having stable mineral salts and low energy value.

For example, when graphitic carbon comes in contact with the modern dioxygen-containing atmosphere, it slowly is oxidized to a core benzenehexacarboxylic acid (mellitic acid) that is metastable with respect to further oxidation. Its aluminum salt is quite stable and appears as the mineral mellite.

Mellite has played a role in the exploration of Mars. Given that meteorites containing graphitic carbon fall to Mars continuously, the 1976 Viking mission that visited and explored Mars fully expected to find some organic material. The instrument designed to find it, however, looked for mass spectrometry signatures, volatile material that would emerge from the Martian soil on heating to temperatures as high as 400 °C. The volatiles from these organic materials were then passed in a stream of dihydrogen gas into a chromatography column; dihydrogen was then removed by a palladium separator and the residual material was injected into a mass spectrometer. The unexpected outcome was that no organic molecules were seen at all.

To explain the missing Mars organics, the immediate hypothesis was that photolysis of water by the ultraviolet light that penetrates to the surface of Mars would lead to hydroxyl radicals that "combust" the graphitic carbon. However, Benner et al. (2000) suggested an alternative explanation, based on the expectation that the graphitic carbon arriving by a meteorite every day to Mars would also be oxidized to the metastable mineral mellite (Boily et al. 2000a, b). Experiments showed that mellite was sufficiently stable so as not to yield substantial amounts of product upon heating under the Viking conditions.

Subsequently, the Mars Polar Lander discovered evidence of perchlorate salts in the Martian soil. Upon heating to 400 °C in the presence of perchlorate, organic matter combust, even in the form of mellite. The combustion products include chlorohydrocarbons; interestingly, these were seen by the Viking 1976 mass spectrometer.

Still other organic minerals on the modern Earth survive due to a combination of instability and lack of metabolic value. Materials are present where the already high level of oxidation makes them metastable with respect to further oxidation. Best known among these are minerals that contain oxalate, which is $C_2H_2O_4$. Oxalate forms an insoluble salt with calcium; in medicine, such salts are commonly found in kidney stones. Further, just one oxidation state away from carbon dioxide, oxalate is not a valuable resource to a microbe even if it were able to metabolize it.

Of course, in the prebiotic world, organic minerals can survive without fear of being eaten by microorganisms. There, organic minerals become possible reservoirs, allowing the accumulation of organic species that would otherwise be transformed to tars over geological periods of time.

3.4.3.1 Organic Carbohydrate Minerals

Carbohydrates are the archetypal examples of organic molecules that react with each other to give unproductive products. This creates problems for prebiotic chemists.

For example, the 2009 model of Powner and Sutherland to form pyrimidine nucleoside phosphates requires large amounts of the C2 carbohydrate glycolaldehyde to await encounter with a large amount of cyanamide to form aminooxazole. Glycolaldehyde needs not wait for cyanamide to arrive, since it can react with itself to form threose and erythrose, and then tar (if the reaction is not stopped by human intervention) (Kim et al. 2011). Next, to manage the reactivity of resulting aminooxazole, Sutherland suggested that it sublimes to condense at a distant location that had substantial amounts of glyceraldehyde (Powner et al. 2009). Glyceraldehyde also reacts with itself (Kim et al. 2011) and may not wait for the arrival of aminooxazole before doing so.

Organic minerals may be exploited as "safe harbors" to allow carbohydrates to accumulate in unreactive form. Borate-ribose (Fig. 3.6) is one such mineral. Here, the stability of the complex arises because ribose forms a cyclic hemiacetal that presents three hydroxyl groups on the same side of the molecule. Borate coordinates tightly to these pre-organized cis-diols, with the five-ring complex stabilized by an internal hydrogen bond.

Smaller carbohydrates such as glycolaldehyde and glyceraldehyde are not as fortunate. Because these species cannot form cyclic hemiacetals that pre-organize cis-diols to present to borate for complexation, their borate complexes are less stable.

Even HCHO is a problem. It is likely to be formed in large amounts in a post-veneer atmosphere (Cleaves 2008). However, if present in sufficient concentrations at high pH, HCHO reacts with itself in the Cannizzaro reaction at a rate that scales with the square of its concentration to yield formate and methanol, neither having much prebiotic value (although MeOH might be a solvent and antifreeze).

To stabilize aldehydes, Powner recently suggested that aminothiazole might react with glycolaldehyde and glyceraldehyde to form imines that form stable crystalline minerals (Islam et al. 2017) (Fig. 3.6). However, mineral species may offer a better solution, if we make a geo-environment more realistically complex. For example, SO_2 is expected to emerge from Hadean volcanoes under the model for planetary history that we have chosen. SO_2 reacts reversibly with HCHO to give hydroxymethanesulfonate (HMS), which crystallizes in an evaporite. HMS is stable against self-reaction but, upon dissolution, slowly releases HCHO into a mixture, where it can react.

A constant source of low HCHO is an effective way to prevent carbohydrates in alkali from becoming “tar.” Here, any enediolate formed is trapped by HCHO (Kim et al. 2011) before it can react with another higher carbohydrate, or suffer beta-elimination, or form saccharinic acids (Fig. 3.7). This is especially true with borate, which moderates formation of enediolates.

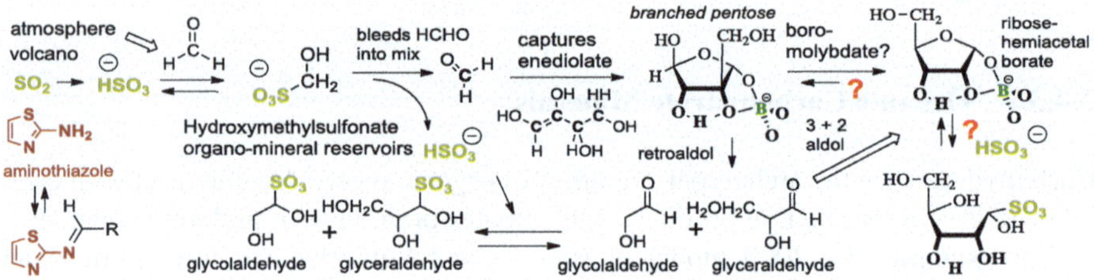

Fig. 3.6 By making the geo-surrounding more complex by including volcanic $\mathbf{SO_2}$, bisulfite addition products (R-(CHOH)-SO_3^-) form with aldehydes, stabilizing these as organic minerals to give reservoirs of carbohydrate adducts for later use. The HCHO bisulfite addition product slowly bleeds HCHO into an alkali mixture of evolving carbohydrates, preventing many classes of reactions that destroy these (Fig. 3.7), while never allowing HCHO to reach levels where Cannizzaro destruction dominates. Thus, Hadean volcanoes mitigate a problem that Powner manages with an organic partner (Islam et al. 2017). Note how bisulfite addition to ribose competes with hemiacetal formation (right)

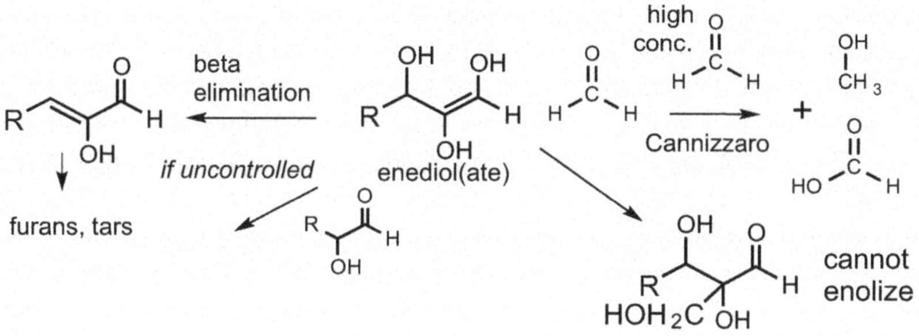

Fig. 3.7 Low levels of HCHO, slowly bled into a mixture from hydroxymethylsulfonate, manage the intrinsic propensity of carbohydrates to form "tars." This arises from the ability of HCHO to capture enediol(ate)s before they have a chance to do undesired reactions (Kim et al. 2011)

3.4.3.2 Organic Minerals Incorporating the Nucleobases

Nucleobases likewise are species that are conceivably formed from hydrogen cyanide, cyanoacetylene, and other small molecules created in a sufficiently reducing Hadean atmosphere. However, their reaction with water generates hydrolysis products. An organic mineral that stabilizes them would be useful. Here, very little work has been done to explore the situation. However, complexes between purines and ferrous iron are known (Richter and Fischer 2003). These two might combine to create organic minerals to allow the accumulation of nucleobases in metastable forms.

Here, it is important to appreciate how choices made in our geological modeling influence how prebiotic chemistry gets driven. For example, Pearce and Pudritz (2015) are troubled from the fact that a Hadean atmosphere dominated by carbon dioxide, dinitrogen, sulfur dioxide, and water would not be very efficient at producing organic molecules. Pearce chose instead to assume that the Hadean atmosphere and lithosphere were in equilibrium with each other, and chose to have purines delivered by meteorite. This made their abundances very low, with the abundance of adenine in a warm little pond being called "negligible." This was the case even if seepage from those ponds was turned off (Pearce et al. 2017).

We exploited recent models for the formation of the late veneer to manage the problem arising from the inefficient formation of purines in oxidized atmospheres. However, various organic minerals might be considered. For example, nucleobases form complexes with various transition metals. These may allow both their concentration and their stabilization.

3.4.3.3 Organic Minerals Incorporating Cyanide

Hydrogen cyanide is a key prebiotic material, serving as a precursor for nucleobases, amino acids and, in some models, even carbohydrates. It is formed best in reducing atmospheres, including those that may have existed above the Earth for about

200 million years following the late veneer. The formation of adenine is especially well studied. However, like formaldehyde, it is difficult to get HCN in high concentrations without itself reacting. The most prominent self-reaction product is the hydrogen cyanide polymer, which is even today poorly defined at a structural level.

Before the end of the last millennium, Keefe, Orgel, and others had noted the incompatibility between hydrogen cyanide and ferrous iron (Keefe and Miller 1996), which was likely to be the dominant oxidation state of iron in a prebiotic ocean, especially under the atmosphere proposed to have been formed during the late veneer. Cyanide reacts with ferrous iron to form ferrous ferrocyanide, well-known as the dye "Prussian blue."

Sutherland recently offered an intriguing proposal that Prussian blue might be a reservoir for hydrogen cyanide, extending the concept to transition metal complexes with other quite unstable prebiotic materials, notably cyanoacetylene (Sutherland 2016). Williams and his group have pointed out that under a reducing atmosphere, ferrous iron (Fe^{2+}) might be a replacement for Mg^{2+}, which is commonly used today under an oxygen-rich atmosphere where iron is scarce due to the poor solubility of many ferric Fe^{3+} salts (Okafor et al. 2017).

This raises the general question of the incompatibility of mineral species and various needed prebiotic organic species. The incompatibility between ferrous iron and cyanide is easily resolved without the formation of Prussian blue. Indeed, depending on the concentration of dissolved ferrous iron, the hydrolysis of cyanide to give formamide and ammonium formate is faster than the formation of Prussian blue. Conversely, Prussian blue will hydrolyze to give ammonium formate. As noted above in the Hud work involving nucleoside phosphorylation, mixtures of formamide and ammonium formate are useful for other prebiotic chemistry problems. Further, as Saladino has shown with his colleagues (Saladino et al. 2003), formamide is as good as a precursor for many nucleobases as hydrogen cyanide. Further, formamide itself is a useful solvent for the discontinuous synthesis model for RNA prebiotic chemistry. This model uses formamide as a solvent to manage the water paradox in the assembly of nucleoside to nucleoside phosphates in the presence of borate minerals and urea.

3.4.4 Minerals as a Way to Stabilize Prebiotic Species That Manage to Be Formed

3.4.4.1 Adsorption onto Phosphate Minerals

Organic minerals allow the storage, one hopes in stable form, of organic species in bulk. They also require the existence of arid desert-like environments (*general geology Questions A and B*). However, an alternative mechanism to stabilize reactive organic species involves their adsorption onto the surfaces of minerals. This allows stabilization of considerably less material in the formation of a bulk organic mineral. However, as one proceeds toward oligomeric RNA, stabilization of even small amounts of material is likely useful.

Precisely because of their low solubility, phosphate minerals, especially apatite, were early candidates for those seeking minerals as to concentrate and catalyze the assembly of scarce organic building blocks to make biological polymers, including RNA, proteins, and even DNA. The more daring researchers considered the apatite group of minerals, which added arsenates and vanadates to phosphates. The resulting mineral collection has a hexagonal or pseudohexagonal monoclinic structure. The general formula is $A_5(BO_4)_3$ (OH, F, Cl), where the A cations can be metal ions such as calcium, barium, sodium, lead, strontium, lanthanum, and/or cerium. The B cations can be phosphorus, vanadium, or arsenic. Carbonate and silicate anions can substitute to a limited extent for the BO_4 groups. This group includes apatite itself, mimetite (a lead arsenate), pyromorphite (a lead phosphate), and vanadinite (a lead vanadate).

The need to concentrate originated in the thermodynamic impossibility of dehydrating building blocks in aqueous environment from dilute solution ("Paradox" 4). Fox had used high temperature for this purpose (Fox 1965). However, the surfaces of soluble minerals may give access to dehydrated products at lower temperatures.

For apatite, much of this was aided by a detailed understanding of the surface of apatite that was developed by Taves (Taves 1963; Taves and Reedy 1969). Thus, Neuman measured small amounts of adsorption of adenosine monophosphate on the surface of apatite. Adenine and adenosine did not adsorb. Therefore, Newman argued that apatite surfaces might allow the concentration of diluted nucleoside monophosphate from aqueous solution. Neuman carried these studies further, heating apatite-adsorbed monophosphates and observing formation of RNA like biopolymers. As the analytical methods were insufficiently advanced to characterize the products to modern standards, these results must be taken as tentative.

3.4.4.2 Adsorption onto Carbonate and Sulfate Minerals

It is easy enough to observe the adsorption of dissolved species onto solid surfaces. It is much more difficult to quantitate that adsorption. When two species interact in aqueous solution, their interaction energy can be determined by the concentration dependence of the two, an interaction that involves two rate processes, a biomolecular association process, and a unimolecular dissociation process.

However, when one of the two interacting components is a mineral, its concentration cannot be determined. Thus, one can easily measure the fraction of a dissolved species that becomes mineral-bound, but this has different meanings in different concentration regimes. Therefore, if the number of dissolved molecules in solution in total is larger than the number of binding sites on the surface of the mineral, then the fraction of solute bound must be less than unity, simply because the surface of the mineral is saturated. That fraction will increase by adding more mineral.

However, in the concentration regime where the amount of solute is less than the number of binding sites, the fraction bound can still be unity, if the affinity of the solute is quite high. Here, the binding rate can be assumed to be limited by the diffusion-controlled encounter of the solute with the surface. The affinity then can be usefully quantitated by measuring the rate of dissociation of the adsorbed solute into a solvent having infinite volume.

However, different solids have different levels of porosity, their surfaces have different levels of smoothness, steps on the surface can greatly alter the number of sites available, and these together make the actual number of sites unknowable. Attempts to determine these parameters by using, for example, noble gas adsorption on powdered minerals are often off point. The number of sites or crevices available to, for example, argon has little relevance to the number of sites on the surface that are available to, for example, an RNA molecule.

Adding to this challenge is the fact that no two natural mineral samples are identical. Natural minerals vary in chemical composition away from their canonical composition from specimen to specimen and certainly from locale to locale. For example, the name of apatite comes from the Greek word meaning "to deceive," which reflects the fact that it comes in many different colors. These are due to different inclusions of atomic species that are not in the canonical formula of the mineral. This, of course, includes substitution of elements within the apatite crystal structures that are discussed above.

In our laboratory, Biondi et al. (2016) recently introduced a general strategy to manage these problems. In this work, we compared the adsorbance of radiolabeled RNA on to binary mineral species obtained in various ways.

In one, the species was precipitated as a synthetic mineral via a double decomposition reaction between the two chemical components. This allowed for an exact (and exacting) level of purity which is the kind of "controlled experiment" that chemists like. However, it may be criticized as being "artificial." Moreover, solid phases with high surface areas, precipitates in particular, are general adsorbents for many molecules, especially macromolecules. Therefore, it is difficult to know, if RNA is seen to adsorb onto a surface, whether the adsorption is in any sense specific or whether it is just another example of a general feature that big molecules adsorb on big surfaces.

Thus, in a second approach, the mineral itself was obtained from a natural source, and the fraction of RNA bound to it was measured.

In a third approach, two precipitated minerals were combined, and the partition of radiolabeled RNA between the two was measured in a competition experiment.

This strategy then asked whether the *trend* in RNA adsorption was consistent across the same minerals in their various forms and presentations, especially within a set of minerals having a common anion (e.g., all carbonates) but differing in their cationic components (e.g., magnesium carbonate, calcium carbonate, strontium carbonate, and barium carbonate). Binary carbonate minerals within Row II (alkaline earth) cations were especially interesting, not only because carbon dioxide is likely to have been an abundant component of an early Earth atmosphere, but also because alkaline earths form a well-known set of binary carbonates that include insoluble magnesium, calcium, strontium, and barium carbonates (magnesite, calcite, strontianite, and witherite, respectively).

In the natural mineral experiments, we observed a periodic table trend. For example, while only a quarter of the radioactivity remained bound to the surface of the specimen of magnesite (with magnesium), ~94% of the reactivity was bound to the surface of the specimen of witherite (with barium). The fraction bound to calcite (calcium) and strontianite (strontium) were intermediate, 47% and 83%. Thus, a periodic table trend was observed with the binding of RNA to the carbonates relatively Ba > Sr > Ca > Mg. We then asked whether the same results could be qualitatively observed with precipitated synthetic minerals, and they could. Here, the percentage adsorption ranged from 95% to 77%, again with the ranking Ba > Sr > Ca > Mg.

To complete the analysis, various pairs of synthetic carbonates were then co-precipitated by mixing aqueous solutions of the alkali metal chlorides with an aqueous solution of sodium carbonate in a 1:1 ratio. These were then separated gravitationally, as the different carbonates have different densities ($CaCO_3$ 2.71 g/cm^3, $MgCO_3$ 2.96 g/cm^3, $SrCO_3$ 3.5 g/cm^3, $BaCO_3$ 4.29 g/cm^3). The partition of RNA between each pair was then observed by pre-equilibration of the radiolabeled RNA in a column with the two minerals, followed by freezing and dissection of the column, counting various slices within it. The labeled RNA partitioned as before; Ba > Sr > Ca > Mg.

As the rationale, if the same trends are observed both in precipitated synthetic minerals and in natural minerals, and if RNA is partitioned consistently between two mineral species precipitated together, the effects cannot easily be nonspecific general adsorption of big biopolymers on to big surfaces. Rather they must be intrinsic statements about the affinity of RNA for the different mineral species.

A series of sulfate minerals was more difficult to obtain as binary alkaline earth minerals, since the first member of the series (magnesium sulfate, epsomite) is quite soluble in water. However, with precipitated synthetic minerals, the same trend was observed, with barium sulfate binding RNA more than strontium sulfate, which bound more RNA than calcium sulfate. The corresponding trend was also observed in the specimens of the natural minerals, with barite > celestine > gypsum (59% > 49% > 20%).

While a periodic table trend is easy to observe in these cases, there is no reason a priori why such a trend should exist. For example, one might speculate that RNA would adsorb better onto a surface if the pattern of anion and cation sites on the surface matched more closely the distances of the anionic sites (phosphates) on the RNA molecule. While one might expect different cations in a mineral would change the spacing of those sites, there is no reason why the heaviest cation would have sites that match RNA the best. Indeed, if this were the mechanism for different surfaces having different affinities for RNA, one might expect a trend to have a mineral that maximally adsorbs somewhere in the middle of the series, rather the end of the series.

This may have been observed. For example, among the precipitated phosphates, the calcium species bound RNA (93%) more than the magnesium phosphate (64%), the strontium phosphate (84%), and barium phosphate (32%). Since "strontium apatite" and "barium apatite" are very seldom found in nature, natural minerals were not available for this study.

Another layer of complexity comes from the fact that same atoms can form different crystal forms. For example, calcium carbonate can precipitate as calcite, aragonite, or vaterite. Calcite crystallizes a trigonal space group; aragonite and vaterite are orthorhombic (Dal Negro and Ungaretti 1971; Maslen et al. 1993). Calcite is the more stable and consequently most common phase, while aragonite is less common, although it does occur in nature as a metastable phase (Fyfe and Bischoff 1965). Vaterite, also known as μ-$CaCO_3$, is a third metastable phase of $CaCO_3$. It occurs much less commonly in nature because it is the least thermodynamically stable. It generally and rapidly transforms itself into one of the other two forms (Grasby 2003). Vaterite is mostly seen when biological systems intervened to precipitate calcium carbonate.

To complete the analysis of the $CaCO_3$ system, natural specimens of aragonite and calcite were examined. Aragonite consistently adsorbed more radiolabeled RNA than calcite. To obtain a synthetic mineral by precipitation, we reasoned that if RNA prefers to bind to aragonite over calcite, then perhaps RNA would nucleate the formation of an aragonite precipitate over a calcite precipitate.

In forming the minerals synthetically, calcite dominates $CaCO_3$ that precipitates upon mixing $CaCl_2$ and Na_2CO_3 in water at near-neutral pH and room temperature and pressure; absent contaminant (McCauley and Roy 1974), aragonite is not formed. We easily reproduced this general result, establishing the structure of the precipitated phases that we obtained by both staining with Feigl stain (silver sulfate and manganese sulfate) (Feigl 1937) and by powder X-ray diffraction.

Initial results were auspicious. Feigl stain suggested that $CaCO_3$ precipitated preferentially as aragonite in the presence of RNA.

We then did powder X-ray diffraction to confirm the crystalline form of the precipitated calcium carbonate. Here, results were variable, but the precipitate formed in the presence of RNA was often identified as being primarily vaterite. We do not have a molecular interpretation of these observations, but this might be a first example of RNA-induced mineral polymorphism.

Interestingly, we also found that RNA bound to aragonite was more thermally stable than the same RNA in aqueous solution. Indeed, ~70% of the RNA bound to aragonite remained full-length after incubation at 95 °C for 2 h, while RNA treated the same way but in aqueous solution showed high levels of degradation.

Thus, we showed that, where it is possible, a comparison of natural minerals, synthetically precipitated minerals, and co-precipitated mineral combinations can be used to drive the conclusion that the adsorbance data collected are relevant to the mineral species themselves and do not merely reflect the adherence of large macromolecules to large surfaces.

The most obvious limitation of this comparative approach comes from nature herself. The rarity of minerals having different elemental compositions determines their availability for these experiments. Some elemental compositions are simply not found in nature at all.

The most striking outcome of these results is the periodic table relationship in the adsorbance of RNA to alkaline earth minerals, both the carbonates and the sulfates. All of these minerals are likely to have been present on early Earth. They are also known on Mars (Ehlmann et al. 2008). Today, most calcium carbonate is the result of biological activity. Where that activity is not present today (perhaps, but perhaps not, on Mars), we might expect to find stabilized RNA formed abiologically.

3.4.4.3 Adsorption onto Opals

Oxygen and silicon are the first and second most abundant elements in Earth's crust (Cox 1989). Therefore, it is not surprising that Earth's near-surface geology is dominated by silicate minerals. Indeed, species containing only silicon and oxygen ("silica") come in multiple forms, including quartz (Agricola 1530), perhaps best known as large crystals forming from supercritical water at high pressures deep in the crust. However, at lower pressures near the Earth's surface, alkaline water dissolves silica in small amounts, from which silica is reprecipitated. In the first phases of reprecipitating silica, the result is "opal."

Thus, it is not surprising that silica has attracted the attention of those interested in the interactions between minerals and organic species during the emergence of the first life on Earth (Kim and Benner 2010; Lambert et al. 2010a, b).

Opal has a very high surface area, raising the possibility that silica might adsorb RNA (Nishiyama et al. 2013). Further, it has long been known that nucleic acids are adsorbed on glass substrates in prep kits that recover small amounts of DNA/RNA from complex biological mixtures (Decher et al. 1994). These, of course, involve chaotropic salts unlikely to have been present on early Earth in abundance.

In our laboratory, we asked, much like we did for carbonates above, whether silica adsorption might have played a role in the stabilization of RNA (Biondi et al. 2016).

Here we showed that synthetic opal does indeed adsorb RNA, surprisingly, much better than to amorphous silica prior to aqueous reworking (ASPAR). Further, RNA adsorbed to opal remains intact, as shown by its ability to give full-length products using reverse transcriptase. Full-length product could also be recovered from ASPAR, but RT-PCR was required to obtain the product in detectable amounts.

For these experiments, opal and amorphous silica were created using a modified Stöber process (Stöber et al. 1968) by acid-catalyzed hydrolysis of tetraethoxysilane. This initially forms fully amorphous SiO_2. Incubation with water over periods of months allows the reworking of the silica to give natural-like opal. These two types of silica are ~ 30% porous, with pore sizes of ~ 20 Å for amorphous silica and ~ 90 Å for opal. For comparison, we note that RNA nucleotide averages ~12 Å in dimension, measured from the Watson-Crick edge of the nucleobase to the phosphate.

Adsorption on SiO_2 surfaces was measured by using a ^{32}P-5′-labeled 83 nt long RNA molecule. For opal, the fraction of RNA tightly bound to the mineral was calculated to be 1.355 pmoles/cm^2 of surface (a calculation that ignored possible movement of RNA from the surface into the bulk solid material). Even a wash treatment with NaF pH 9.5 released only ~5.4% total of the bound radiation after 2 days incubation. As the label is still associated with macromolecular RNA, this was consistent with the protection of RNA by silica against alkaline degradation. For amorphous SiO_2 on the other hand, adsorption was only 0.32 pmoles/cm^2, about a quarter of what was seen with the synthetic opal. This lower adsorption might be ascribed to the smaller pore size of this mineral compared to the opal version of SiO_2. This would also indicate either that the (main) RNA-mineral interaction is based on occupation of mineral pores, or that the total surface was inaccessible by way of the small pores (20 Å) to large molecules such as RNA, compared to the pore size in structured opal of ~ 90 Å. Considering that the 83 nt long RNA molecule used in this study might range from 228 Å in size for the fully relaxed (but mostly double-stranded) form to 48 Å in diameter in a fully folded form [calculated from the Flory law of gyration radius: RG = 5.5x(number of nts)1/3 (Hyeon et al. 2006)], the RNA could only partially enter the 20 Å pores of SiO_2 rod through its axial single- or double-stranded edges (10–20 Å), while it could gain access to the pores of opal in various degrees of folding.

The hypothesis that RNA in SiO_2 rod is adsorbed via the small pores of the mineral was also supported by results of reverse transcription of the adsorbed RNA. This was used to demonstrate that full-length material was adsorbed. Reverse transcripts were in good yields, even after 2 days incubation in NaF pH 9.5, indicating that the adsorption mechanism did not compromise the integrity of the molecule, but rather stabilized the RNA against alkaline degradation.

On the other hand, with amorphous SiO_2, full-length reverse transcripts could not be obtained in detectable amounts, even though shorter amplified molecular fragments could be observed. Full-length cDNA could only be detected after 25 cycles of PCR amplification. This indicated that very little intact RNA survived.

These results suggested that as silica progresses from its completely amorphous state (the state presumably created the instant that a tetraalkoxysilane derivative hydrolyzes) to more organized (but still bulk amorphous) states, RNA oligonucleotides are better able to be adsorbed and stabilized. These experiments do not allow us to clearly distinguish, as an explanation of these results, between the possibility that the greater crystalline order after months of aqueous reworking is directly responsible for the improved adsorption/stabilization, or whether it is indirectly responsible as a result of greater pore size, greater accessible surface area, or both.

Interestingly, very little RNA adsorbed on the surface of a fully grown natural quartz crystal (here, a natural "Herkimer diamond") or a natural agate. Natural obsidian (a volcanic glass) adsorbed some RNA better. However, among the natural silica species, natural opals adsorbed the best (Biondi et al. 2017).

In the "discontinuous synthesis model" for the origin of RNA (Neveu et al. 2013), RNA emerged in dry valleys fed intermittently by aqueous runoff from basaltic high lands, such as valleys currently found on Mars or Death Valley in California. It is in exactly these types of dry valleys that silica minerals would be reworked to give opals. Indeed, much of the gem opals found widely in jewelry are created in these environments. Through results reported here that show the stabilization of RNA in the environments on these minerals, additional steps might be added to this model, providing stabilized oligoribonucleotides that might be useful as starting points for molecular Darwinism. Indeed, the length of the 83 nt RNA molecule reported in this study, which is stabilized on opal, is comparable to the length of the RNA that Holliger, Joyce, and others suggest might be able to catalyze its own self-replication.

3.5 The Big Picture: The Discontinuous Synthesis Model for the Prebiotic Formation of RNA

We are now prepared to put this together in a "big picture," which we have called the "discontinuous synthesis model" (DSM) for the prebiotic formation of RNA in a mineral environment (Fig. 3.8) (Neveu et al. 2013). Because it incorporates ideas and data from many labs, the DSM has no particular ownership. It requires these elements:

1. A basaltic watershed region containing serpentinizing olivines, igneous apatites, and igneous tourmalines that are eroded by rain falling from a post-veneer atmosphere containing HCHO, HNCNH, HCN, HCCCN, and trace glycolaldehyde ($HOCH_2CHO$). The liquid is alkaline (pH 10–11) and contains borate at 35–100 mM concentrations; it supports the aqueous processes in the DSM (blue in Fig. 3.8), including the borate-moderated formose process (Kim et al. 2011) that creates borate-carbohydrate complexes. These conditions also support hydrolysis of HCN and $NCNH_2$, generating formamide, urea, ammonium formate, and carbonate.
2. The effluent emerges into an arid evaporite environment with intermittent water. When water evaporates, formamide, ammonium formate, and urea remain behind. Formamide boils with decomposition at over 200 °C at 1 atm; it is conceivably a liquid on the surface of Mars today.
3. Temperatures in the arid environment range from −20 to 60 °C. However, when encountering local geothermal environments, higher temperatures allow formamide, ammonium formate, and other products derived from HCN (e.g., diaminomaleonitrile, DAMN) to form nucleobases (Saladino et al. 2003). Borate minerals assist this (Saladino et al. 2011). These environments also allow chemistry of Becker et al. (2016) and Kim and Benner (2017) to form nucleosides and nucleoside phosphates.
4. The pH in the arid environment is buffered by atmospheric CO_2 (Sleep et al. 2011) but is intermittently raised as alkaline solutions are reintroduced. At pH ~6, molybdate can complete the formation of ribose, which is stabilized as an organic mineral.
5. The arid environment is also exposed to volcanic gasses, including sulfur dioxide. These may stabilize simple carbohydrates like glyceraldehyde and glycolaldehyde, preserving them in reservoirs in sufficient amounts to rescue the Powner et al. (2009) model for the formation of pyrimidine nucleoside phosphates.

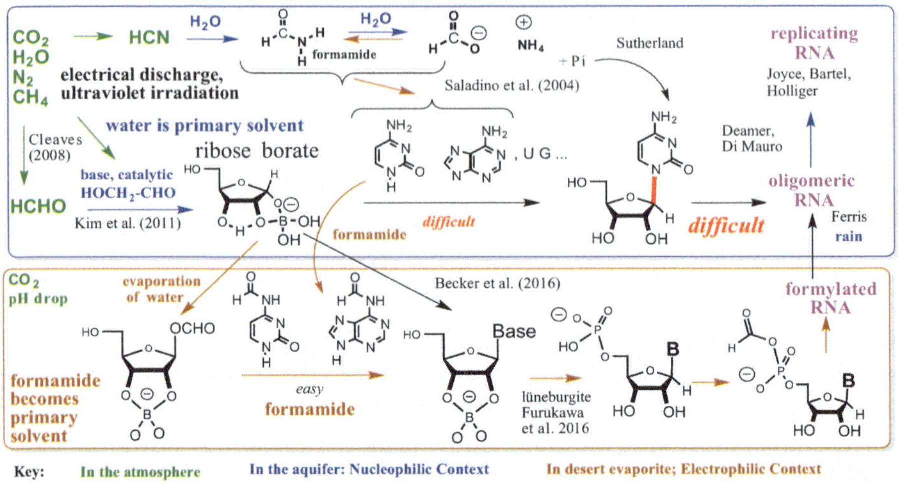

Fig. 3.8 The Discontinuous Synthesis Model (DSM) (Neveu et al. 2013) for RNA formation, based on results from labs worldwide, with atmospheric steps (green), steps in water environments (blue), and steps in nonaqueous "exotic" solvents such as formamide and urea. Each step has experimental support, with key papers indicated. First, HCHO, glycolaldehyde, $NCNH_2$ (cyanamide), and HCN are formed in a CO_2-CH_4-N_2-H_2O post-veneer atmosphere by UV, electrical discharge, silent discharge (Löb 1913), and cosmic energy (Pinto et al. 1980; Holland 1984). These are rained into aquifers in contact with serpentinizing basalts containing peridotite and igneous tourmalines, which create aqueous alkali that convert HCN to formamide and ammonium formate, cyanamide to urea, and formaldehyde to carbohydrates, the last being stabilized as their borate complexes. The effluent is then delivered to an arid evaporite at −20 to 60 °C, where atmospheric CO_2 lowers the pH to ~ 6.5 (Sleep et al. 2011), allowing formation of polyphosphates and phosphate-formate anhydrides and, with Mo^{6+}, in turn allowing Bilik reactions for carbohydrate processing (Kim et al. 2011; Petrus et al. 2001). Desert dehydration enriches exotic solvents such as formamide and urea, also precursors for nucleobases in local geothermal environments (Anumukonda et al. 2011; Yuasa et al. 1984). Nucleosides arise by Becker et al. (2016), Powner et al. (2009), or Kim and Benner (2017) routes. Next, borophosphates (e.g., lüneburgite) form 5′-nucleoside phosphates. Later, mixed anhydrides (formate, phosphate, thiophosphate) generate oligomeric RNA, adsorbing in homochiral form on mineral surfaces and/or stabilized by adsorption by opals (Biondi et al. 2016). Flooding of evaporite basin returns water, leading to the hydrolysis of formate esters and N-formylated nucleobases (half-life ~1 month at 80 °C, pH 6). Subsequent steps involving catalysis by oligomeric RNA rely on work from Lehman, Szostak, and Joyce (Lehman 2003; Lincoln and Joyce 2009)

6. Dehydration followed by rehydration is frequent. This gives polyphosphates, thiophosphates (if volcanic H_2S is available, or with a combination of phosphite and S_8), and phosphate-formate anhydrides. These condense to form linear RNA, which is stabilized and concentrated by adsorption on mineral surfaces.

We titled DSM as "discontinuous" to illustrate why it is not yet (and should not yet be) accepted as a solution to the "origins of Darwinism" problem. Too many steps are disconnected, requiring a chemist to clean up the output of the first step to prepare it for input into the second step. *However, we know when we are succeeding in this framework when this cleanup becomes less severe or goes away entirely.*

References

Abramov O, Mojzsis SJ (2009) Microbial habitability of the Hadean Earth during the late heavy bombardment. Nature 459:419–422

Agricola G (1530) Quarzum. In: Bermannus, Sive De Re Metallica, in Aedibus Frobenianis. Basileae, p 129

Allègre CJ, Manhès G, Göpel G (2008) The major differentiation of the Earth at ~4.45 Ga. Earth Planet Sci Lett 267:386–398

Andreani M, Munoz M, Marcaillou C et al (2013) μXANES study of iron redox state in serpentine during oceanic serpentinization. Lithos 178:70–83

Anet FA (2004) The place of metabolism in the origin of life. Curr Opin Chem Biol 8:654–659

Anumukonda LN, Young A, Lynn DG et al (2011) Adenine synthesis in a model prebiotic reaction: connecting origin of life chemistry with biology. J Chem Educ 88:1698–1701

Appayee C, Breslow R (2014) Deuterium studies reveal a new mechanism for the formose reaction involving hydride shifts. J Am Chem Soc 136:3720–3723

Bach W, Paulick H, Garrido CJ et al (2006) Unravelling the sequence of serpentinization reactions: petrography, mineral chemistry, and petrophysics of serpentinites from MAR 15_N (ODP Leg 209, Site 1274). Geophys Res Lett 33:L13306

Bada JL, Chalmers JH, Cleaves HJ (2016) Is formamide a geochemically plausible prebiotic solvent? Phys Chem Chem Phys 18:20085–20090

Becker H (2006) Highly siderophile element composition of the Earth's primitive upper mantle: constraints from new data on peridotite massifs and xenoliths. Geochim Cosmochim Acta (17):4528–4550

Becker S, Thoma I, Deutsch A et al (2016) A high-yielding, strictly regioselective prebiotic purine nucleoside formation pathway. Science 352(6287):833–836

Bell EA, Boehnke P, Harrison TM et al (2015) Potentially biogenic carbon preserved in a 4.1 billion-year-old zircon. Proc Natl Acad Sci U S A 112(47):14518–14521

Benner SA (2009) The life, the universe and the scientific method. FfAME Press, .Gainesville, 312 pp.

Benner SA (2013) Synthesis as a route to knowledge. Biol Theory 8:357–367

Benner SA (2017a) Detecting Darwinism from molecules in the Enceladus plumes, Jupiter's moons, and other planetary water lagoons. Astrobiology 17(9):840–851

Benner SA (2017b) Uniting natural history with the molecular sciences. The ultimate multidisciplinarity. Acc Chem Res 50:498–502

Benner SA, Kim HJ (2015) The case for a Martian origin for Earth life. In: Hoover RB, Levin GV, Rozanov, AY, Wickramasinghe NC (eds) Instruments, methods, and missions for astrobiology XVII, SPIE Optical Engineering+ Applications, 9606, 96060C

Benner SA, Ellington AD, Tauer A (1989) Modern metabolism as a palimpsest of the RNA world. Proc Natl Acad Sci U S A 86:7054–7058

Benner SA, Devine KG, Matveeva LN et al (2000) The missing organic molecules on Mars. Proc Natl Acad Sci U S A 97:2425–2430

Benner SA, Caraco MD, Thomson JM et al (2002) Planetary biology. Paleontological, geological, and molecular histories of life. Science 293:864–868

Benner SA, Bains W, Seager S (2013) Models and standards of proof in cross-disciplinary science: the case of arsenic DNA. Astrobiology 13:510–513

Benner SA, Karalkar NB, Hoshika S et al (2016) Alternative Watson-Crick synthetic genetic systems. Synthetic Biology. Cold Spring Harb Perspect Biol 8(11). doi: https://doi.org/10.1101/cshperspect.a023770

Bernal JD (1951) The physical basis of life. Routledge and Kegan Paul, London

Berndt ME, Allen DE, Seyfried WE (1996) Reduction of CO_2 during serpentinization of olivine at 300 °C and 500 bar. Geology 24(4):351–354

Biondi E, Branciamore S, Maurel MC et al (2007a) Montmorillonite protection of an UV-irradiated hairpin ribozyme. Evolution of the RNA world in a mineral environment. BMC Evol Biol 7 (Suppl 2):S2
Biondi E, Branciamore S, Fusi L et al (2007b) Catalytic activity of a hammerhead ribozyme in a clay mineral environment: implications for the RNA World. Gene 389:10–18
Biondi E, Howell L, Benner SA (2016) Opal adsorbs and stabilizes RNA – a hierarchy of prebiotic silica minerals. Syn Lett 27:A–E
Biondi E, Furukawa Y, Kwai J et al (2017) Adsorption of RNA on mineral surfaces and mineral precipitates. Beilstein J Org Chem 13:393–404
Blöchl E, Keller M, Wächtershäuser G et al (1992) Reactions depending on iron sulfide and linking geochemistry with biochemistry. Proc Natl Acad Sci U S A 89(17):8117–8120
Boehnken P, Harrison TM (2016) Illusory late heavy bombardments. Proc Natl Acad Sci U S A 113 (39):10802–10806
Boily JF, Persson P, Sjöberg S (2000a) Benzenecarboxylate complexation at the goethite-water interface: I. A mechanistic description of pyromellitate surface complexes from the combined evidence of infrared spectroscopy, potentiometry, adsorption data and surface complexation modeling. Langmuir 16:5719–5729
Boily JF, Persson P, Sjöberg S (2000b) Benzenecarboxylate surface complexation at the goethite (α-FeOOH)/water interface: II. Linking IR spectroscopic observations to mechanistic surface complexation models for phthalate, trimellitate, and pyromellitate. Geochim Cosmochim Acta 64(20):3453–3470
Bonner WA, Kavasmaneck PR, Martin FS et al (1974) Asymmetric adsorption of alanine by quartz. Science 186(4159):143–144
Bonner WA, Kavasmaneck PR, Martin FS et al (1975) Asymmetric adsorption by quartz: a model for the prebiotic origin of optical activity. Orig Life 6(3):367–376
Borisov AA (2016) Mutual interaction of redox pairs in silicate melts: equilibria involving metallic phases. Petrology 24(2):117
Brandes JA, Boctor NZ, Cody GD et al (1998) Abiotic nitrogen reduction on the early Earth. Nature 395(6700):365–367
Brasser R, Mojzsis SJ (2017) A colossal impact enriched Mars' mantle with noble metals. Geophys Res Lett. https://doi.org/10.1002/(2017GL074002
Brasser R, Mojzsis SJ, Werner SC et al (2016) Late veneer and late accretion to the terrestrial planets. Earth Planet Sci Lett 455:85–93
Bregestovski PD (2015) "RNA World", a highly improbable scenario of the origin and early evolution of life on Earth. J Evol Biochem Physiol 51:72–84
Burcar BT, Barge LM, Trail D et al (2015) RNA oligomerization in laboratory analogues of alkaline hydrothermal vent systems. Astrobiology 15:509–522
Burcar BT, Pasek M, Gull M et al (2016) Darwin's warm little pond: a one-pot reaction for prebiotic phosphorylation and the mobilization of phosphate from minerals in a urea-based solvent. Angew Chem Int Ed 55(42):13249–13253
Burton FG, Neuman MW, Neuman WF (1969) On the possible role of crystals in the origins of life. I. The adsorption of nucleosides, nucleotides and pyrophosphate by apatite crystals. Biosystems 3(1):20–26
Butlerov A (1861) Bildung einer zuckerartigen Substanz durch Synthese. Ann Chem 120:295–298
Cairns-Smith AG (1977) Takeover mechanisms and early biochemical evolution. Biosystems 9 (2–3):105–109
Cairns-Smith AG (1982) Genetic takeover and the mineral origins of life. Cambridge University Press, Cambridge
Cairns-Smith AG (2005) Sketches for a mineral genetic material. Elements 1:157–161
Calvin M, Calvin GI (1964) Atom to Adam. Am Sci 52:163
Cech TR (2000) The ribosome is a ribozyme. Science 289(5481):878–879
Cech TR, Bass BL (1986) Biological catalysis by RNA. Annu Rev Biochem 55:599–629

Chapman CR, Cohen BA, Grinspoon DH (2007) What are the real constraints on the existence and magnitude of the late heavy bombardment? Icarus 189(1):233–245

Cherniak DJ, Hanchar JM, Watson EB (1997) Rare-earth diffusion in zircon. Chem Geol 134:289–301

Cleaves HJ (2008) The prebiotic geochemistry of formaldehyde. Precambrian Res 164:111–118

Cody GD (2004) Transition metal sulfides and the origins of metabolism. Geophysical Laboratory, Carnegie Institution of Washington, Washington, DC, pp 569–599

Cody GD (2005) Geochemical connections to primitive metabolism. Elements 1:139–143

Cody GD, Boctor NZ, Filley T et al (2000) The primordial synthesis of carbonylated iron-sulfur clusters and the synthesis of pyruvate. Science 289:1339–1339

Cody GD, Boctor NZ, Hazen RM et al (2001) Geochemical roots of autotrophic carbon fixation: Hydrothermal experiments in the system citric acid, H_2O-($\pm$FeS)-($\pm$NiS). Geochim Cosmochim Acta 65(20):3557–3576

Cody GD, Boctor NZ, Brandes JA et al (2004) Assaying the catalytic potential of transition metal sulfides for abiotic carbon fixation. Geochim Cosmochim Acta 68(10):2185–2196

Cohen BA, Swindle TD, Kring DA (2000) Support for the lunar cataclysm hypothesis from lunar meteorite impact melt ages. Science 290:1754–1756

Condie KC (2018) A planet in transition: the onset of plate tectonics on Earth between 3 and 2 Ga. Geosci Front 9(1):51–60

Cox PA (1989) The elements: their origin, abundance, and distribution. Oxford University Press, Oxford

da Silva JAL, Holm NG (2014) Borophosphates and silicophosphates as plausible contributors to the emergence of life. J Colloid Interface Sci 431:250–254

Dal Negro A, Ungaretti L (1971) Refinement of the crystal structure of aragonite. Am Mineral 56:768–772

Danielson LR, Righter K, Newville M et al (2011) Molybdenum valence in basaltic silicate melts: effects of temperature and pressure. In: 42nd Lunar and planetary science conference, 2609

Dauphas N, Schauble EA (2016) Mass fractionation laws, mass-independent effects, and isotopic anomalies. Annu Rev Earth Planet Sci 44:709–783

Dauphas N, Chen J, Papanastassiou D (2015) Testing Earth–Moon isotopic homogenization with calcium-48. In: Lunar and planetary science conference XXXXVI, 2436

Day JM, Walker RJ (2015) Highly siderophile element depletion in the Moon. Earth Planet Sci Lett 423:114–124

Decher G, Lehr B, Lowack K et al (1994) New nanocomposite films for biosensors – layer-by-layer adsorbed films of polyelectrolytes, proteins or DNA. Biosens Bioelectron 9:677–684

Decker P, Schweer H, Pohlamnn R (1982) Bioids: X. Identification of formose sugars, presumable prebiotic metabolites, using capillary gas chromatography/gas chromatography—mas spectrometry of n-butoxime trifluoroacetates on OV-225. J Chromatogr A 244:281–291

Dyson F (1985) Origin of life. Cambridge University Press, Cambridge

Ehlmann BL, Mustard JF, Murchie SL et al (2008) Orbital identification of carbonate-bearing rocks on Mars. Science 322:1828–1832

Ertem G, Ferris JP (1996) Synthesis of RNA oligomers on heterogeneous templates. Nature 379 (6562):238–240

Ertem G, Ferris JP (1997) Template-directed synthesis using the heterogeneous templates produced by montmorillonite catalysis. A possible bridge between the prebiotic and RNA worlds. J Am Chem Soc 119(31):7197–7201

Ertem G, Ertem MC, McKay CP et al (2017) Shielding biomolecules from effects of radiation by Mars analogue minerals and soils. Int J Astrobiol 16(03):280–285

Eschenmoser A (1997) Towards a chemical etiology of nucleic acid structure. Orig Life Evol Biosph 27(5–6):535–553

Evgenii K, Wolfram T (2000) The role of quartz in the origin of optical activity on earth. Orig Life Evol Biosph 30(5):431–434

Feely RA, Trefry JH, Lebon GT et al (1998) The relationship between P/Fe and V/Fe ratios in hydrothermal precipitates and dissolved phosphate in seawater. Geophys Res Lett 25:2253–2256
Feigl F (1937) Qualitative analysis by spot tests. Nordemann, New York, p 400
Ferris JP (2005) Mineral catalysis and prebiotic synthesis: montmorillonite-catalyzed formation of RNA. Elements 1(3):145–149
Ferris JP, Ertem G (1992) Oligomerization of ribonucleotides on montmorillonite: reaction of the 5′-phosphorimidazolide of adenosine. Science 257(5075):1387–1389
Ferris JP, Ertem G (1993) Montmorillonite catalysis of RNA oligomer formation in aqueous solution. A model for the prebiotic formation of RNA. J Am Chem Soc 115(26):12270–12275
Ferris JP, Sanchez RA, Orgel LE (1968) Studies in prebiotic synthesis: III. Synthesis of pyrimidines from cyanoacetylene and cyanate. J Mol Biol 33(3):693–704
Feynman R (1974) Cargo cult science. Caltech commencement address. Reproduced in “Surely You’re Joking, Mr. Feynman”. Norton, New York
Fiore M, Strazewski P (2016) Prebiotic lipidic amphiphiles and condensing agents on the early Earth. Life 6. https://doi.org/10.3390/life6020017
Fox SW (1965) A theory of macromolecular and cellular origins. Nature 205:328
Fuchs LH (1969) The phosphate mineralogy of meteorites. In: Meteorite research. Springer, Dordrecht, pp 683–695
Fyfe WS, Bischoff JL (1965) The calcite-aragonite problem. In Pray LC, Murray RC (eds), Dolomitization and limestone diagenesis: a symposium. Society of Economic Paleontologists and Mineralogists, Special Publication, 13, pp 3–13
Galison PL (1987) How experiments end. University of Chicago Press, Chicago
Genda H, Fujita T, Kobayashi H et al (2017) Impact erosion model for gravity-dominated planetesimals. Icarus 294:234–246
Goldschmidt VM (1952) Geochemical aspects of the origin of complex organic molecules on the earth, as precursors to organic life. New Biol 12:97–105
Grasby SE (2003) Naturally precipitating vaterite (μ-$CaCO_3$) spheres: Unusual carbonates formed in an extreme environment. Geochim Cosmochim Acta 67:1659–1666
Grosjean M, Geyh MA, Messerli B et al (1995) Late-glacial and early Holocenelake sediments, groundwater formation and climate in the Atacama altiplano 22–241S. J Paleolimnol 14:241–252
Guerrier-Takada C, Gardiner K, Marsh T et al (1983) The RNA moiety of ribonuclease-P is the catalytic subunit of the enzyme. Cell 35:849–857
Gull M, Mojica MA, Fernández FM et al (2015) Nucleoside phosphorylation by the mineral schreibersite. Sci Rep 5:17198. https://doi.org/10.1038/srep17198
Hamano K, Abe Y, Genda H (2013) Emergence of two types of terrestrial planet on solidification of magma ocean. Nature 497:607–610
Hanczyc MM, Fujikawa SM, Szostak JW (2003) Experimental models of primitive cellular compartments: encapsulation, growth, and division. Science 302(5645):618–622
Harman CE, Kasting JF, Wolf ET (2013) Atmospheric production of glycolaldehyde under hazy prebiotic conditions. Orig Life Evol Biosph 43:77–98
Hartmann WK (1975) Lunar “cataclysm”: a misconception? Icarus 24(2):181–187
Hazen RM (2005) Genesis. The scientific quest for life’s origins. Joseph Henry Press, Washington, DC
Hazen RM, Papineau D, Bleeker W et al (2008) Mineral evolution. Am Mineral 93 (11–12):1693–1720
Hedenquist JW, Arribas A, Gonzalez-Urien E (2000) Exploration for epithermal gold deposits. Rev Econ Geol 13:45–77
Herdewijn P (2001) TNA as a potential alternative to natural nucleic acids. Angew Chem Int Ed 40 (12):2249–2251
Holland H (1984) The chemical evolution of the atmosphere and oceans. Princeton University Press, Princeton, 598 p

Holm NG, Ertem G, Ferris JP (1993) The binding and reactions of nucleotides and polynucleotides on iron oxide hydroxide polymorphs. Orig Life Evol Biosph 23(3):195–215
Huber C, Wächtershäuser G (1997) Activated acetic acid by carbon fixation on (Fe,Ni)S under primordial conditions. Science 276(5310):245–247
Huber C, Wächtershäuser G (1998) Peptides by activation of amino acids with CO on (Ni,Fe)S surfaces: implications for the origin of life. Science 281(5377):670–671
Huber C, Eisenreich W, Hecht S et al (2003) A possible primordial peptide cycle. Science 301 (5635):938–940
Huber C, Kraus F, Hanzlik M et al (2012) Elements of metabolic evolution. Chem Eur J 18 (7):2063–2080
Hud NV, Jain SS, Li X et al (2007) Addressing the problems of base pairing and strand cyclization in template-directed synthesis. Chem Biodivers 4(4):768–783
Hyeon C, Dima RI, Thirumalai D (2006) Size, shape, and flexibility of RNA structures. J Chem Phys 125(19):194905
Ichida JK, Zou K, Horhota A et al (2005) An in vitro selection system for TNA. J Am Chem Soc 127(9):2802–2803
Inoue T, Orgel LE (1982) Oligomerization of (guanosine 5′-phosphor)-2-methylimidazolide on poly (C): an RNA polymerase model. J Mol Biol 162(1):201–217
Islam S, Bučar DK, Powner MW (2017) Prebiotic selection and assembly of proteinogenic amino acids and natural nucleotides from complex mixtures. Nat Chem 9:584–589
Izgu EC, Oh SS, Szostak JW (2016) Synthesis of activated 3′-amino-3′-deoxy-2-thio-thymidine, a superior substrate for the nonenzymatic copying of nucleic acid templates. Chem Commun 52 (18):3684–3686
Javoy M, Kaminski E, Guyot F et al (2010) The chemical composition of the Earth: enstatite chondrite models. Earth Planet Sci Lett 293:259–268
Jermann TM, Opitz JG, Stackhouse J et al (1995) Reconstructing the evolutionary history of the artiodactyl ribonuclease superfamily. Nature 374:57–59
Jones LC, Rosenbauer R, Goldsmith JI et al (2010) Carbonate control of H_2 and CH_4 production in serpentinization systems at elevated P-Ts. Geophys Res Lett 37:L14306
Joyce GF, Orgel LE (1999) Prospects for understanding the origin of the RNA world. In: Gestland RF, Cech RTR, Atkins JF (eds) The RNA World, 2nd edn. Cold Spring Harbor Press, Cold Spring Harbor, NY, pp 49–78
Keefe AD, Miller SL (1996) Was ferrocyanide a prebiotic reagent? Orig Life Evol Biosph 26 (2):111–129
Kim HJ, Benner SA (2010) Comment on “The silicate-mediated formose reaction: bottom-up synthesis of sugar silicates”. Science 329(5994):902-a
Kim HJ, Benner, SA (2017) Prebiotic stereoselective synthesis of purine and noncanonical pyrimidine nucleotide from nucleobases and phosphorylated carbohydrates. Proc Natl Acad Sci U S A (on line)
Kim HJ, Ricardo A, Illangkoon HI et al (2011) Synthesis of carbohydrates in mineral-guided prebiotic cycles. J Am Chem Soc 133:9457–9468
Kim HJ, Furukawa Y, Kakegawa T et al (2016) Evaporite borate-containing mineral ensembles make phosphate available and regiospecifically phosphorylate ribonucleosides: borate as a multifaceted problem solver in prebiotic chemistry. Angew Chem 55:15816–15820
Kring DA, Cohen BA (2002) Cataclysmic bombardment throughout the inner solar system 3.9–4.0Ga. J Geophys Res 107(E2):4-1–4-6
Kruijer TS, Kleine T, Fischer-Gödde M et al (2015) Lunar tungsten isotopic evidence for the late veneer. Nature 5:534–537
Lahav N (1994) Minerals and the origin of life – hypotheses and experiments in heterogeneous chemistry. Heterog Chem Rev 1:159–179
Lahav N, White D, Chang S (1978) Peptide formation in the prebiotic era: thermal condensation of glycine in fluctuating clay environments. Science 201(435 O):67–69

Lambert JB, Gurusamy-Thangavelu SA, Ma KBA (2010a) The silicate-mediated formose reaction: bottom-up synthesis of sugar silicates. Science 327:984–986
Lambert JB, Gurusamy-Thangavelu SA, Ma KBA (2010b) Response to comment on "The silicate-mediated formose reaction: bottom-up synthesis of sugar silicates". Science 329(5994):902-b
Lapen TJ, Righter M, Brandon AD et al (2010) A younger age for ALH84001 and its geochemical link to shergottite sources in Mars. Science 328:347–351
Larralde R, Robertson MP, Miller SL (1995) Rates of decomposition of ribose and other sugars. Implications for chemical evolution. Proc Natl Acad Sci U S A 92:8158–8160
Lehman N (2003) A case for the extreme antiquity of recombination. J Mol Evol 56:770–777
Leman LJ, Orgel LE, Ghadiri MR (2006) Amino acid dependent formation of phosphate anhydrides in water mediated by carbonyl sulfide. J Am Chem Soc 128(1):20–21
Levy M, Miller SL (1998) The stability of the RNA bases: implications for the origin of life. Proc Natl Acad Sci U S A 95:7933–7938
Levy M, Miller SL, Oró J (1999) Production of guanine from NH_4CN polymerizations. J Mol Evol 49:165–168
Li L, Prywes N, Tam CP et al (2017) Enhanced nonenzymatic RNA copying with 2-aminoimidazole activated nucleotides. J Am Chem Soc 139(5):1810–1813
Lincoln TA, Joyce GF (2009) Self-sustained replication of an RNA enzyme. Science 323:1229–1232
Löb W (1913) Uber das Verhalten des Formamids unter der Wirkung der stillen Entladung Ein Beitrag zur Frage der Stickstoff-Assimilation. Ber Dtsch Chem Ges 46:684–697
Lohrmann R (1972) Formation of urea and guanidine by irradiation of ammonium cyanide. J Mol Evol 1:263–269
Maher KA, Stevenson DJ (1988) Impact frustration of the origin of life. Nature 331:612–614
Malaterre C (2013) Synthetic biology and synthetic knowledge. Biol Theory 8:346–356
Mann U, Frost DJ, Rubie DC et al (2012) Partitioning of Ru Rh Pd Re Ir and Pt between liquid metal and silicate at high pressures and high temperatures – Implications for the origin of highly siderophile element concentrations in the Earth's mantle. Geochim Cosmochim Acta 84:593–613
Marchi S, Bottke WF, Cohen BE et al (2013) High-velocity collisions from the lunar cataclysm recorded in asteroidal meteorites. Nat Geosci 6:303–307
Markovitch O, Lancet D (2014) Multispecies population dynamics of prebiotic compositional assemblies. J Theor Biol 357:26–34
Maslen EN, Streltsov VA, Streltsova NR (1993) X-ray study of the electron density in calcite, $CaCO_3$. Acta Cryst B49:636–641
McCauley JW, Roy R (1974) Controlled nucleation and crystal growth of various $CaCO_3$ phases by the silica gel technique. Am Mineral 59:947–963
McCollom TM (2013) Miller-Urey and beyond: What have we learned about prebiotic organic synthesis reactions in the past 60 years? Annu Rev Earth Planet Sci 41:207–229
McCollom TM, Seewald JS (2001) A reassessment of the potential for reduction of dissolved CO2 to hydrocarbons during serpentinization of olivine. Geochim Cosmochim Acta 65 (21):3769–3778
Miller SL (1953) A production of amino acids under possible primitive earth conditions. Science 117:528–529
Miyakawa S, Cleaves HJ, Miller SL (2002) The cold origin of life: B. Implications based on pyrimidines and purines produced from frozen ammonium cyanide solutions. Orig Life Evol Biosph 32:209–218
Mojzsis SJ, Arrhenius G, McKeegan KD et al (1996) Evidence for life on Earth before 3,800 million years ago. Nature 384(6604):55–59
Molster FJ, Lim TL, Sylvester RJ et al (2001) The complete ISO spectrum of NGC 6302. Astron Astrophys 372:165–172
Morbidelli A, Marchi S, Bottke WF et al (2012) A sawtooth-like timeline for the first billion years of lunar bombardment. Earth Planet Sci Lett 355-356:144–151

Müller J, Fabricius F (1978) Lüneburgite [$Mg_3(PO_4)_2B_2O(OH)_4 \times 6\ H_2O$] in Upper Miocene sediments of the Eastern Mediterranean Sea. Init Rep DSDP 42:661–664. https://doi.org/10.2973/dsdp.proc.42-1.127.1978
Mutschler H, Wochner A, Holliger P (2015) Freeze-thaw cycles as drivers of complex ribozyme assembly. Nat Chem 7:502–508
Neveu M, Kim HJ, Benner SA (2013) The "Strong" RNA World hypothesis. Fifty years old. Astrobiology 13:391–403
Nishiyama T, Kagami Y, Yamauchi T et al (2013) Assembly of stimulus-sensitive gel particles with DNA-dye complexes. Polym J 45:659–664
Okafor CD, Lanier KA, Petrov AS et al (2017) Iron mediates catalysis of nucleic acid processing enzymes: support for Fe (II) as a cofactor before the great oxidation event. Nucleic Acids Res 45:3634–3642
Orgel LE (2000a) A simpler nucleic acid. Science 290(5495):1306–1307
Orgel LE (2000b) Self-organizing biochemical reactions. Proc Natl Acad Sci U S A 97:12503–12507
Oró JJ (1965) Investigation of organo-chemical evolution. In: Mamikunian G, Briggs MH (eds) Current aspects of exobiology. Pergamon Press, Oxford, pp 13–39
Parsons I, Lee MR, Smith JV (1998) Biochemical evolution II: origin of life in tubular microstructures on weathered feldspar surfaces. Proc Natl Acad Sci 95(26):15173–15176
Pasek MA (2016) Schreibersite on the early earth: scenarios for prebiotic phosphorylation. Geosci Front 8(2):329–335
Pearce BKD, Pudritz RE (2015) Seeding the pregenetic Earth: meteoritic abundances of nucleobases and potential reaction pathways. Astrophys J 807:85–94
Pearce BKD, Pudritz RE, Semenov DA et al (2017) Origin of the RNA world: the fate of nucleobases in warm little ponds. Proc Natl Acad Sci U S A. https://doi.org/10.1073/pnas.1710339114
Petrus L, Petrusová M, Hricovíniová Z (2001) The Bilik reaction. In: Stutz AE (ed) Topics in current chemistry: glycoscience, epimerisation, isomerisation and rearrangment reactions of carbohydrates, vol 215. Springer, Berlin, pp 15–41
Piccoli PM, Candela P (2002) Apatite in igneous systems. Rev Miner Geochem 48:255–292
Pinto J, Gladstone G, Yung Y (1980) Photochemical production of formaldehyde in Earth's primitive atmosphere. Science 210:183–185
Powner MW, Gerland B, Sutherland JD (2009) Synthesis of activated pyrimidine ribonucleotides in prebiotically plausible conditions. Nature 459:239–242
Puchtel IS, Walker RJ, Touboul M et al (2014) Insights into early Earth from the Pt–Re–Os isotope and highly siderophile element abundance systematics of Barberton komatiites. Geochim Cosmochim Acta 125:394–413
Qin L, Alexander CMD, Carlson RW (2010) Contributors to chromium isotope variation of meteorites. Geochim Cosmochim Acta 74:1122–1145
Ricardo A, Carrigan MA, Olcott AN et al (2004) Borate minerals stabilize ribose. Science 303:196
Rich A (1962) On the problems of evolution and biochemical information transfer. In: Kasha M, Pullmann B (eds) Horizons in biochemistry. Academic Press, NY, pp 103–126
Richter Y, Fischer B (2003) Characterization and elucidation of coordination requirements of adenine nucleotides complexes with Fe (II) ions. Nucleosides Nucleotides Nucleic Acids 22 (9):1757–1780
Righter K, Yang H, Costin G et al (2008) Oxygen fugacity in the Martian mantle controlled by carbon: New constraints from the nakhlite MIL 03346. Meteorit Planet Sci 43:1709–1723
Righter K, Danielson LR, Pando KM et al (2015) Highly siderophile element (HSE) abundances in the mantle of Mars are due to core formation at high pressure and temperature. Meteorit Planet Sci 50:604–631
Rosing MT (1999) ^{13}C-Depleted carbon microparticles in >3700-Ma sea-floor sedimentary rocks from West Greenland. Science 283:674–676

Rubie DC, Jacobson SA, Morbidelli A et al (2015a) Accretion and differentiation of the terrestrial planets with implications for the compositions of early-formed Solar System bodies and accretion of water. Icarus 248:89–108
Rubie DC, Nimmo F, Melosh HJ (2015b) Formation of the Earth's core. In: Schubert G (ed) Treatise on geophysics, vol 9: Evolution of the Earth, 2nd edn. Elsevier, Oxford, pp 43–79
Russell MJ, Hall AJ (1997) The emergence of life from iron monosulphide bubbles at a submarine hydrothermal redox and pH front. J Geol Soc 154(3):377–402
Russell MJ, Martin W (2004) The rocky roots of the acetyl-CoA pathway. Trends Biochem Sci 29 (7):358–363
Russell MJ, Daniel RM, Hall AJ et al (1994) A hydrothermally precipitated catalytic iron sulphide membrane as a first step toward life. J Mol Evol 39(3):231–243
Sahai N, Kaddour H, Dalai P (2016) The transition from geochemistry to biogeochemistry. Elements 12(6):389–394
Saladino R, Ciambecchini U, Crestini C et al (2003) One-pot TiO_2-catalyzed synthesis of nucleic bases and acyclonucleosides from formamide: implications for the origin of life. Chembiochem 4:514–521
Saladino R, Barontini M, Cossetti C et al (2011) The effects of borate minerals on the synthesis of nucleic acid bases, amino acids and biogenic carboxylic acids from formamide. Orig Life Evol Biosph 41:317–330
Sanchez RA, Orgel LE (1970) Studies in prebiotic synthesis: V. Synthesis and photoanomerization of pyrimidine nucleosides. J Mol Biol 47:531–543
Sanchez RA, Ferris JP, Orgel LE (1966) Cyanoacetylene in prebiotic synthesis. Science 154:784–785
Santos AR, Agee CB, McCubbin FM et al (2013) Apatite and merrillite from Martian meteorite NWA 7034. In: Lunar and planetary science conference 44, 2601
Schoonen M, Smirnov A (2016) Staging life in an early warm 'seltzer' ocean. Elements 12 (6):395–400
Schoonen M, Smirnov A, Cohn C (2004) A perspective on the role of minerals in prebiotic synthesis. Ambio 33(8):539–551
Sekimoto K, Takayama M (2012) Formation of hydrogen cyanide HCN under limited discharge conditions in non-reduced ambient air. ESCAMPIG XXI, Viana do Castelo, Portugal, 10–14 July
Shapiro R (1995) The prebiotic role of adenine: a critical analysis. Orig Life Evol Biosph 25:83–98
Shapiro R (1999) Prebiotic cytosine synthesis: a critical analysis and implications for the origin of life. Proc Natl Acad Sci U S A 96:4396–4401
Shapiro R (2007) A simpler origin for life. Sci Am 296:46–53
Sleep NH, Meibom A, Fridriksson T et al (2004) H_2-rich fluids from serpentinization: geochemical and biotic implications. Proc Natl Acad Sci U S A 101:12818–12823
Sleep NH, Bird DK, Pope EC (2011) Serpentinite and the dawn of life. Philos Trans R Soc B 366:2857–2869
Smith JV (1998) Biochemical evolution. I. Polymerization on internal, organophilic silica surfaces of dealuminated zeolites and feldspars. Proc Natl Acad Sci U S A 95(7):3370–3375
Smith JV, Arnold FP, Parsons I et al (1999) Biochemical evolution III: polymerization on oganophilic silica-rich surfaces, crystal-chemical modeling, formation of first cells, and geological clues. Proc Natl Acad Sci U S A 96(7):3479–3485
Stephenson JD, Hallis LJ, Nagashima K et al (2013) Boron enrichment in Martian clay. PLoS One 8 (6):e64624
Stephenson JD, Popović M, Bristow TF et al (2016) Evolution of ribozymes in the presence of a mineral surface. RNA 22:1893–1901
Stober W, Fink A, Bohn E (1968) Controlled growth of monodisperse silica spheres in micron size range. J Colloid Interface Sci 26:62–69
Summers DP, Chang S (1993) Prebiotic ammonia from reduction of nitrite by iron (II) on the early Earth. Nature 365(6447):630–633

Sutherland JD (2016) The origin of life. Out of the blue. Angew Chem Int Ed 55(1):104–121
Taves DR (1963) Similarity of octacalcium phosphate and hydroxyapatite structures. Nature 200 (4913):1312–1313
Taves DR, Reedy RC (1969) A structural basis for the transphosphorylation of nucleotides with hydroxyapatite. Calcif Tissue Int 3(1):284–292
Tera F, Papanastassiou DA, Wasserburg GJ (1974) Isotopic evidence for a terminal lunar cataclysm. Earth Planet Sci Lett 22(1):1–21
Touboul M, Puchtel IS, Walker RJ (2012) 182W evidence for long term preservation of early mantle differentiation products. Science 335:1065–1069
Touboul M, Liu J, O'Neil J et al (2014) New insights into the Hadean mantle revealed by 182 W and highly siderophile element abundances of supracrustal rocks from the Nuvvuagittuq greenstone belt, Quebec Canada. Chem Geol 383:63–75
Trail D, Watson EB, Tailby ND (2011) The oxidation state of Hadean magmas and implications for early Earth's atmosphere. Nature 480(7375):79–82
Van Vleck JH (1928) The correspondence principle in the statistical interpretation of quantum mechanics. Proc Natl Acad Sci U S A 14(2):178–188
Wächtershäuser G (1988a) Before enzymes and templates: theory of surface metabolism. Microbiol Rev 52(4):452–484
Wächtershäuser G (1988b) Pyrite formation, the first energy source for life: a hypothesis. Syst Appl Microbiol 10(3):207–210
Wächtershäuser G (1990a) Evolution of the first metabolic cycles. Proc Natl Acad Sci U S A 87 (1):200–204
Wächtershäuser G (1990b) The case for the chemoautotrophic origin of life in an iron-sulfur world. Orig Life Evol Biosph 20(2):173–176
Wächtershäuser G (1993) The cradle chemistry of life: on the origin of natural products in a pyrite-pulled chemoautotrophic origin of life. Pure Appl Chem 65(6):1343–1348
Warren PH (2011) Stable-isotopic anomalies and the accretionary assemblage of the Earth and Mars: a subordinate role for carbonaceous chondrites. Earth Planet Sci Lett 311:93–100
Weber AL (1995) Prebiotic polymerization: oxidative polymerization of 2,3-dimercapto-l-propanol on the surface of iron(III) hydroxide oxide. Orig Life Evol Biosph 25(1–3):53–60
Willbold M, Elliot T, Moorbath S (2011) The tungsten isotopic composition of the Earth's mantle before the terminal bombardment. Nature 477:195–198
Willbold M, Mojzsis SJ, Chen HW et al (2015) Tungsten isotope composition of the Acasta Gneiss Complex. Earth Planet Sci Lett 419:168–177
Young ED, Kohl IE, Warren PH et al (2016) Oxygen isotopic evidence for vigorous mixing during the Moon-forming giant impact. Science 351:493–496
Yuasa S, Flory D, Basile B et al (1984) Abiotic synthesis of purines and other heterocyclic compounds by the action of electrical discharges. J Mol Evol 21:76–80
Zhang J, Dauphas N, Davis AM et al (2012) The proto-Earth as a significant source of lunar material. Nat Geosci 5:251–255
Zhang S, Blain JC, Zielinska D et al (2013) Fast and accurate nonenzymatic copying of an RNA-like synthetic genetic polymer. Proc Natl Acad Sci U S A 110(44):17732–17737

Chapter 4
From the Dawn of Organic Chemistry to Astrobiology: Urea as a Foundational Component in the Origin of Nucleobases and Nucleotides

César Menor-Salván

Abstract Urea is formed in significant quantities in classic prebiotic model reactions and simply by hydrolysis of cyanide. It is a very interesting molecule, with chemical properties that make it a potential precursor of nucleobases and related molecules, as well as a promoter of phosphorylation. In addition, urea's physicochemical properties allow it to form a range of viscous eutectic solutions by simple evaporation or freezing. Thus, urea is the basis of a potential prebiotic environment that forms "little ponds." This chapter provides a historical perspective on the prebiotic chemistry of urea, from Wohler's synthesis in the early nineteenth century to the most recent works.

4.1 Introduction: What Is Prebiotic Chemistry?

The origin of life is one of the fundamental questions of humanity. Humans have wondered about the origin of themselves and all living beings since the emergence of abstract thinking. In trying to find answers, religions arose, philosophers debated, and scientists have been seduced by the challenges of this question in the frontier of knowledge.

The pursuit of a reasoned answer for the origin of life is as old as philosophy. According to Roman and Greek doxographers (because nothing written by his own hand has reached us), Anaximander of Miletus (c. 610–540 BCE) was the first western philosopher who envisioned a proto-evolutionary model (Kočandrle and

C. Menor-Salván (✉)
NSF-NASA Center for Chemical Evolution, School of Chemistry and Biochemistry, Georgia Institute of Technology, Atlanta, GA, USA

Department of Systems Biology, Universidad de Alcala, Madrid, Spain
e-mail: cesar.menor@uah.es

© Springer International Publishing AG, part of Springer Nature 2018

C. Menor-Salván (ed.), *Prebiotic Chemistry and Chemical Evolution of Nucleic Acids*, Nucleic Acids and Molecular Biology 35,
https://doi.org/10.1007/978-3-319-93584-3_4

Fig. 4.1 Anaxagoras of Clazomenae was one of promoters of philosophy in Athens. His rudimentary visions of evolution and the atomic nature of matter were the first attempt to give an elaborated, non-strictly mythological explanation of the origin of life

Kleisner 2013). According to the *Refutations of All Heresies* by Hippolytus of Rome:

> Living creatures arose from the moist being evaporated by the Sun. In the beginning man was similar to a different kind of animal, namely a fish

An idea about the origin of life is related to moisture and mud, with further diversification that led to different animals and the importance of the alternation of cold and hot periods. This theory was elaborated by Anaxagoras of Clazomenae (c. 510–428 BCE), who studied the philosophy of Anaximander through his teacher Anaximenes and his disciple Archelaus, who was said to be the teacher of Socrates. According to the important doxographer Diogenes Laertius (third century AD; Bredlow 2010), Anaxagoras taught that

> living beings were formed from the moisture, heat and earthy substances; later the species were propagated by generation from one another.

Anaxagoras also envisioned the rudiments of an atomic theory of matter in his book *Peri Physeos*, of which only some fragments have been preserved (Curd 2007). Anaxagoras coined the term *spermata*, which refers to the myriad of infinitesimal particles that compose all things through the separation and combination of particles from an original chaotic mixture. This concept is, rather incorrectly, regarded as the origin of the term and theory of panspermia. However, the term *panspermia* was never used by Anaxagoras or his doxographs, as far as can be inferred from the bibliography studied for this chapter (Fig. 4.1).

The ideas of Anaximandro and Anaxagoras were early non-religious and reason-based approaches to the question of the origins of life. Due to the ulterior dominance of Abrahamic religions, which condemned any idea that seemed to contradict the Scriptures, questions about abiogenesis (i.e., the chemical roots of the origins of life) did not resume until the nineteenth century and the famous letter of Charles Darwin of 1871 to his friend, the botanist Joseph Dalton Hooker, where he mused about a "warm little pond, with all sorts of ammonia and phosphoric salts... that a protein compound was chemically formed ready to undergo still more complex changes (Peretó et al. 2009)."

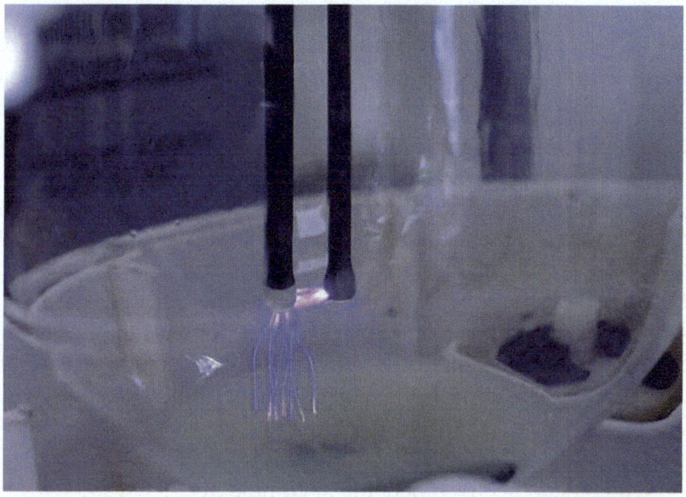

Fig. 4.2 A simplified setup based on the classic Stanley Miller experiment is shown: spark discharges simulating thunderbolts in a primordial, anoxic atmosphere containing methane lead to the synthesis of organic molecules, including amino acids. These organic substances accumulate in the form of a brown slick on the small water pond at the bottom of the reactor

Darwin was not actually interested in abiogenesis, however, and concluded that it was "mere rubbish, thinking at present of the origin of life." This is not really surprising, as the nineteenth century was a period of intense development of organic chemistry. Inadvertently, the foundations of the chemistry of life's origin—that is, prebiotic chemistry and chemical evolution—were set up. However, deep knowledge of biochemistry and molecular biology, together with the rise and development of the field of geochemistry, were necessary to build a meaningful chemical approach to the origins of life. Hence, it is not surprising that the study of the chemical origin of nucleic acids and life imply, in part, the rediscovery of chemistry from the nineteenth century to the mid-twentieth century, revised with the most advanced chemical technology and modern knowledge and then applied in geochemically plausible ways.

Thanks to the latest advances in biochemistry, geochemistry, astrochemistry, supramolecular chemistry, and molecular biology, we are entering into a golden age in the study of the origins of life—more than 60 years after the seminal work of Stanley Miller in 1953 (Fig. 4.2). These advances are making it possible to uncover the origins of life, using a bottom-up approach (prebiotic chemistry) together with a biological top-down approach toward the first minimal genomes and metabolisms.

The chemistry of the origins of life, or *prebiotic chemistry* is currently an important area of the science of astrobiology (Des Marais et al. 2008). One of the main goals of astrobiology is to understand how key prebiotic molecules, which were potentially relevant for the emergence of life, could be synthesized or delivered in planetary environments and how geochemical and environmental conditions lead to the emergence of life. Prebiotic chemistry therefore could be considered the link between geochemistry and astrochemistry on one hand, and chemical evolution and biochemistry on the other. Our role as researchers in the field of the chemical roots of life is to provide meaningful scenarios for the synthesis of raw organic materials and the origin of a biochemical system capable of evolution. Those scenarios are not necessarily mutually incompatible, and several mechanisms could lead to complementary or similar chemical spaces.

As Leslie Orgel wrote in 2004, it is not wise to try to define prebiotic chemistry too tightly, because it is a very elastic term that is highly dependent on interactions with the considered geological setting and feedback from other areas in the transdisciplinary field of astrobiology. Without forgetting the warning of one of the major luminaires of the field, prebiotic chemistry can be roughly defined as the abiotic organic chemistry of nature—that is, all the naturally occurring, or naturally plausible, chemical processes leading to the formation and diversification of an organic chemical inventory, from inorganic carbon sources and without intervention of living organisms, which ultimately lead to the origin of biopolymers, self-organization, and emergence of biochemical complexity.

Prebiotic chemistry differs from classic organic chemistry in its goals and approaches. It should be considered in tandem with realistic geological and environmental conditions—or, at least, as providing a narrative that is extrapolatable to a natural setting. Nature does not have glassware, exotic solvents, and pure reagents. Instead, it provides minerals and changing environments (Cleaves et al. 2012). Aside from the approximation of the origins of life on early Earth, prebiotic chemistry and chemical evolution (which could be considered a global field that includes prebiotic chemistry in the classic sense through the emergence of the first living system to undergo Darwinian evolution), are very valuable tools to predict and understand the chemistry and habitability of other worlds, inside (e.g., Europa and Enceladus) or outside of our solar system (e.g., new exoplanets). Essential feedback can be provided through the direct analysis of extraterrestrial materials, where active prebiotic chemistry is occurring, or from a frozen image of the organic chemistry that preceded life (Goesmann et al. 2015; Burton et al. 2012).

This chapter provides an overview of the prebiotic chemistry of the building blocks of nucleic acids, with special emphasis on the role of urea. Urea is a special solution concentrated by environmental cycles, such as drying or freezing of "little ponds"; it is a precursor and promoter of the prebiotic synthesis of nucleobases and nucleotides. The general idea behind urea as a prebiotic precursor is that the key reactions of interest for the origins of life are condensation and polymerization, requiring dehydration and concentration. The drying-wetting and freeze-thaw cycle scenarios, together with the concentration of prebiotic urea and its derived chemical space, can provide the right conditions for the origins of life.

4.2 Open Questions on the Origin of Nucleic Acids

The easy spontaneous formation of amino acids from atmospheric precursors, in a variety of plausible prebiotic conditions, was first demonstrated by the German biochemist Walther Löb in 1913 in what was possibly the first prebiotic chemistry experiment. Löb designed his experiment with the explicit objective of understanding some aspects of the origin of life. Löb observed the formation of glycine after silent discharges in an atmosphere composed of carbon dioxide, water, and ammonia; A common mistake in the citations of Löb's work is that he used carbon monoxide, a translation error further propagated by citing the English papers without reading the original German publication. Löb clearly stated his goal: "The question of natural nitrogen fixation is especially interesting, in that it presents the source of the first organic nitrogen-containing product for the formation of the first albumin bodies." The author later concluded:

> Here, succeeding for the first time, an amino acid has been produced artificially from the input products of the natural synthesis, which, in the simplest phase, plays a role in the origin of natural protein as the final products of the reaction of carbonic acid, ammonia and water, without application of other materials, purely through supplying a special energy that remains in close connection with the natural radiation. (Löb 1913, translated from the German original text)

It is interesting to note that Löb considered formamide as the fundamental intermediate in the formation of glycine and performed some experiments using formamide as a starting point, obtaining glycine (among other products). During Löb's time, the analytical technology that is necessary to analyze complex mixtures was not yet developed. Thus, Löb was unable to elaborate on the chemistry behind his experiments by identifying the amino acids and nitrogen heterocycles, including nucleobases, that possibly formed in his experimental conditions. Almost one century after the pioneering work of Löb, formamide is now regarded as one of the key precursors in prebiotic chemistry. (For readers who are interested in formamide chemistry, I recommend reading the works of Raffaele Saladino and Ernesto di Mauro, including Saladino et al. 2009.)

The experimental work of Löb is usually not regarded in studies on the history of the origins of life, in part due to the common belief that he used carbon monoxide instead of carbon dioxide in his experiments. The seminal work of Stanley Miller, synthesizing both proteinaceous and non-proteinaceous amino acids in simulated prebiotic conditions, was the starting point of the experimental study of abiogenesis (Miller 1953). Miller centered his analysis on the product created in the formation of amino acids. However, his reactor and experimental designs have been tested, both as in the original study and with several variations, in a variety of experimental models and for the synthesis of molecules other than amino acids. The products have been examined with modern analytical tools, revealing the complexity of the chemical mixtures formed.

Samples obtained by Miller during his experiments were analyzed in recent years to determine the details of their composition (Johnson et al. 2008). The dominant products in the small-molecule chemical space formed during Miller-type

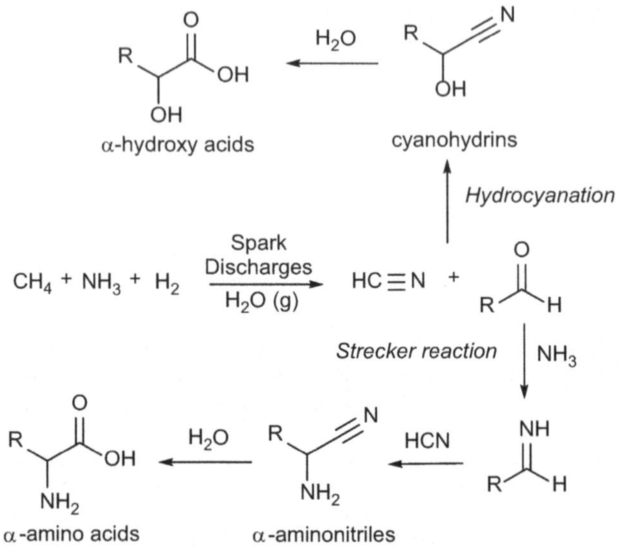

Fig. 4.3 The formation of amino acids and hydroxy acids by Strecker and hydrocyanation reactions in the classic Miller experiment are shown. It is interesting to note that the formation of aldehydes did not lead to sugars by homologation reactions under the conditions of spark discharge experiments

experiments were amino acids (in particular, glycine, alanine, and β-alanine) and hydroxy acids. The formation of hydrogen cyanide (HCN) and subsequent Strecker and hydrocyanation reactions of aldehydes lead, respectively, to amino acids and hydroxy acids (Fig. 4.3). This reaction was discovered by Adolph Strecker in 1850; its importance in prebiotic chemistry was predicted as early as 1914 by the Spanish chemist and pharmacist Jose Rodriguez Carracido. Carracido described this reaction 39 years before Miller's experiment, highlighting its role in the origin of amino acids, which is currently well established (Bada 2004; Pascal et al. 1980). Carracido wrote that in

> the epochs prior to the appearance of life, [the atmospheric composition] would allow a higher incidence of ultraviolet radiation on the earth, which in turn would make possible the synthesis of the aldehyde-forming cyanohydrin that generates the first amino acid (Viqueira 2010).

In 1920, Carracido published in Spanish a theory on the origin of life entitled "The Chemical Phylogeny of the Albuminoid Molecule", in which he provided the first definition of Chemical Evolution: "The chemical phylogenetic series that, from the initial member of a simple carbon-nitrogen combination (HCN), gradually grows to proteins and proteinaceous compounds of greater molecular magnitude, articulating the pieces in the very complex mosaic of life." In the same text, Carracido proposed that imidazole may be an important prebiotic reagent that could be formed from atmospheric acetylene (as we showed recently in Menor-Salván and Marín-Yaseli 2013). He also proposed that the origin of nucleic acids could be traced to a combination of the phosphate minerals and urea, which would be the precursor of bases.

Fig. 4.4 The Spanish biochemist and pharmacist Jose Rodriguez Carracido (1856–1928), a forgotten pioneer of prebiotic chemistry, defined the concept of chemical evolution and proposed, in 1920, the key roles of HCN and urea in the origins of life

Carracido wrote his theory on the origin of life in 1919–1920. Given the state of the art at that time, it could be considered visionary, esspecially because that we recently demonstrated that phosphate minerals could phosphorylate nucleosides in urea-rich environments (Burcar et al. 2016) and that urea is an efficient precursor of bases (Menor-Salván et al. 2009). In 1965, Juan Oró recognized the visionary theory of Carracido: "I have noticed the prophetic ideas of Dr. Carracido and the surprising similarity of his theoretical considerations with the experimental results that have been obtained in present times" (Juan Oró, letter to Angustias Sanchez-Moscoso, 1965). Unfortunately, soon after publication of his theory, Carracido retired and ended his investigations on prebiotic chemistry (which he called *chemical phylogeny*), as he lamented in his *Confessions*, published after he passed away in 1928 (Fig. 4.4).

A number of factors led to the consideration of proteins as the first biopolymers of life: the easy formation of amino acids in a variety of simulated prebiotic conditions (Zaia et al. 2008), the difficulties of forming the precursors of nucleic acids in simple spark experiments, and increasing knowledge of molecular biology (which began during the 1950s and 1960s). However, without nucleic acids, the evolution of a protein-first world seemed to be unlikely. During the late 1960s, Leslie Orgel, Francis Crick, and Carl Woese each independently proposed that RNA was more likely than polypeptides to be the first polymer of life (Woese 1967; Crick 1968; Orgel 1968). A common inspiration for these proposals, crystallized in the RNA-world hypothesis, is that RNA is more likely to be an ancient polymer than other biopolymers.

In the early 1980s, the research groups of Altman and Cech discovered that different RNA molecules performed catalytic processes in living cells, including the self-splicing of the RNA, (Guerrier-Takada et al. 1983; Kruger et al. 1982). This discovery influenced decisively the strategies for the study of origins of life. The catalytic RNA molecules were named ribozymes, emphasizing the fact that these are enzyme-like molecules composed of RNA. Prior to the discovery of ribozymes, it was generally accepted that only proteins catalyzed chemical reactions in living cells. With the discovery of ribozymes, it seemed that the paradox concerning which came first—proteins or nucleic acids—was solved. Because RNA could both store information and catalyze chemical reactions, it had the two basic features believed to be necessary for self-replication and evolution. This prompted Walter Gilbert in 1986 to coin the phrase "RNA world," referring to a time in early evolution when life used RNA for information storage and chemical catalysis (Guilbert 1986). Later, Noller and coworkers would also demonstrate that the essential component of the ribosome is RNA and that ribosomal proteins play a more supportive role (Noller et al. 1992). These results were taken as evidence that RNA "invented" protein synthesis—a scenario that fits perfectly with *RNA world.*

In 1990, the laboratory of Jack Szostak invented techniques for the selection of RNA molecules with specific properties from pools of RNA molecules containing random sequences (Ellington and Szostak 1990). Since then, many laboratories have produced an impressive array of RNA molecules with catalytic abilities. Many thinkers cited this as evidence that RNA can accomplish almost any task that is now carried out in life by proteins, and thus provides support for the RNA-world hypothesis.

Despite the huge amount of publications generated—64 years since Miller's experiment and 134 since from the synthesis of purines from HCN (Johnson and Nicolet 1914)—and the general enthusiasm that still exists for the RNA-world hypothesis, thanks to some recent studies that were very well received in the field (Wagner and Blackmond 2016; Patel et al. 2015; Powner et al. 2009), major gaps remain in our knowledge about the connection between geochemistry and the abiotic production of small molecules, the processes that give rise for the first RNA or RNA-like molecules, the coevolution with proteins and the subsequent processes of organization, the rise of the cellularization, and the origin of a complex metabolic system.

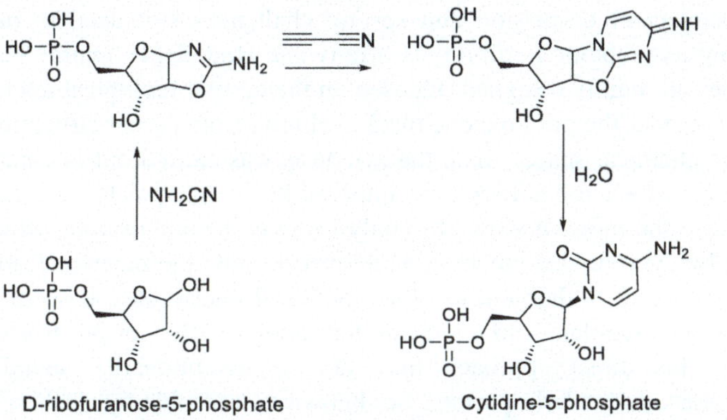

Fig. 4.5 Bypassing the problem of glycosidation in the synthesis of CMP through a pentose aminooxazoline, by Sanchez and Orgel (1970). The work of Shutherland and coworkers largely improved this route, solving the problem of ribose formation in the origin of cytidine

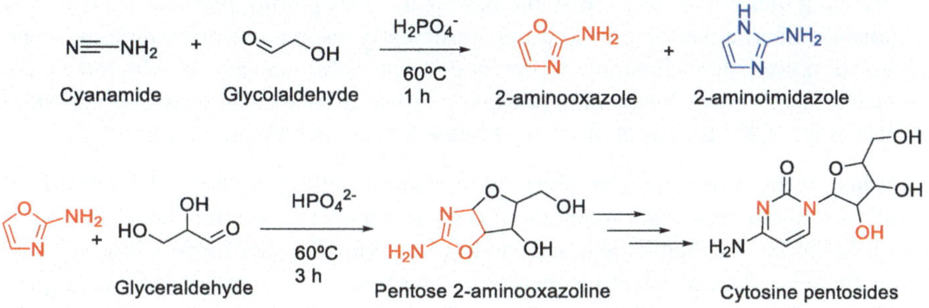

Fig. 4.6 Route to cytidine from cyanamide and glycolaldehyde precursors

The overall main questions regarding the origin of RNA could be classified as follows:

1. **The nucleosidation problem**: The formation of a glycosidic bond between ribose (or other alternative sugars) and bases to produce beta-nucleosides was one of the most significant problems for the RNA-world hypothesis. The problem could be divided in three subordinated parts: (1) the origin of sugars and the selection of ribose; (2) the prebiotic formation of nucleosides; and (3) the formation of a glycosidic bond with canonical RNA bases. The canonical-ribocentric model of Sutherland and coworkers, which followed the idea of the synthesis of cytidine proposed by Sanchez and Orgel in 1970 (Fig. 4.5), elegantly bypasses the problem of the prebiotic ribose formation; it was a partial solution to the nucleoside problem. This route is based in the formation of 2-aminooxazole by the condensation of cyanamide and glycolaldehyde, and further formation of pentose aminooxazoline by a reaction with glyceraldehyde. Recently, it has been shown that the formation of the 2-aminooxazole precursor and 2-aminoimidazole, which could activate the phosphate in nucleotides and promote polymerization, is possible in the same reaction (Fahrenbach et al. 2017, Fig. 4.6). However, this model

is the subject of discussion and several challenges still need to be addressed regarding its prebiotic feasibility (Cafferty and Hud 2014), mainly because these syntheses are highly designed, focused on the current biological nucleotides, and do not consider the possible chemical evolution from earlier structures in a more complex chemical space. Also, the synthesis was carried out in clean multi-step processes, which are difficult or impossible to extrapolate to real geological settings; furthermore, it worked mainly for cytidine and uridine, which could be formed by cytidine deamination. Moreover, despite the intense efforts to extend the chemistry of Sutherland to other canonical nucleotides, it seems to be very difficult to extrapolate it to a realistic formation of RNA or pre-RNA molecules. Overall, this direct approach may not be evolutionarily sound. Although nucleobases, particularly purines, are known to assemble by stacking in aqueous solutions (Ts'o et al. 1963) and nucleic acid double helices are held together in large part by hydrogen-bond interactions between purine and pyrimidine bases, few of the intermediates in Sutherland's synthesis possess aromatic stacking surfaces; most of them also appear to be incapable of hydrogen bonding or orienting themselves to form stable base pairs. This pairing problem leads one to consider alternative solvents to water, particularly viscous eutectic solutions; when based in urea, they constitute a prebiotically plausible scenario in which the easy synthesis of nucleobases and phosphorylation of nucleosides is also possible (He et al. 2017; Burcar et al. 2016; Menor-Salván and Marín-Yaseli 2012).

When using prebiotic chemistry in a natural setting without the controlled syntheses, instruments, and methods of organic chemistry, the likelihood of assembly in Sutherland's chemistry-based models into complex structures is low; furthermore, there is a lack of selective pressure to preserve these compounds in a complex prebiotic environment. The formation of a rich nucleobase chemical space in prebiotic model reactions (Menor-Salván et al. 2009) and the occurrence in natural samples (Callahan et al. 2011) suggests an improbability that the nucleotides present in today's biochemistry were the only species present during the origins of life; a more likely scenario is a nucleic acid-like polymer that contained a variety of easy-to-synthesize and easy-to-assemble elements, which evolved into canonical nucleotide cofactors and polymers. This idea is supported by the straightforward high-yield synthesis of a non-canonical C-nucleoside from 2,4,6-triaminopyrimidine and ribose (Chen et al. 2013) and the self-assembly of the nucleoside with cyanuric acid, forming hexad rosettes. Also, the equivalent formation of barbituric acid C-nucleotide by direct ribosylation (which shows a preference for the beta-nucleotide) and its assembly in stacked hexads with melamine (Fig. 4.7, Cafferty et al. 2016), all prebiotically available bases, suggests that the formation of assemblies of nucleosides and bases is also a selection mechanism, which could preserve, concentrate, and facilitate the backbone binding, leading to a pre-RNA molecule containing canonical and non-canonical bases (Hud et al. 2013). The C-nucleotide of barbituric acid, moreover, could be considered an analog of pseudouridine, which is incorporated by post-translational modification to RNA and is known to stabilize the

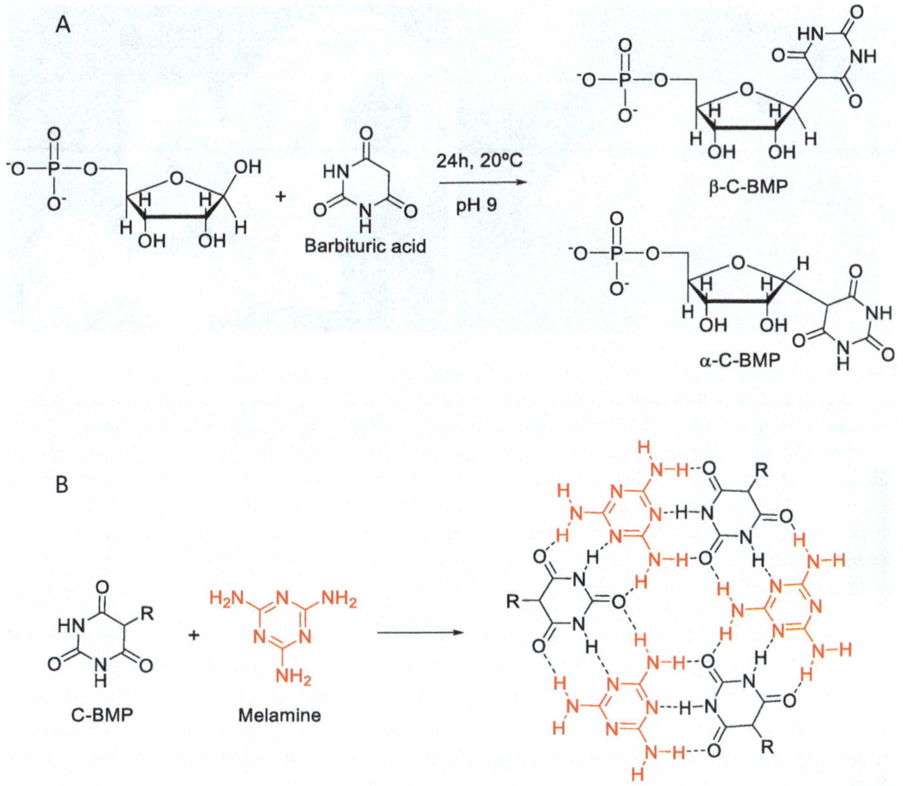

Fig. 4.7 Direct ribosylation of barbituric acid (**a**) and formation of hexad rosettes by base pairing between barbituric acid nucleotides and melamine (**b**). The BMP is an analog of pseudouridine and could have played a role in the chemical evolution of pre-RNA molecules

secondary structure (Chawla et al. 2015). The selection mechanism not only works for the bases but also for the sugars in the backbone, assuming a complex mixture of aldoses, ketoses, sugar acids, and alcohols derived from formaldehyde (i.e., a formose reaction scenario or glyoxylate scenario; Benner et al. 2012). This approach also could solve the base-pairing paradox—that is, if the nucleobases of RNA do not pair as monomers, then how and why would they have been selected for incorporation into polymers instead of the many other non-pairing heterocycles that otherwise have similar chemical and physical properties? In the words of Nicholas Hud: "Without a mechanism that selects for pairing at the monomer level, it stands to reason that the first nucleic acids would have been horribly compromised by the inclusion of non-pairing nucleobases and other undesirable reactive molecules" (Cafferty and Hud 2014).

2. **The phosphorylation problem**: The phosphorylation problem, named by Alan Schwartz (2006), reflects the unlikely prebiotic formation of organophosphates due the low availability of soluble and mobile phosphate in a realistic

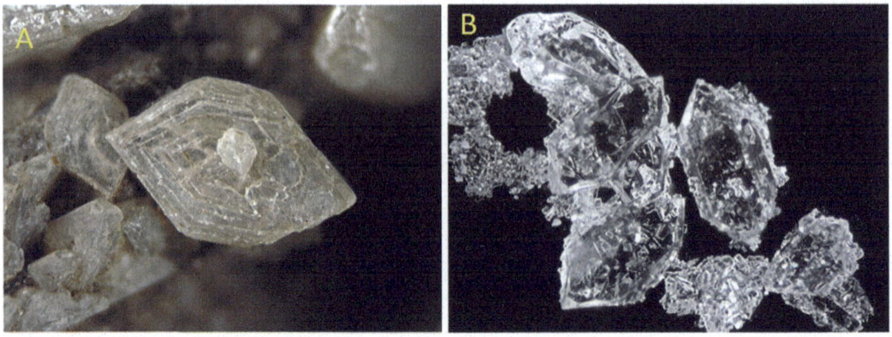

Fig. 4.8 Struvite could be formed in natural conditions and manure ponds, in urea-rich solutions and environments with low redox potential. In these conditions, struvite is formed in almost pure crystals even in the presence of calcium (**a**). In small, evaporating water ponds containing a set of plausible organic compounds under simulated prebiotic conditions, struvite can be formed abiotically also in presence of calcium. Abiotic struvite is usually accompanied by newberyite, which are both good phosphate-transfer agents (**b**). It seems that urea content of the solution, in both modern or prebiotic scenarios, determines the phosphate mineral formation and the formation of organophosphates

geochemical setting. Leslie Orgel discovered in 1973 that the mineral struvite could be an efficient prebiotic phosphorylating agent (Handschuh and Orgel 1973). Struvite is uncommon on modern Earth and is always associated with the decay of putrescent organic matter or a biomineral. The prebiotic availability of struvite (Fig. 4.8) was questioned since Orgel's finding (Gull and Pasek 2013). Recently, a significant advance was proposed, in which urea plays a key role, thanks to the previously undescribed mobilization of primary phosphate from insoluble sources by urea + Mg^{2+} in an evaporite setting and the transformation of primary apatite in secondary gypsum + struvite, demonstrating a novel, previously undescribed mineral transformation that could have worked in prebiotic times (Burcar et al. 2016). This result suggests that the phosphorylating agent could have been present in the prebiotic era, possibly being the first plausible connection between geochemistry and the formation of organophosphates and useful 5′-nucleotides, cyclic, and 3′-nucleotides. The finding of phosphate minerals that could constitute a source of mobile phosphate in Mars (although is richer in total phosphorus than Earth) is a support for a phosphorylation model based on P (V) minerals (Adcock et al. 2013). The "warm little pond model" therefore extends the prebiotic roles of urea within the inorganic realm by modulating the precipitation of magnesium and calcium salts, favoring the formation of evaporite minerals consistent with a plausible and straightforward prebiotic phosphorylation. The presence of significant urea concentrations increases the remobilization of insoluble phosphate and shows a preference for the accumulation of magnesium phosphates, instead of the more insoluble calcium phosphate. This effect is now observed in reducing environments that are rich in urea, phosphorus, and organics, where struvite accumulates. Prebiotic chemistry could have formed similar environments on early Earth, in which urea and other molecules (e.g., cyanide, ammonia, organic acids) are determinant factors for the associated supergenic mineralogy.

It is interesting that this "little water pond" approach is closer to current ideas about geothermal scenarios in the connection between geochemistry and the origins of life (Damer 2016). Moreover, another solution to the phosphorylation problem based in the same principle, but with different mineralogy, was published recently, validating the idea that prebiotic processes could be robust enough to lead to nucleic acid precursors in evaporitic conditions (Kim et al. 2016). The phosphorylation problem, together with the difficulty of explaining sugar availability, may be alternative foundations for primordial genetic polymers. This problem is treated extensively in other chapters of this book.

3. **The polymerization problem**: Given the formation of nucleotides, their polymerization in realistic prebiotic conditions, without artificial activators, is difficult. Attempts for the non-enzymatic formation of polynucleotides result in short polymers and are difficult to reproduce, despite some enthusiastic claims (Saladino et al. 2012). Aside from the formation of the polymer itself, the problems of a lack of stability and a strong sensitivity to depolymerization, dephosphorylation, and nucleobase alteration of the nascent oligonucleotide need to be overcome. Difficulties with the prebiotic formation of polynucleotides with catalytic activity and the generation of a population able to evolve have occurred not only in the laboratory, but also in the theoretical considerations. A population of replicators lacks enough functional diversity to be self-sustaining, leading to the enunciation of alternatives on the primary origin of nucleic acids, such as the metabolism-first hypotheses (Francis 2015; Martin and Russell 2007; Russell and Martin 2004; Holm 2003; Wächtershäuser 1988). The problem with metabolism-first hypotheses is the lack of experimental results that constitute real advances in the formation of chemical evolution models. We consider the duality of metabolism-first vs. RNA-first to be outdated. Researchers on the origins of life field began to consider a scenario in which the three subsystems of life—metabolism, informational and functional polymers, and membranous compartmentalization—coevolved. The chemical evolution for the formation of far-from-equilibrium proto-metabolic systems alone, without the formation of polymers, is unlikely; in addition, the polymerization requires dehydration-friendly settings. The recent findings on the formation of depsipeptides (mixed ester-amide polymers) in a simple evaporitic environment could solve the polymerization problem in the origin of peptides (Forsythe et al. 2015) and constitutes a major finding for the field of the origins of life. The evolution of depsipeptides could link the the prebiotic chemistry of amino acids and peptides with nucleobases and nucleotides, as both take place in similar environments (the concentration of solutes in evaporated water ponds) and could lead to the origin of polymers with catalytic activity, the binding of nucleotides to peptide-rich polymers, and ribosomal origins (see Chaps. 9 and 10 in this book).

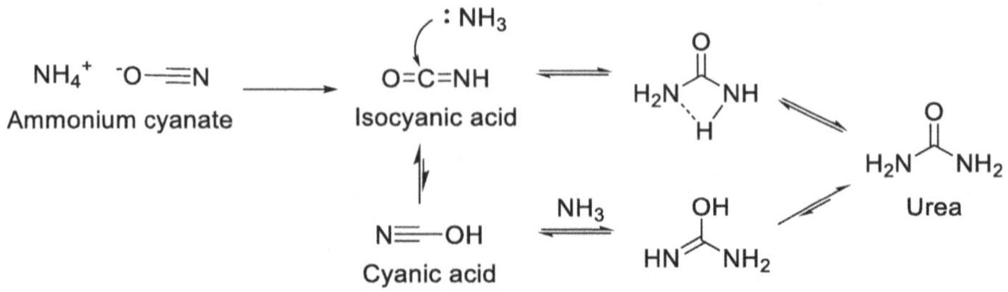

Fig. 4.9 Wöhler synthesis of urea from cyanate salts. The equilibrium in solution between urea and cyanic-isocyanic acid is key in the prebiotic formation of several nucleobases

4.3 The Prebiotic Origin of Urea

One of the most famous experiments in the history of chemistry was the serendipitous synthesis of Friedrich Wöhler, published in 1828. Wöhler wanted to study the properties of ammonium cyanate salt and tried to obtain it by treating lead cyanate with ammonia, to precipitate lead hydroxide and isolate the ammonium salt by evaporation of the solution. Instead of the desired salt, he obtained an unexpected white solid, whose properties matched those of urea isolated from urine (Wöhler 1828). This reaction, together with the synthesis of oxalic acid from cyanamide (also by Wöhler, in 1824), could be considered the first prebiotic chemistry experiments.

Contrary to popular belief, Wöhler's synthesis did not signal the end of vitalism. Wöhler recognized, in a letter to his friend Berzelius, that the synthesis of urea was not counterevidence of the *vitalism* hypothesis, because an organic source was still necessary to obtain the cyanate salts (Ramberg 2000). It was necessary to wait a few more decades, for the development of the new science of organic chemistry, to accumulate evidence against vitalism. Although the fate of vitalism has been sealed since the mid-nineteenth century, the birth of prebiotic chemistry and the beginning of experimental approaches to the abiogenesis hypothesis ended to the idea of *élan vital.* The mechanistic details of the Wöhler synthesis have not been completely established (Tsipis and Karipidis 2003). The reaction implies the rearrangement of ammonium cyanate by proton transfer from ammonium cation to cyanate, followed by a nucleophilic attack of ammonia to either isocyanic acid (the most stable tautomer), yielding urea, or cyanic acid followed by tautomerization (Fig. 4.9).

The plausible availability of urea as a prebiotic precursor has been well established. It could have been formed efficiently in the solar system's icy bodies via mechanisms similar to the Wöhler synthesis, as well as by radical mechanisms (Nuevo et al. 2010), and delivered to early Earth during comet bombardment. The formation of urea (along with one of its most important derivatives, hydantoin; see Sect. 4.6) by photochemistry in icy bodies lead affirms that "one can reasonably assume that molecules such as hydantoin, urea, and α-amino acids seeded the oceans of primitive Earth" (De Marcellus et al. 2011). If we think of small water ponds instead of oceans, the concentration of urea by evaporation or freezing of those water ponds creates a very interesting scenario for chemical evolution. Extraterrestrial delivery is not the only option for the accumulation of urea in the primitive Earth. The efficient formation of urea (and its related compound guanidine) from cyanide or cyanamide has been known since the nineteenth century (Oró and Kimball 1962). In addition, urea was formed in the spark discharge experiments (Miller 1957), where cyanide was formed in significant concentrations from a methane-containing atmosphere. In fact, as Jeffrey Bada told us personally in a delightful story, Stanley Miller was surprised by the high amount of urea formed in his experiments, which was far superior to that of glycine.

Although it is an open question, if prebiotic Earth's atmosphere contained even low or local concentrations of methane, it could have constituted an efficient source of cyanide in the presence of molecular nitrogen. The cyanide could have accumulated in the form of metallic complexes, such as Prussian blue (iron ferrocyanide). The hydrolysis of cyanide complexes could have been a way for the sustained release and concentration of urea in small water ponds (Ruiz-Bermejo et al. 2009a, b). Urea is a strong hydrogen bond donor that could form a variety of eutectic solutions (Grover et al. 2015; Parnica and Antalik 2014; Smith et al. 2014). The urea molecules in water solutions form hydrogen bonds with the neighboring water molecules in both the amino and the carbonyl groups (Soper et al. 2003). The number of hydration water molecules per molecule of urea has been reported to be approximately 2 molecules at concentrations of less than 5 M urea (Hayashi et al. 2007). Neutron diffraction measurements at 25 °C on aqueous 10 M urea showed that approximately 4.3 water molecules are hydrogen bonded to the carbonyl oxygen atom (Kameda et al. 2006). Infrared and dielectric spectroscopy studies showed that the two predominant interactions are those of urea–urea, observed at urea concentrations higher than 1 M, and water–urea. At a concentration of 11 M, nearly all the urea molecules have other urea molecules as their nearest neighbors, because few water molecules remain to hydrate the urea molecules at high concentrations. Concentration-dependence studies showed that this is due to the aggregation of urea molecules in dimers (approximately 18% of urea in 0.1 M solution is in the form of non-covalent dimers; Stokes 1965) and/or oligomers at higher concentrations (Grdadolnik and Maréchal 2002). A considerable amount of urea dimers or clusters is present in a water solution; the described properties of urea solutions allow the formation of highly viscous, dense solutions that could be separated by evaporation or freezing of a diluted solution in the form of a low-melting-point viscous solution in which urea is present as hydrogen-bonded dimers or higher association clusters, randomly distributed within the ice.

On the other hand, the evaporation of a water pool containing urea would lead to its crystallization if pure. However, urea forms eutectic solvents with other prebiotic precursors, such as ammonium formate (Burcar et al. 2016), acetate salts, and glycerol. These eutectics extend the liquid range of concentrated organic solutions while allowing the solubilization of organic solutes, dehydration, and hence polymerization reactions. The physicochemical properties of urea, together with its relative stability and its nucleophilic character (which gives urea a role as a chemical precursor of nitrogen heterocycles), results in urea-based eutectics for a plausible and very interesting scenario on the origin of nucleobases and subsequent processes, such as phosphorylation (Burcar et al. 2016).

4.4 Pioneering Works on the Synthesis of Nucleobases Applied to the Prebiotic Chemistry of Nucleic Acids

Many chemists of the nineteenth century and first half of the twentieth century practiced prebiotic chemistry without realizing it. In fact, throughout the history of prebiotic chemistry, including very recent times (Becker et al. 2016; Menor-Salván et al. 2018), a significant number of works on chemical approaches to Abiogenesis were based on the application of chemical reactions studied during the last two centuries. In this section, we will review some early works relevant to the modern studies of the origins of life.

In 1864, the Nobel-prizewinning German chemist Adolf von Baeyer published a study on the synthesis of 2,4,6-trihydroxypyrimidine or barbituric acid, whose prebiotic importance was discussed in Sect. 4.2. The heterocycle was obtained after a relatively complex process during his studies on the chemistry of uric acid, in a paper in which also described the formation of hydantoin and parabanic acid, another relevant prebiotic chemistry process (Carter 1951; von Baeyer 1864). It is commonly believed that the common name of barbituric acid comes from the festivity of Saint Barbara. However, this story does not fit with the temporal sequence of his works. The name instead comes from the fact that Baeyer (who, at that time, was only 29 years old) was "charmed by a youthful beauty Barbara X"; in her honor, he coined the name "barbituric" (Carter 1951).

The direct synthesis of barbituric acid was accomplished a few years later by the French chemist Edouard Grimaux in 1879, using the quantitative reaction of urea and malonic acid in chloroform in the presence of phosphoryl chloride (Fig. 4.10a). The Grimaux synthesis introduced the use of barbituric acid as a precursor of a complete family of drugs and organic precursors by the use of substituted malonates and urea or by Knoevenagel condensation between aldehydes and the nucleophilic C5 of the pyrimidine ring. This property of the pyrimidine ring of barbituric acid also made possible the synthesis of C-nucleosides and C-nucleotides in very mild, prebiotically plausible, conditions (Cafferty et al. 2016).

The synthesis of barbituric acid could be applied in a prebiotically plausible way (Fig. 4.10b). Malonic acid, which is formed in classic cyanide-based model prebiotic reactions and also by photoalteration of precursors as hydantoins (see Sect. 4.6), react with urea in a urea-eutectic solution at low temperatures in dehydrating conditions (Menor-Salván et al. 2017, 2018). An interesting variant of this reaction uses aminomalonic acid, which forms at a good yield in classic spark reactions (Ruiz-Bermejo et al. 2006) and in cyanide-based model prebiotic reactions (Oró and

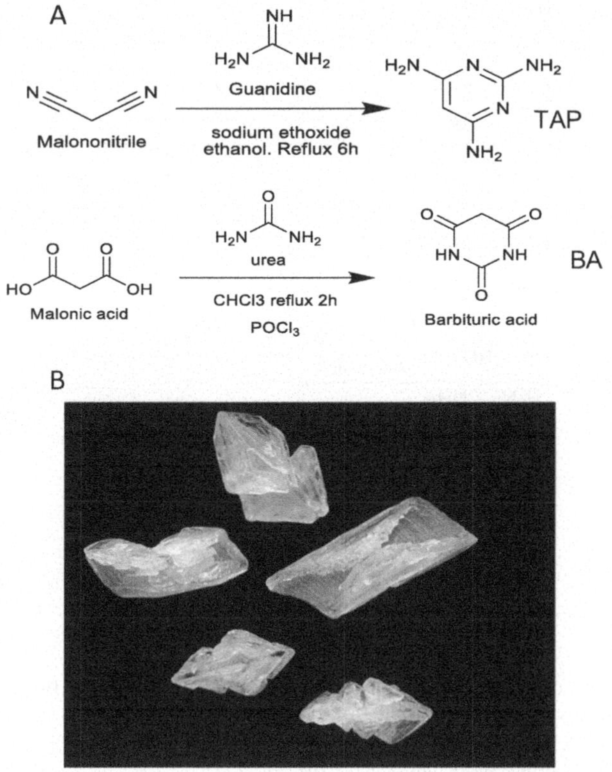

Fig. 4.10 (**a**) Traube synthesis of 2,4,6-triaminopyrimdine (TAP) and Grimaux synthesis of malonylurea or barbituric acid (BA). (**b**) Crystals of barbituric acid formed in possible prebiotic conditions, by simple evaporation-drying-dilution cycles of an urea and malonic acid solution at 65 °C during 3 days

Kimball 1962). The aminomalonic acid condenses with urea, forming 5-aminobarbituric acid or uramil, which could react with urea and lead to uric acid. This process is not new at all; its origins can be traced to the nineteenth century.

In 1884, the French chemist M. Gautier reported the formation of xanthine by the gentle heating of ammonium cyanide in the presence of acetic acid in a sealed tube (Hayem 1888). During these times, adenine was believed to be a polymer of HCN (Hayem 1888; Johnson and Nicolet 1914); in addition, in a series of papers in 1900 and 1904, the German chemist Wilhelm Traube connected the chemistry of pyrimidines with purines. In this very important reaction, the preferred formylation of the C5 amino group in 5,6-diaminopyrimidines by formic acid, formate salts, or formamide, followed by ring closure, lead to the synthesis of guanine and adenine (among other purine derivatives) in water and mild conditions (Fig. 4.11). This synthesis was applied and improved upon in subsequent years, sometimes with the goals of understanding the possible origin of purines in plants, synthesizing labelled

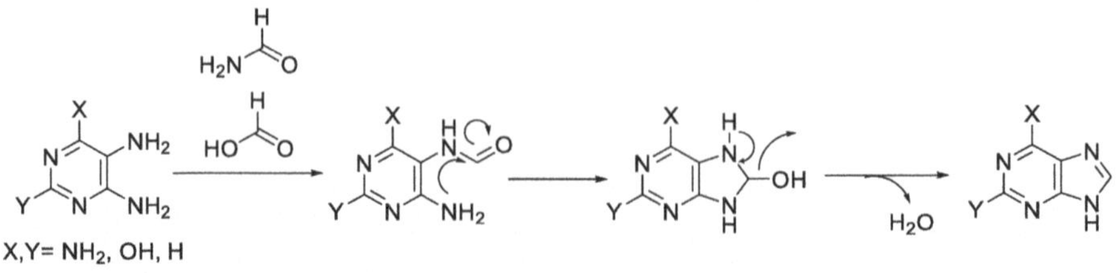

Fig. 4.11 Traube synthesis of purines

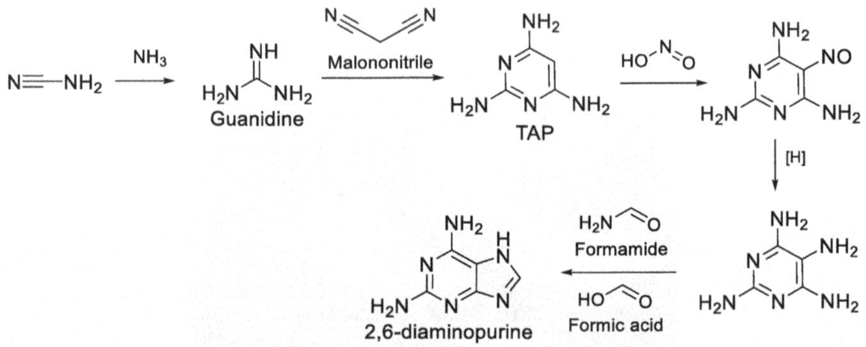

Fig. 4.12 Synthesis of a purine ring from cyanamide and malononitrile, by Bendich et al. (1950). This sequence of reactions may be of interest for the prebiotic chemistry of nucleobases after the proposal that nitrogen oxides, particularly NO, could have been essential components in the atmosphere of early Earth (Airapetian et al. 2016)

purines, and elucidating nucleobase metabolic pathways (Hurst 1979; Cavalieri and Brown 1949).

The work of the American chemist Aaron Bendich is especially remarkable. He developed several useful syntheses of nucleobases with a clear application in the forthcoming prebiotic chemistry field. Bendich et al. (1950) synthesized diaminopurine from cyanamide and malononitrile, in a pathway mediated by the formation of tetraaminopyrimidine by nitrosylation of 2,4,6-triaminopyrimidine (Fig. 4.12). The nitrosylation pathway, using nitrous acid, could become a plausible prebiotic pathway after the discovery that nitrogen oxides could have been essential products of the atmospheric chemistry of early Earth (Airapetian et al. 2016).

Another prebiotically interesting work, performed by Bendich et al. (1949), was the synthesis of cytosine from urea and cyanoacetaldehyde (Fig. 4.13). Although the synthesis performed by Bendich was not formally prebiotic (he used the diethylacetal form of cyanoacetaldehyde in sodium butoxide/*n*-butanol), the reaction principle has prebiotic application and could explain the high-yield formation of cytosine in urea-based model prebiotic reactions (Menor-Salván et al. 2009).

During the first years of the twentieth century, a remarkable American chemist, the Yale University professor Treat B. Johnson, thoroughly studied the chemistry of pyrimidines, purines, and hydantoins. The main motivation for his studies was to

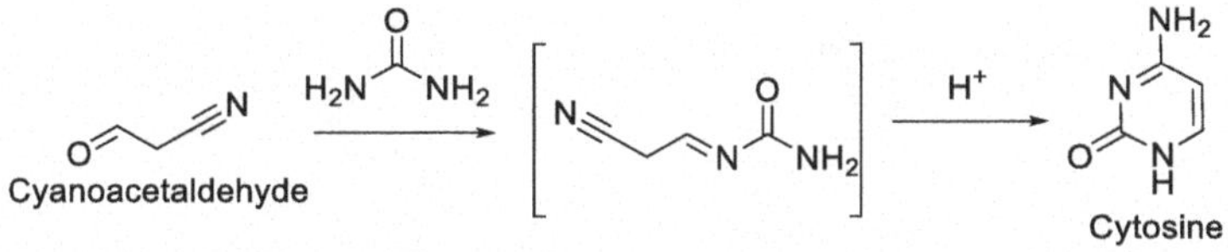

Fig. 4.13 Bendich synthesis of cytosine. This synthesis was tested unsuccessfully, probably because of the lesser reactivity of the secondary amine, for the direct formation of cytidine using urea riboside as the reagent

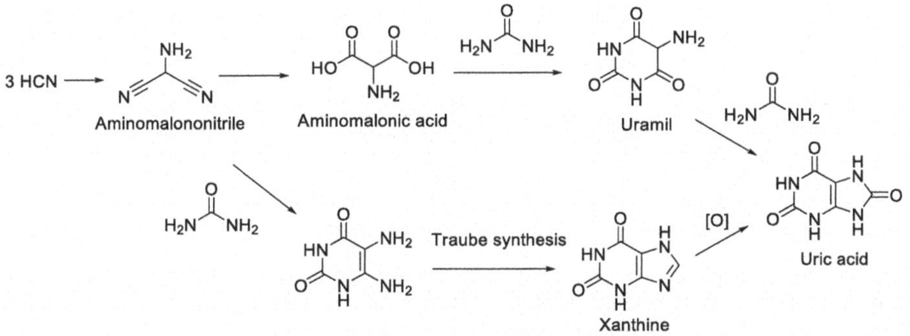

Fig. 4.14 Synthetic scheme proposed by Treat B. Johnson and Ben H. Nicolet in 1914 to explain the origin of uric acid from hydrogen cyanide. The formation of aminomalononitrile from HCN was discovered in 1873 and a very simple synthesis of purines from HCN was published by the French chemist Armand Gautier in 1884. We have shown that this reaction sequence could be a prebiotic pathway to the purine ring

gain an understanding of the origin of these compounds in plants. It is interesting that, during these times, they associated chemical synthesis in water and mild conditions with the biological origin of bases, thinking that the biochemistry followed the easiest chemical pathway. (Today, we use the term *chemomimetic* for a biochemical pathway that follows a prebiotically plausible pathway and *biomimetic* for the chemical synthesis inspired by biochemical pathways.) Johnson, knowing that adenine could be easily formed by the polymerization of HCN, proposed that the formation of pyrimidines and purines from HCN is a route that explains the origin of bases in living organisms (Fig. 4.14; Johnson 1914; Johnson and Nicolet 1914). Hence, Johnson's experiments can be considered as the first experimental prebiotic chemistry approach to the origin of nucleobases.

Treat Johnson was the first scientist to observe the nucleosidation problem—that is, the difficulty in forming direct ribose derivatives of RNA nucleobases and their polymerization. Johnson was interested in the chemical synthesis of an RNA polymer. In 1930, he wrote: "The problem of nucleoside synthesis resolves itself into one of finding a practical method of coupling a sugar with a pyrimidine (uracil, thymine or cytosine) at the 3-position of the ring" (Hilbert and Johnson 1930). Finding a naturally plausible pathway for the origin of the canonical nucleosides

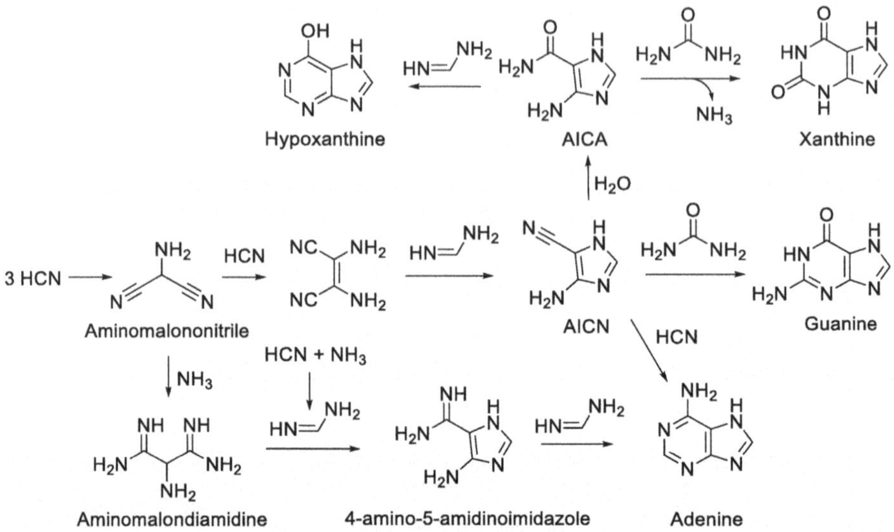

Fig. 4.15 Possible prebiotic formation of purine nucleobases by polymerization of HCN. A key step, the formation of AICN, takes place at room temperature by ultraviolet irradiation. Urea is also easily formed if HCN and ammonia are present as precursors

would become, a few decades later, one of the most refractory problems of the chemical science.

The prebiotic role of HCN, envisioned by Treat Johnson, was elaborated in the seminal works of Joan Oró in the years 1960–1964 and the works of Ferris and Orgel (Eschenmoser 2007; Orgel 2004a, b; Ferris and Hagan 1984; Ferris and Orgel 1966; Lowe et al. 1963; Oró 1960; Oró and Kimball 1961–1962). Oró and coworkers studied the formation and mechanisms of the synthesis of adenine from cyanide in prebiotic conditions. Ferris et al. (1968) observed the formation of guanine and xanthine and the Ferris–Orgel photochemical synthesis of the cyanoimidazole 4-amino-5-cyano-imidazole (AICN, also 4-aminoimidazole-5-carbonitrile) from the HCN tetramer. The synthesis of hypoxanthine by condensation of 4-aminoimidazole-5-carboxamida (AICA), the hydrolysis product of AICN, and urea (Fig. 4.15) and the relation of AICA with cyano compounds was studied by Elliot Shaw (1950) and rediscovered for prebiotic chemistry one decade later. These works established the importance of HCN and urea as prebiotic precursors of nucleobases and confirmed, nearly a century later, that the pioneering observations of Gautier, doubted by eminent chemists during his time, were correct.

The prebiotic relevance of early Earth and atmospheric HCN in its chemical evolution have been discussed (Menor-Salván and Marín-Yaseli 2012; Orgel 2004a, b). Recent publications support an atmospheric origin for HCN (and, hence, other atmospheric carbon and nitrogen sources), showing the compatibility between the young Sun, the presence of liquid water in prebiotic Earth, and the *seeding* of water pools by carbon and nitrogen compounds produced by atmospheric chemistry (Airapetian et al. 2016).

The HCN polymerization pathway to adenine and guanine proposed by Joan Oró resembled the biochemical de novo pathway for the synthesis of inosine monophosphate and adenosine monophosphate (Fig. 4.16). This resemblance lead to the suggestion that the biochemical pathway is a chemomimetic of the prebiotic pathway. Some studies on the abiotic formation of adenosine and inosine following the same principle were done, with very limited success (Ferris and Hagan 1984). This lack of success lead to a reinterpretation of the chemomimetic biochemical pathway as an example of chemical determinism: the possible synthetic pathways, biochemical or abiotic, are limited by the chemistry of the heterocycles involved.

Atmospheric HCN also could lead also to the formation of cytosine and uracil through the precursor cyanoacetylene. Cyanoacetylene shows great reactivity in water, even at low temperatures (Menor-Salván and Marín-Yaseli 2012). The potential relevance of cyanoacetylene in the origin of nucleobases was pointed out by Ferris and coworkers in a key experiment performed in 1968. They observed that the reaction of cyanoacetylene with aqueous sodium cyanate or urea gave cytosine at up to 5% yield; this is a high yield from a prebiotic chemistry point of view (Fig. 4.17). The mechanism of this reaction could be explained by cyanoacetaldehyde, which was generated by aqueous hydrolysis of cyanoacetylene.

Stanley Miller and coworkers followed the study of the cyanoacetaldehyde pathway with its reaction with urea in an ice matrix at 253 K to give cytosine and uracil at 0.005% and 0.02% yields, respectively (Nelson et al. 2001). In the same experimental approach, cyanoacetaldehyde reacted with guanidine at 253 K to give cytosine at 0.05% yield, uracil at 10.8% yield, and isocytosine and 2,4-diaminopyrimidine after 2 months (Cleaves et al. 2006). The mechanism of this reaction could proceed through the cyanoacetaldehyde dimer derivative 4-(hydroxymethylene) pentenedinitrile (Fig. 4.17).

The synthetic connection between the pyrimidines and purines, which was discovered by Traube, was extensively studied in a variety of conditions and with a variety of substituents to obtain substituted and alternative purines for pharmacological use. A great advantage of the Traube synthesis, exploited for its synthetic potential, is in the condensation of 5,6-diaminopyrimidines with one carbon fragment (formic acid, formamide, dithioformic acid, or urea); the course of the reaction is usually an attack at the 5-amino group, followed by cyclization to give the purine. On the other hand, complex aldehydes (e.g., aldose sugars) show preference for attacking the 6-amino group. One application derived by this observation, developed by Kenner and coworkers in 1949, is especially interesting because it could be a clue for a solution of the nucleosidation problem of purines: the synthesis of adenosine by formylation or thioformylation of 5-amino group and ring closure of a 4,5,6-triaminopyrimidine previously ribosylated in the 6-amino position (Fig. 4.18).

Researchers in the laboratory of Thomas Carell in Germany have recently shown that this reaction could be plausible for the origins of life, if the formylation in the 5-amine group occurs previous to the glycosidation (Becker et al. 2016). This work demonstrated that tetrosides and pentosides of 5-formamido-6-aminopyrimidines could be formed directly by a Formose reaction. Further ring closure leads to the canonical beta-furanoside anomer of adenosine, among other isomers and

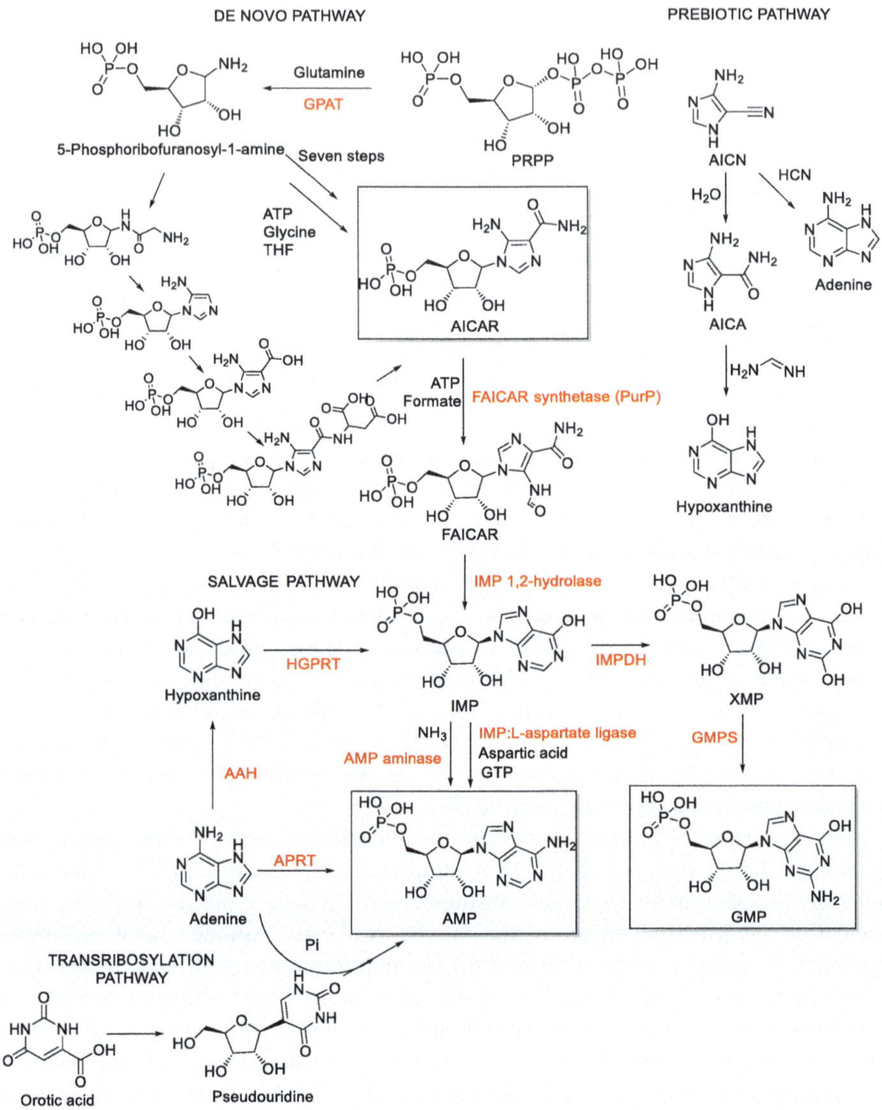

Fig. 4.16 Overview of the purine anabolism. Despite the coincidence of the 4-aminoimidazole-5-carboxamide (AICA) intermediate in both the de novo synthesis of inosine and in the prebiotic synthesis of purines from HCN, it is not possible to reproduce it including ribose. This apparent coincidence between the biochemical and prebiotic pathways is due to the chemical determinism in the synthetic pathways rather than an evolutionary trait. Also, the direct ribosylation of adenine and hypoxanthine, which is highly efficient in the salvage pathway, is not possible without enzymes (although this statement may be outdated soon, as we are working on a plausible direct ribosylation). The transribosylation pathway, the transfer of a ribosyl group from a C-nucleoside to adenine, is a rare metabolic route identified in some prokaryotes and in the ciliated protozoos of the genus *Tetrahymena* (Kusama et al. 1966). This route could be another interesting case of the chemomimetic route; our group is studying the possible prebiotic transfer of ribosyl to canonical purines

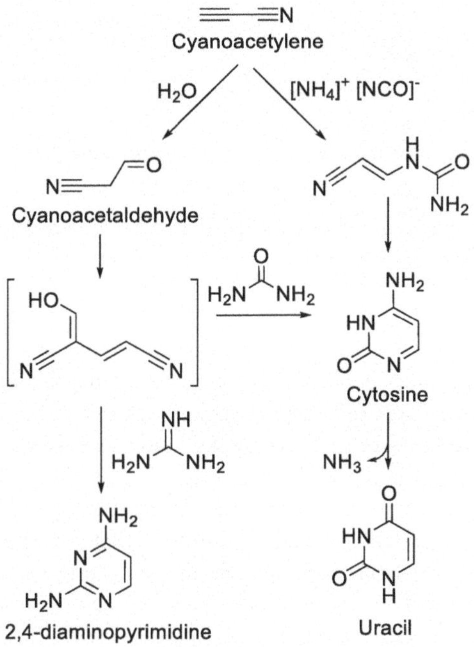

Fig. 4.17 Possible prebiotic formation of pyrimidine nucleobases from atmospheric cyanoacetylene, formed photochemically in an atmosphere containing methane and hydrogen cyanide

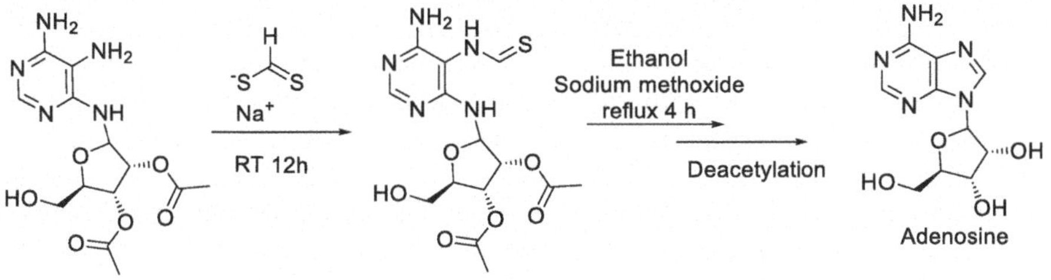

Fig. 4.18 Synthesis of adenosine by application of the Traube reaction to a ribosylated aminopyrimidine, developed by in 1949 by Kenner and coworkers. The reaction also works with formic acid/formamide. The higher nucleophilicity of the 5-amino group complicates the synthesis, which requires several steps because it is necessary to incorporate the sugar in the 6-amido group. Despite this problem, the reaction has had recent applications in the prebiotic chemistry of nucleosides

byproducts (Fig. 4.19). It is interesting that we were working in a similar line on the synthesis of guanosine by formylation and ribosylation of 6-hydroxy-2,4,5-triaminopyrimidine, which could be synthesized in prebiotic conditions from urea and hydantoin (Menor-Salván et al. 2018, see Sect. 4.7).

A problem of the synthesis by Becker and coworkers, which also complicated the adenosine synthesis by Kenner et al. in 1949, is that it is necessary to carry out the formylation *before* the ribosylation in a separate step, as the aldehydes compete for

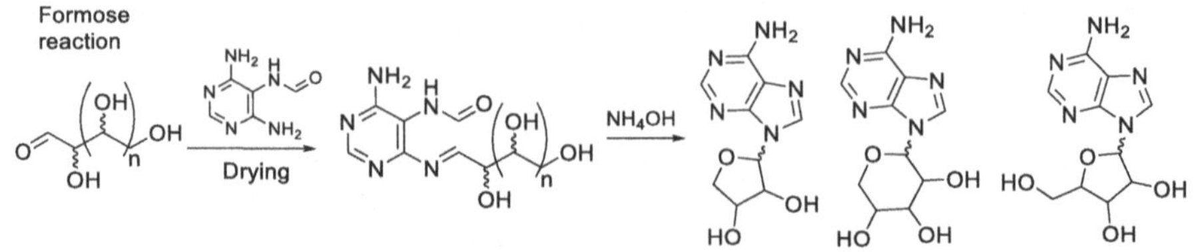

Fig. 4.19 Synthesis of adenine nucleosides by application of the Traube reaction, developed by Becker et al. (2016). The aldoses generated in the sugar mixture by a formose reaction attack the 6-amino group of previously 5-amino formylated pyrimidine. A ring closure in basic conditions leads to a mixture of nucleosides, including canonical adenosine

the more reactive 5-amino position. We think that the plausibility of multistep reactions that require the separation of intermediates in a natural setting is unlikely, and we tested the synthesis of guanine in the same urea-rich eutectic used for the synthesis of pyrimidines and for the subsequent phosphorylation of nucleosides (Burcar et al. 2016). In these conditions, aldehydes compete for the more reactive 5-amino group, resulting in the formation of pterins that constitute the main process, particularly the synthesis of neopterin and isoneopterin if ribose is used as sole sugar in the experiment (Sect. 4.7). In any case, the formation of guanosine is possible, although in prebiotic conditions there is competition with the formation of pterins (this is not really a problem, given the central role of pterins as biochemical cofactors) and the formation of strong orange and red pteridine-based dyes. The presence of other nucleophiles, such as urea, in high concentrations could be an obstacle for the formation of nucleosides.

The model reactions for the prebiotic synthesis of nucleosides, along with the possibility of chemical evolution from a diverse nucleobase chemical space to the canonical nucleobases in the RNA, offer possible solutions for the question of the origin of nucleic acids. The described model reactions are not problem free, are performed in relatively clean conditions, and even involve several steps. However, nature does not have glassware; it is unlikely that the prebiotic milieu was formed by small pools of water solutions of pure reagents, mixed at the right timing by *deus ex machina* events. Hence, it is necessary to make some assumptions. For example, in the case of the nucleoside formation by Traube synthesis, it is necessary to assume the concomitant origin of sugars; a model for the prebiotic origin of aldoses, and specifically ribose, would support the models based on direct glycosidation of nucleobases, even if aldoses are formed in low yields. This is possible because the nucleosides could undergo processes of selection by aggregation in supramolecular structures. The recent finding of ribose and other sugars in ultraviolet-irradiated ice in interplanetary conditions supports the direct glycosidation model (Meinert et al. 2016)

However, if we consider the essence of biochemistry, the described prebiotic model reactions can be viewed as consistent with the known biochemical systems:

- The prebiotic synthesis of nucleobases is very easy and robust in several conditions and includes canonical and non-canonical bases. Specifically, the prebiotic availability of adenine is possibly high. In biochemistry, adenosine plays a central role as a cofactor component, as a nucleic acid component, and in energy transfer and signaling. This raises the question of whether there is a correlation between the prebiotic availability of adenine and the evolution of a system based in adenine nucleotides as an essential molecule.

- The prebiotic formation of nucleobases and pteridines is closely linked. Pterin cofactors are essential and probably played a role in the origins of biochemical systems.
- Despite the difficulties of the sugar phosphate backbones, it is likely that nucleosides and phosphate as ionized linker were present from the beginning, as the prebiotic plausibility of alternatives is possibly very low.
- The model reactions, including synthesis and phosphorylation, gradually point to a specific environment: small, subaerial water ponds where concentration by evaporation and polymerization promoted by dehydration is possible. This environment favors the formation of bases, nucleosides, lipids vesicles, and peptide-containing polymers (Damer 2016; Forsythe et al. 2015).

A frequent criticism of the prebiotic synthesis of bases and nucleosides is the low yield of relevant products, such as the formation of alpha anomers of adenosine, mixtures of furanose and pyranose nucleosides, and a variety of secondary adducts with other sugars or electrophiles present in the mixture. This criticism only make sense if we focus prebiotic chemistry on the origin of the extant biomolecules, applying the methodologies and ways of thinking of synthetic organic chemistry (identification of the structure we want, retrosynthetic analysis and synthesis) without considering other possibilities and the whole chemical space, with its potential interactions. However, a rich prebiotic nucleobase chemical space offers the possibility of the selection of the right structures: the most stable glycosides and polymers in repeated cycles of hydrolysis and drying in a water pond, for example, will organize and survive, whereas less stable structures or molecules unable to combine in non-covalent associations will remain in a monomeric state. Also, the formation of supramolecular structures by base-pairing could be a mechanism of selection and accumulation of relevant molecules, formed even at low yields. In other words, after the prebiotic synthesis of a given chemical space begins a process of chemical evolution, in which the selection of the right structures leads to the next level of complexity.

4.5 s-Triazines as Alternative Nucleobases

The 1,3,5-triazines, symmetrical triazines, s-triazines, and the family of cyanuric acid (2,4,6-trihydroxytriazine) are good examples of non-biological molecules that are traditionally overlooked by prebiotic chemistry but that could have played a role in the prebiotic chemical evolution.

The prebiotic formation of triazines depends on the availability of urea and cyanide; it could be found in all experiments involving the two precursors. Together with hydantoins, the amino acids glycine and alanine, and glycolic, acetic, lactic, formic, succinic, and oxalic acids, the triazines are the most common and abundant prebiotic molecules to be found in a variety of conditions. Consider the direct analysis of the products generated by spark discharges in the experiment depicted in Fig. 4.2, which was performed at room temperature. The atmosphere is composed of molecular nitrogen and enriched with methane. As expected, a significant part of the carbon is fixed in the form of brown macromolecular materials, or *tholins*; the remaining carbon is formed of classic Miller's experiment products: amino acids (mainly glycine and alanine), urea, hydantoins, and diverse carboxylic acids, with no evidence of canonical nucleobases. The presence of water in the form of aerosol largely improves the result, limiting the formation of *tholins*, increasing the yields of amino acids and other products, and even increasing the formation of nucleobases (Ruiz-Bermejo et al. 2007).

If we consider the same experiment but with a water pool enriched in urea, it could be delivered, in a natural scenario, by cometary bombardment (de Marcellus et al. 2015). HCN and organics produced in the prebiotic atmosphere could dissolve and concentrate. This scenario, which considers both the extraterrestrial supply of small organics (the so-called soft panspermia, in contrast with the hard panspermia hypothesis of direct delivery of living cells to the planet) and the formation of precursors by atmospheric chemistry, may be more plausible than a scenario that considers only one of the options. The enrichment of urea reduces the carbon fixed in form of *tholins* and the analysis shows that the dominant components are hydantoins (Fig. 4.20).

The most interesting finding comes after repeating the experiment (same atmosphere, same energy source, water-urea solution) but under temperature cycles that creates freeze-melt cycles in the solution, maintaining a low temperature (maximum 5 °C). In these conditions, the chemistry of the system changes completely, showing the formation of an interesting set of nucleobases—mainly cytosine, barbituric acid, and s-triazines (Fig. 4.21). The one-pot synthesis of triazines and pyrimidines—particularly the barbituric acid family, which forms nucleosides spontaneously in presence of aldoses—supports the idea of the origin of a prebiotic self-assembled supramolecular structure, formed by non-canonical nucleosides and triazines. The triazines could play a role in prebiotic chemical evolution in two ways: as an early component of proto-RNA molecules or as a selector and concentrator of nucleosides formed in low yields in prebiotic conditions.

The prebiotic model synthesis of the s-triazine series is consistent with its finding in meteorites (melamine, ammelide, and cyanuric acid were found in the Murchison meteorite). The aminotriazines could spontaneously form exocyclic ribosides by nucleophilic attack of the aldose to the amine group (Fialho et al. 2018; Kaur et al. 2017; Hysell et al. 2005). The exocyclic ribosides of triazines could mimic both purine or pyrimidine nucleosides, forming base pairs with canonical nucleosides or between the triazine nucleosides themselves (Hysell et al. 2005). An interesting difference between s-triazines and pyrimidines of the barbituric acid family is that the ring nitrogen of the former cannot be attacked by electrophilic sugars, such as ribose. Otherwise, the C5 of the pyrimidine ring in the barbituric acid family is strongly nucleophilic; it is the preferential position for electrophilic attack by

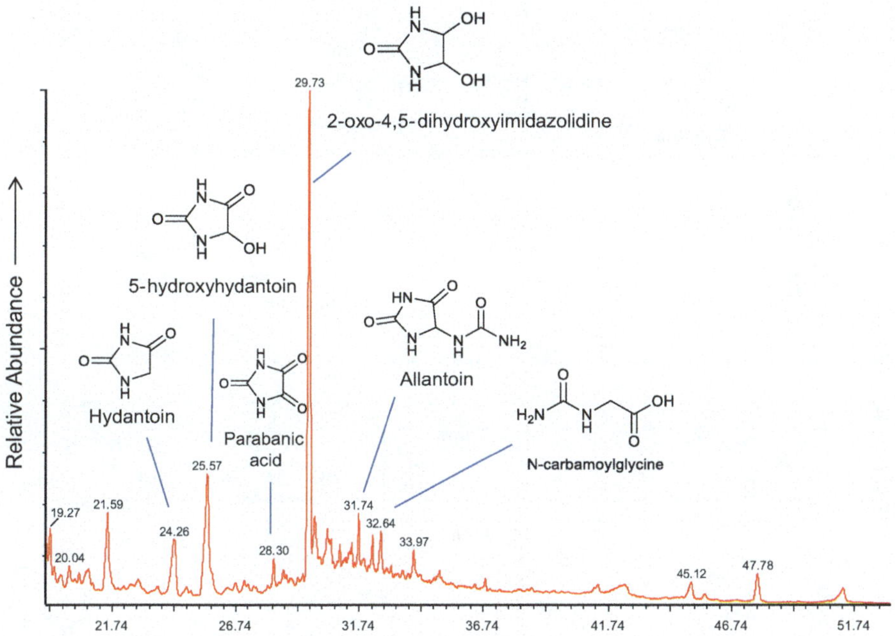

Fig. 4.20 Gas chromatography–mass spectrometry chromatogram of the products formed after sparking an N_2 + CH_4 atmosphere over a urea-enriched water pool (0.1 M urea) at room temperature. The analysis was revised and actualized from Menor-Salván et al. (2009). Along with hydantoins, there are formations of glycine, alanine, and *N*-formylglycine. It is necessary to concentrate the urea (by freezing or evaporation) to get a nucleobase-enriched chemical space

aldoses, even in amine-substituted pyrimidines such as 2,4,6-triaminopyrimidine. To date, the prebiotically plausible formation of endocyclic nucleosides of s-triazines have not been reported. The potential formation of an s-triazine bond to a backbone (sugar or other type) would have important implications regarding the structure and stability of the potential pre-RNA molecules. Regardless of the stability of potential prebiotic triazine nucleosides or nucleoside-like molecules, the free triazine bases may be useful molecular scaffolding components that provide strong self-assembling supramolecular complexes, such as rosettes (Li et al. 1996; ten Cate et al. 2005). In neutral or acidic water solutions, the keto form is the favored resonance structure of cyanuric acid. This form constitutes a versatile scaffold with both a donor and acceptor of hydrogen bonding; the sp3 hybridized nitrogen atoms of its triazine ring provide three N–H hydrogen bond donors and six pairs of unshared electrons for the carbonyl oxygens, oriented within the triazine plane, which act as hydrogen bond acceptors (Fig. 4.22). Also, aromatic π–π ring stacking in triazines contributes to the structure and stabilizes the binding, allowing the formation of stacked rosettes. The melamine exhibits exactly complementary hydrogen bonding abilities. Uracil, barbituric acid, uric acid, thymine, and xanthine (bases

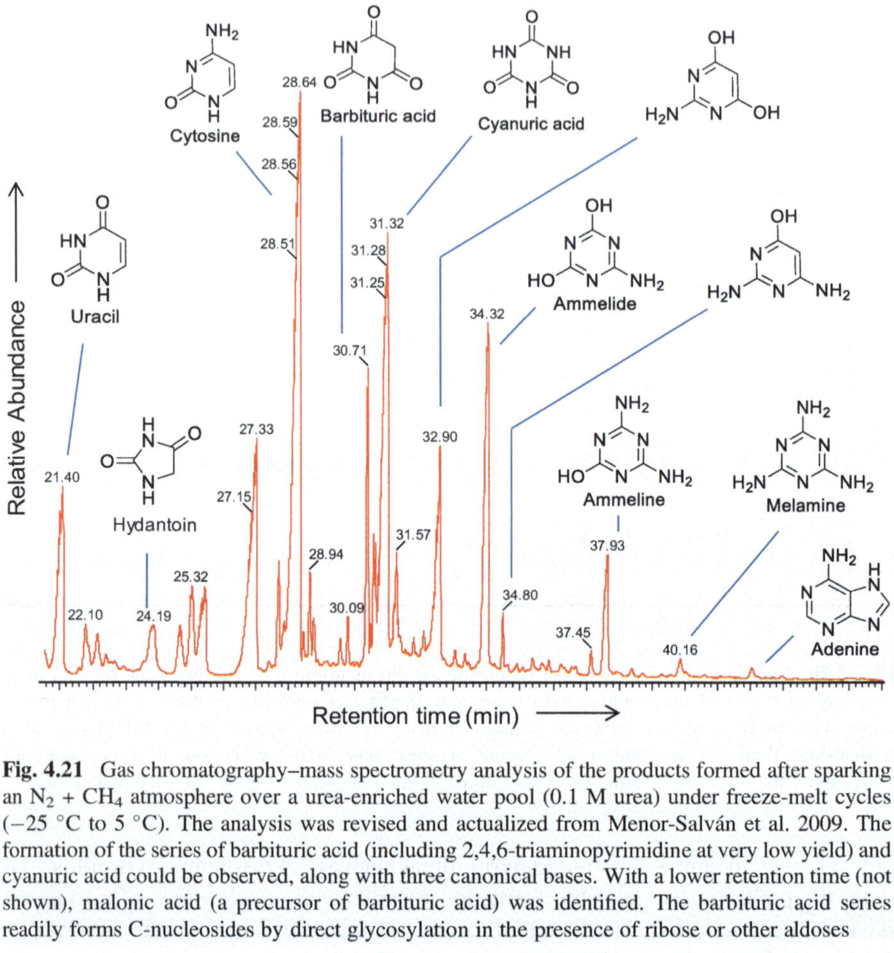

Fig. 4.21 Gas chromatography–mass spectrometry analysis of the products formed after sparking an N_2 + CH_4 atmosphere over a urea-enriched water pool (0.1 M urea) under freeze-melt cycles (−25 °C to 5 °C). The analysis was revised and actualized from Menor-Salván et al. 2009. The formation of the series of barbituric acid (including 2,4,6-triaminopyrimidine at very low yield) and cyanuric acid could be observed, along with three canonical bases. With a lower retention time (not shown), malonic acid (a precursor of barbituric acid) was identified. The barbituric acid series readily forms C-nucleosides by direct glycosylation in the presence of ribose or other aldoses

and nucleosides) interact with triazines, particularly melamine, forming self-assemblies (Tolleson et al. 2009).

The base-pairing properties of s-triazines could add a selection mechanism to the prebiotic milieu; the strong base pairing between triazines and pyrimidine or purine nucleobases or nucleosides could help in the selection, concentration, and auto-organization of pre-RNA building blocks (Fig. 4.23). The triazine-nucleobase or triazine-nucleoside assemblies have low solubility even in diluted solutions; after precipitation of triazine assemblies, an increase of urea concentration leads to the complete redissolution of the precipitate by disruption of the hydrogen bonds. This could have provided a very simple concentration, preservation in the form of insoluble assemblies, and a redissolution mechanism through dry-wet environmental cycles. This potential triazine role in prebiotic evolution through base-pairing is strongly dependent on the pK_a of the bases and pH of the solution (Krishnamurthy 2012).

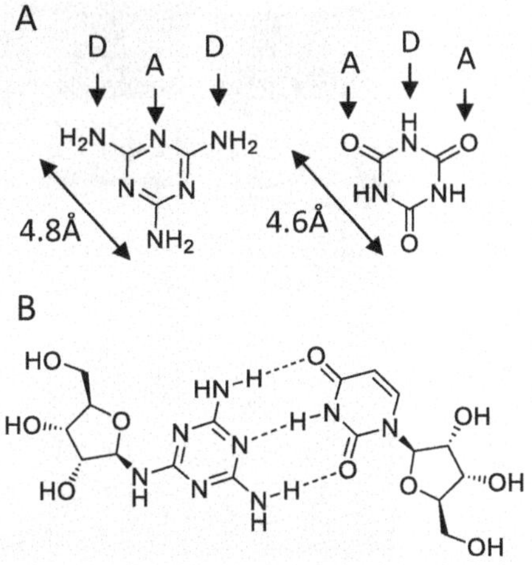

Fig. 4.22 (**a**) The hydrogen bonding capabilities of melamine (left) and cyanuric acid (right). Each molecule has three binding orientations, forming a triangle with three possible hydrogen bonding positions, both donor (D) and acceptor (A), by each side (figure based on Tolleson et al. 2009). This symmetry in base pairing converts the triazine bases in extraordinary pieces in potential pre-RNA scaffolds. (**b**) Base-pairing between uridine and exocyclic melamine riboside

Urea is the precursor for the synthesis of s-triazines in both industrial synthesis and prebiotic chemistry. The industrial process is a high-temperature and high-pressure process in which urea is converted to cyanuric acid (the cyclic trimer of cyanic acid) or melamine (trimer of guanidine) through the decomposition of urea to cyanic acid and its polymerization (Tolleson et al. 2009). The first synthesis of cyanuric acid was studied by Liebig and Wöhler, who determined its composition at the beginning of nineteenth century. The first synthesis of cyanuric acid was performed by Scheele after oxidative pyrolysis of uric acid (Smolin and Rapoport 1959). Later, the synthesis was performed simply by heating a urea melt until the emission of ammonia ceased. This reaction could be relevant from a prebiotic point of view, because gently heating urea in dehydrating conditions leads to the formation of biuret by a reaction of urea with cyanic acid formed by its decomposition (Kurzer 1956). Biuret further reacts with another molecule of urea or with cyanic acid, producing the hypothetical (as we have not yet observed it) intermediate triuret (Fig. 4.24). Triuret could decompose readily in two ways, producing cyanuric acid and ammelide. This process is consistent with our observations in model prebiotic reactions using urea eutectic formation (Menor-Salván et al. 2009).

In prebiotic conditions, we observed the formation of triazines after the concentration of urea (by freezing) and energization by means of spark discharges in a methane-enriched atmosphere (Menor-Salván et al. 2009). This work demonstrated that a high-temperature scenario is not necessary to form triazines; they could be formed even at low temperatures if a source of urea and/or cyanic acid is present. Later, we showed that urea under dehydrating conditions also could lead to triazines by mild heating and confirmed the biuret pathway in its formation (Menor-Salván et al. 2018). The polymerization of cyanic acid could lead, depending on the conditions, to insoluble and stable cyamelide, which is regarded as a linear polymer (Smolin and Rapoport 1959). Also, cyanuric acid and melamine form insoluble

Fig. 4.23 Precipitation (**a**) and redissolution by disruption of base pairing after adding concentrated urea (**b**) of an insoluble complex between solutions of the triazine melamine and the nucleoside uridine. Precipitation could have helped to concentrate and preserve prebiotic nucleic acid building blocks and drying-wetting cycles could have induced molecular selective pressure for the chemical evolution of proto-nucleic acid

resins by a methyolation reaction with formaldehyde. Those polymers may be incorporated to the *tholin* pool and could constitute a burden in chemical evolution models based on the prebiotic availability of triazines. The concentration of urea eutectic solutions may help to form triazines, thus avoiding its loss in form of intractable polymers.

In the formation of the amine-rich terms of the s-triazine series, the mechanism is not straightforward. Cyanuric acid reacts with ammonia and the heating of biuret yields melamine, but both processes require high temperatures (Kurzer 1956) and/or high pressures (Kinoshita 1953). If cyanuric acid could be regarded as a cyanic acid or urea trimer, melamine is the equivalent cyclic trimer of cyanamide or guanidine. The formation of melamine by heating guanidine carbonate has been well known since the beginning of the twentieth century (Davis 1921). Several plausible

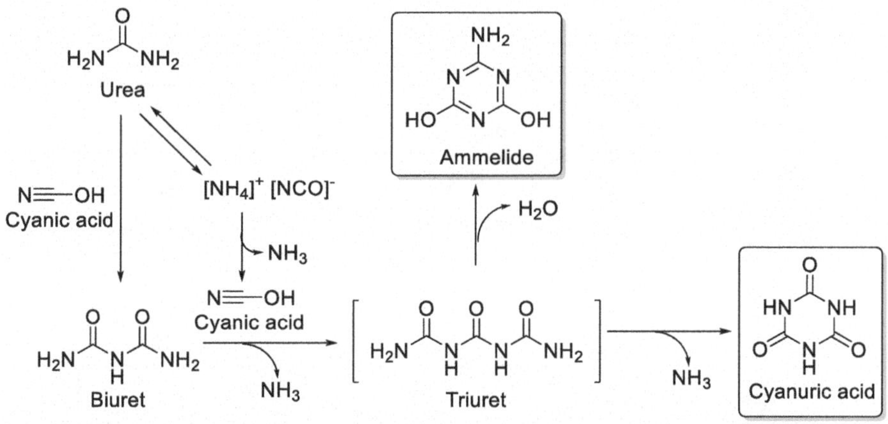

Fig. 4.24 Synthesis of ammelide and cyanuric acid from urea through cyanic acid formation. In prebiotic conditions, the formation of urea eutectics and their dehydration could lead to triazines in a pathway comparable to the high-temperature synthesis

prebiotic conditions could promote the formation of cyanamide, such as spark discharges in a methane-enriched nitrogen atmosphere or urea concentration and dehydration (Fenton 1882). If cyanamide is formed, even at a low or moderate temperature, the formation of triazines is plausible in prebiotic conditions, which are milder than the usual high-temperature synthetic methods (Fig. 4.25). This could explain the significant formation of melamine and ammeline in experiments using spark discharges through nitrogen-methane, in which reactive species formed in the atmosphere (cyanamide and cyanic acid) will react with urea solutions, leading to the formation of triazines.

In our model prebiotic reactions (Menor-Salván et al. 2009), it is interesting to note that triazines were formed significantly only in cold environments, with the formation of ice within the range of temperature of liquid urea-water eutectic solutions. Experiments at room temperature with liquid, diluted solutions did not show significant triazine formation, whereas experiments using urea solutions in an inert atmosphere showed the formation of triazines. This could be explained by several hypotheses: (1) the ice trap stabilizes reactive species, diverting the synthesis to the formation of triazines thanks to the formation of interstitial urea eutectic solutions; or (2) the spark discharges through the eutectic urea solution lead to the formation of a reactive species (cyanic acid, cyanamide) from the urea, hence following the classic synthetic pathway (Fig. 4.25). The atmospheric reactions at low temperatures could favor triazine synthesis through HCN and cyanamide formation, whereas higher temperatures may lead to hydantoin precursors (see Sect. 4.6). The change of energy source to ultraviolet irradiation in similar conditions shows very low yield formation of cyanuric acid, with negligible formation of other triazines (Menor-Salván and Marín-Yaseli 2013), suggesting that, together with urea concentration and/or temperature, the nature of the energy source determines the final composition of the prebiotic mixture.

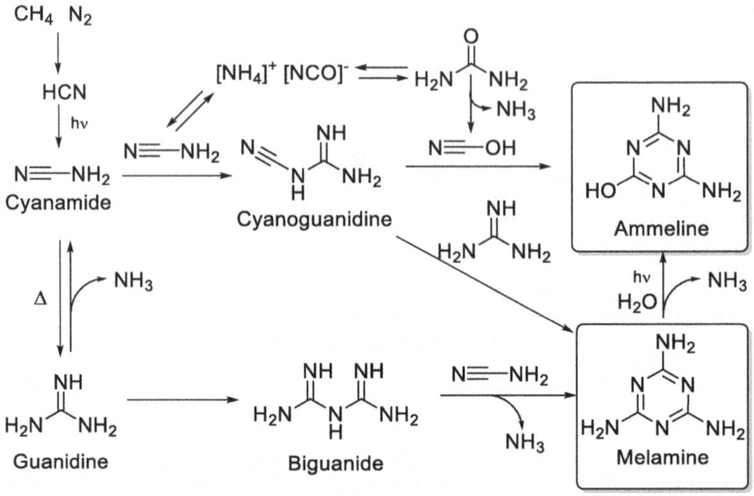

Fig. 4.25 Synthesis of ammeline and melamine through guanidine or cyanamide formation

A possible explanation for our results was provided by Yassin Jeilani, Thom Orlando, and their coworkers in 2014, who suggested a free radical mechanism for the origin of triazines in low-temperature conditions. The irradiation of urea or ammonia could lead to release of aminyl radical, which activates the urea by hydrogen abstraction, leading eventually to the formation of a guanylurea intermediate and melamine or to a biuret intermediate and cyanuric acid (Fig. 4.26). The final triazine assemblage and yield depends on the energies and probability of several possible free radical reactions. Regardless of the final mechanism (as Jeilani et al. pointed out, "prebiotic systems are dynamic and speculative"), the plausible formation of free triazine rings in relatively stable conditions could be an important process for chemical evolution. We hope that further experimental and theoretical works with model pre-RNA molecules will soon show us whether the triazines played a relevant role in prebiotic chemical evolution.

4.6 Hydantoins: A Prebiotic Nitrogen Reservoir for the Origin of Nucleic Acids?

Traditional prebiotic chemistry was mainly RNA-centric. The effort was focused on the elucidation of the origin and organization of the canonical components of extant biopolymers, overlooking a plethora of organic molecules that are not involved, or play a small part, in known biochemistry. One of the molecular families traditionally overlooked or that has not received a lot of attention is the hydantoin family. Hydantoins are two-nitrogen, five-membered rings derived from the imidazolidine-

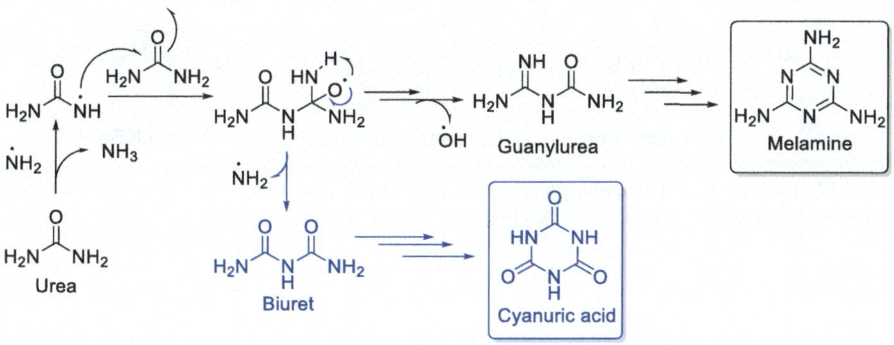

Fig. 4.26 Free radical mechanism proposed to explain the synthesis of s-triazines by irradiation or spark discharges in prebiotic conditions. Depending on the path followed by radical reactions, cyanuric acid (blue path) or melamine (black path) is formed. Based on Jeilani et al. (2014)

2,4-dione; they are the main nitrogen heterocycles formed in all prebiotic simulations based on irradiation of methane-containing atmospheres. Together with urea and amino acids, hydantoins could have been the most abundant organics on primitive Earth and other prebiotic environments (De Marcellus et al. 2011) and could have formed in interstellar ice (Bernstein et al. 2002). Hydantoins were found in carbonaceous chondrites and are precursors of glycine and alanine (Sarker et al. 2013). In fact, as we observed in our laboratory, the simplest hydantoin yields glycine, diglycine, and carbamoylglycine by simple drying-wetting cycles at 50 °C of a solution containing hydantoin and urea (Fig. 4.27); also, hydantoin could form glycine-glycolic depsipeptides by drying-wetting a solution of hydantoin and glycolic acid.

Hydantoin formation explain the large excess of glycine and alanine relative to other amino acids. How is hydantoin formed in prebiotic conditions? The ultimate precursors of hydantoin are methane/CO_2 and nitrogen, and a key factor is the formation of HCN. Calculations made by Airapetian et al. (2016) showed that a nitrogen-dominate atmosphere containing carbon dioxide and methane would be able to produce a HCN concentration higher than previously thought. If we consider an atmosphere containing methane and nitrogen, it is reasonable to expect the formation of HCN and acetylene (Fig. 4.28, Ruiz-Bermejo et al. 2009a, b). As we reviewed in this and other chapters in this book, HCN is the precursor of urea, among other molecules, including glyoxylic acid (Marín-Yaseli et al. 2016). Acetylene, which could be formed in atmosphere and has been detected in cometary ice (Brooke et al. 1996), is a very efficient source of glyoxylic acid and glyoxal, among other carboxylic acids of prebiotic interest, such as glycolic and lactic acids (Menor-Salván and Marín-Yaseli 2013).

Considering the atmospheric chemistry of methane, acetylene, and hydrogen cyanide (and not disregarding extraterrestrial sources, locally or globally enriched atmospheres, or water ponds), we could propose a "warm little pond" model in which, by means of dry-wet and/or freeze-melt environmental cycles, the synthesis

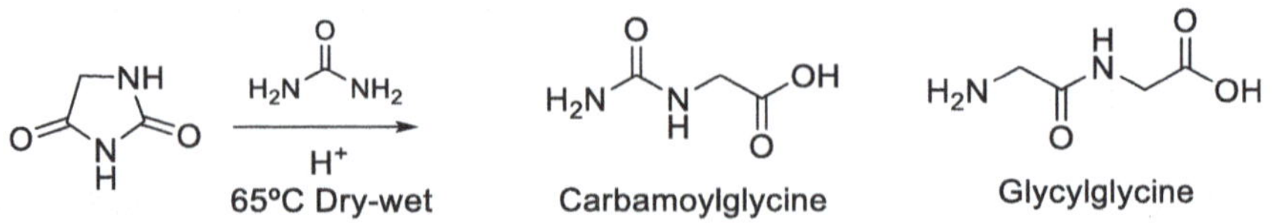

Fig. 4.27 Hydrolysis of hydantoins yields amino acids. Drying-wetting cycles of a solution of urea and hydantoin lead to a product enriched in diglycine and carbamoylglycine

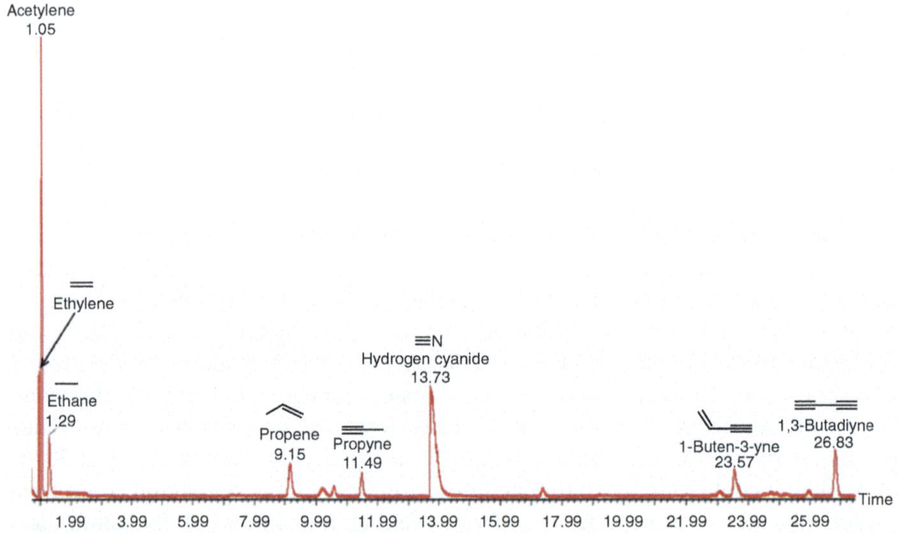

Fig. 4.28 Nitrogen-based atmospheres containing methane lead, after irradiation, to a mixture rich in HCN and acetylene (Adapted from Ruiz-Bermejo et al. 2009a, b)

of relevant raw materials and subsequent chemical evolution occurs (Fig. 4.29). This water pond could easily accumulate organics, especially urea, given the colligative properties of the urea-water system. This allows the formation of eutectic solutions and the accumulation of assemblies.

The glyoxylic acid reacts readily with urea to form 5-hydroxy-hydantoin, which is a compound commonly found in model prebiotic chemistry experiments. The simple drying-rewetting at 65 °C of a urea and glyoxylic acid solution leads to a mixture of 5-hydroxy-hydantoin, dihydroorotic acid, and uracil (Fig. 4.30). The reaction is dependent on the pH and relative concentrations of reactants, resulting in variable yields of the hydantoin and dihydroorotic acid. This could be consistent with the formation of dihydroorotic and orotic acid observed in prebiotic experiments (Menor-Salván and Marín-Yaseli 2013). The photochemical oxidation of dehydroorotic acid yields orotic acid, whose decarboxylation in water is thermodynamically favored, leading to uracil; this is a parallel pathway to the biochemical, enzyme-driven counterpart, which can be interpreted as evidence of the chemical

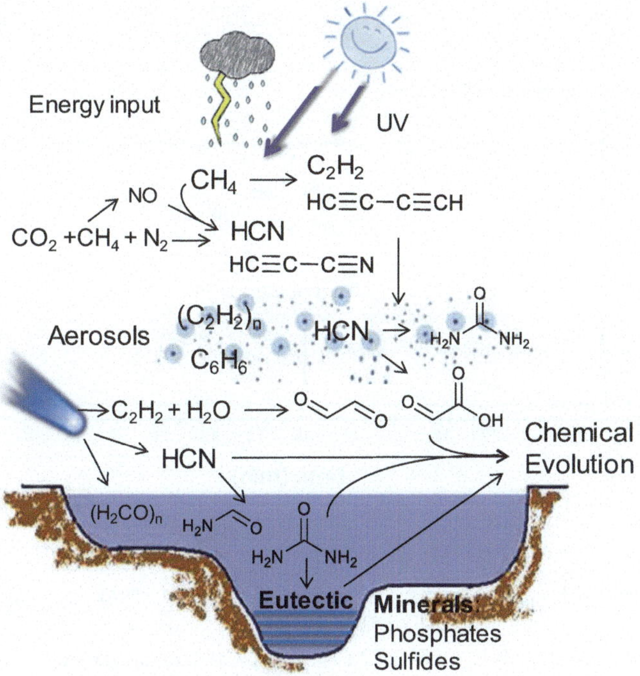

Fig. 4.29 A general "warm little pond" model for the origin of chemical evolution based on the atmospheric and extraterrestrial input of hydrogen cyanide and acetylene. Overall, the carbon triple bond is the source of urea, glyoxylic acid, and other components that would be essential for chemical evolution

determinism in chemical evolution (Lazcano 2009; Yamagata et al. 1990). The mechanism of formation of dihydroorotic and orotic acids and uracil from glyoxylate and urea is yet to be explained, but probably goes through a hydantoin or hydantoic acid intermediate; in addition, it could be a pathway for the origin of these pyrimidines in a prebiotic setting. Uracil formation could also lead to the synthesis of barbituric acid by hydroxyl radical addition (Menor-Salván and Marín-Yaseli 2013).

Another route from hydantoin to orotic acid and uracil could be the condensation of oxalacetic acid with urea, which proceed in gentle conditions (simply by evaporation at 50 °C, as we observed in our laboratory) to give 5-carboxymethylidene-hydantoin (Mitchell and Nyc 1947). This derivative rearranges by more vigorous heating to form orotic acid (Fig. 4.31). Another possible source of 5-carboxymethylidene-hydantoin is the condensation of hydantoin with glyoxylic acid, given the nucleophilic character of the C5 position of the hydantoin ring (Ware 1950).

If the reaction between glyoxylic acid and urea occurs in a medium with an excess of urea or in urea/guanidine eutectics, the main product formed is allantoin (Fig. 4.32). This is a prebiotic chemistry version of a reaction first observed by the

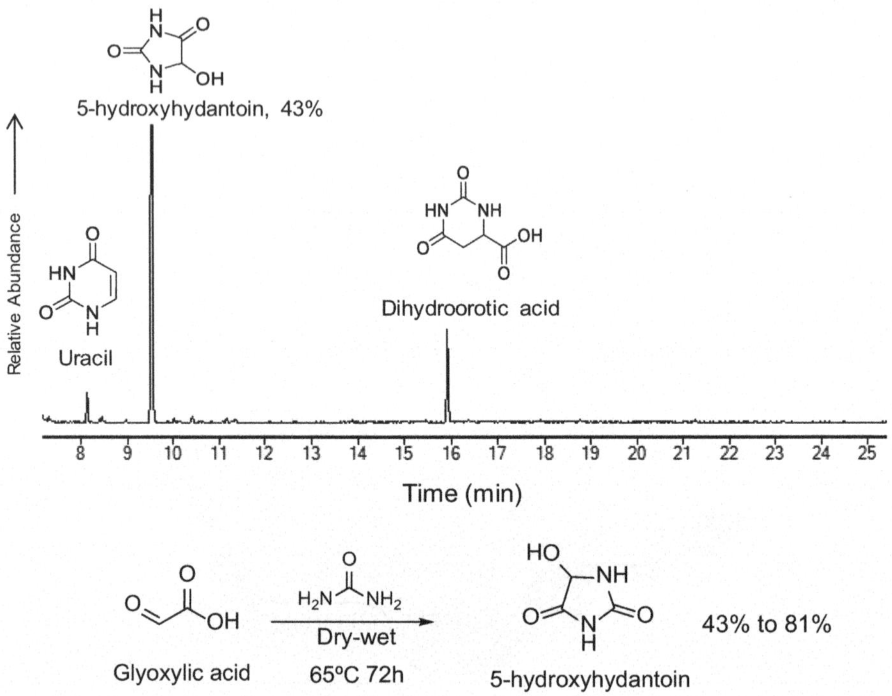

Fig. 4.30 Gas chromatography–mass spectrometry chromatogram of the reaction product resulting from drying/rewetting a solution containing urea and glyoxylic acid at 65 °C (Menor-Salván, unpublished results). This reaction is consistent with the results published in model prebiotic reactions (Menor-Salván and Marín-Yaseli 2013) and suggest that a hydantoin ring could be a precursor/intermediate in the prebiotic formation of orotic acid and uracil

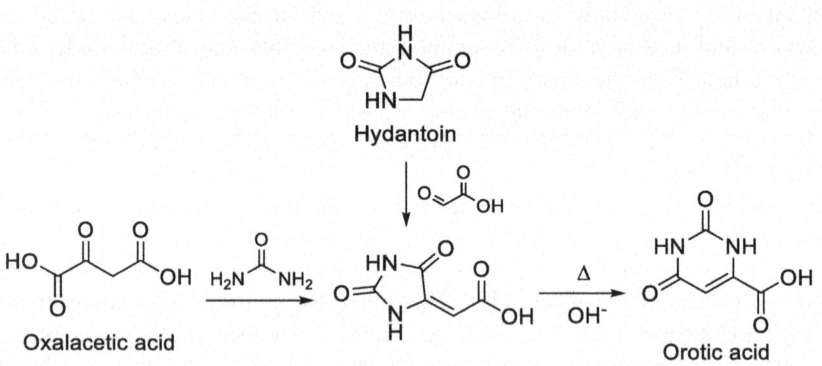

Fig. 4.31 Synthesis of orotic acid involving hydantoin. The hydantoin-glyoxal reaction could explain the orotic acid found in prebiotic model reactions

great chemist Edouard Grimaux (Ben-Ishai et al. 1977). Overall, the reactions described above could explain the formation of allantoin (Fig. 4.20) and orotic acid (Menor-Salván and Marín-Yaseli 2013) in model prebiotic reactions.

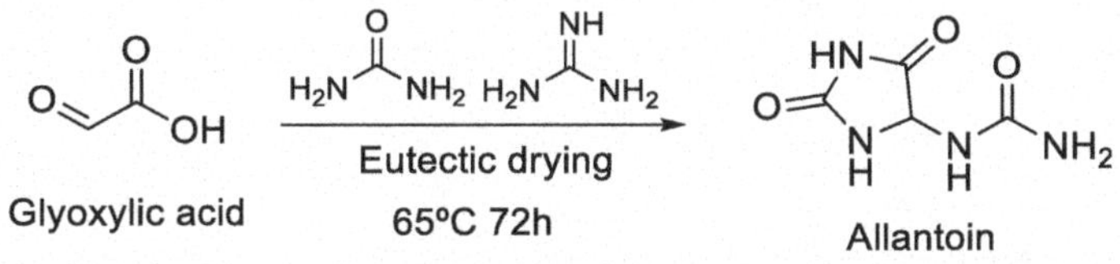

Fig. 4.32 Prebiotic synthesis of allantoin from glyoxylic acid in urea eutectic solution

The reaction between glyoxylic acid and urea could be a drawback for the glyoxylate-world models. It is not likely that glyoxylic acid is formed in a prebiotic milieu completely free of urea, which is formed in significant proportions by hydrolysis of hydrogen cyanide. As an example of the possible prebiotic role of glyoxylate in the emergence of a proto-metabolism is the synthesis of tartaric acid by the glyoxoin reaction of glyoxylate catalyzed by cyanide (Butch et al. 2013). The glyoxoin reaction could explain the tartaric acid observed in model prebiotic reactions (Ruiz-Bermejo et al. 2006). In the presence of urea, however, the efficiency of glyoxylate as a source of the Krebs cycle, or other proto-metabolic molecules, is largely reduced. For example, the synthesis of tartaric (or other possible related products of glyoxylate) by glyoxoin reaction is impaired, even in presence of cyanide, with a reduction of yield from 80% in DL-tartaric acid reported by Butch et al. (2013) to less than 6% (Fig. 4.33). Hence, the prebiotic formation and accumulation of urea in small water ponds depends on the availability of prebiotic products, diverting active carbon compounds to the synthesis of heterocycles.

Another important prebiotic reactant is the glyoxal, which could be formed in a variety of reactions and from precursors including formaldehyde (Delidovich et al. 2011; Weber 2001), glycolaldehyde (Perri et al. 2009), simulations of precometary interstellar ices (de Marcellus et al. 2015), browning reactions of sugars (Martins et al. 2001; Powrie et al. 1986), and acetylene (Menor-Salván and Marín-Yaseli 2013). The reaction between glyoxal and urea yields hydantoin in a reaction that is dependent on the pH (Menor-Salván et al. 2018, Fig. 4.34). The pH dependence could confirm that the formation of hydantoin follows a mechanism mediated by a pinacol-pinacolone rearrangement (Fig. 4.35), as we suggested previously after experiments performed using deuterated glyoxal.

The formation of glyoxal may be a source of prebiotic imidazole if ammonia is present (Fig. 4.36). The irradiation of an acetylene-bearing atmosphere over an ammonia solution leads to the formation of imidazole in a mechanism apparently mediated by the glyoxal diamine, with formation of the methanediol derivative of imidazole as a byproduct.

Possible prebiotic reactions involving the formation of hydantoins and related compounds are summarized in Fig. 4.37. Overall, the depicted substances are expected in prebiotic environments in which urea, glyoxylic acid, and glyoxal are formed. Of course, these reactions are not exclusive. The photochemistry of purines would lead to the formation of hydantoin or hydantoin derivatives (in another example of parallelism between biochemical pathways and prebiotic chemistry).

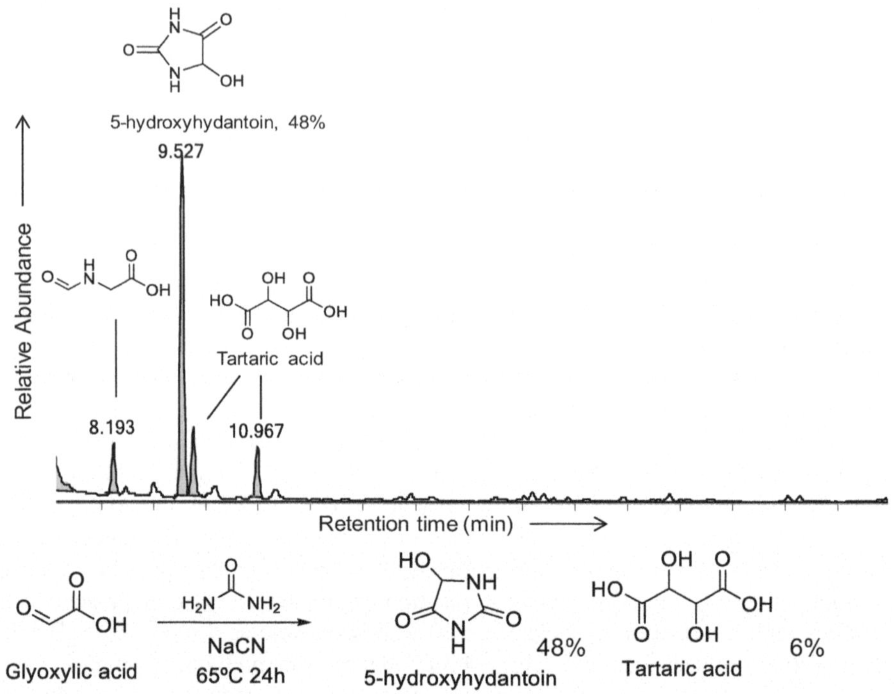

Fig. 4.33 Glyoxoin reaction (glyoxylate dimerization catalyzed by cyanide) in the presence of urea. The yield of tartaric acid is significantly reduced due to competition with the hydantoin ring formation in the presence of urea. Reaction performed by the author, based on Butch et al. (2013)

Also, the hydantoins could follow the reverse path and favor the formation of nucleobases.

Another possible reaction of potential prebiotic interest involving hydantoin is the formation of glycouril. This compound is formed by a slow reaction of glyoxal with excess urea at room temperature (Fig. 4.38). In a prebiotic setting, urea concentration by the evaporation of water ponds could absorb glyoxal by the formation of glycouril. The glycouril readily reacts with formaldehyde, forming methylol glycouril (Fig. 4.39). When heated in acidic conditions, this leads to a gel formed by the cucurbituril macrocycle, a hexameric structure that could self-assemble to form supramolecular structures (Day et al. 2001):

When methylol glycouril is left in water or gently heated, it slowly releases formaldehyde; when it is heated in basic conditions with an adequate catalyst (calcium hydroxide), it could form sugars. Thus, the tetramethylol glycouril could have been one of the possible formaldehyde reservoirs (Cleaves 2008) because of its insolubility and formaldehyde storage capacity (4 mol formaldehyde per mol of glycouril). The tetramethylol glycouril could release formaldehyde in a sustained manner and lead to the formation of sugars by a formose reaction in a urea-compatible scenario.

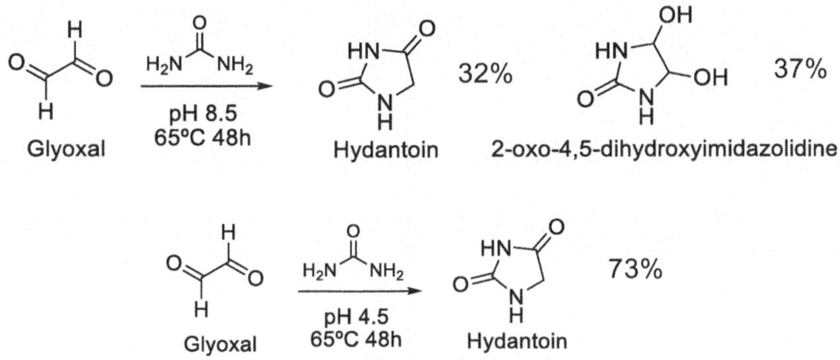

Fig. 4.34 pH dependence of the reaction between urea and glyoxal. Together with isotopic labelling experiments, this demonstrates the mechanism of prebiotic formation of hydantoin from urea

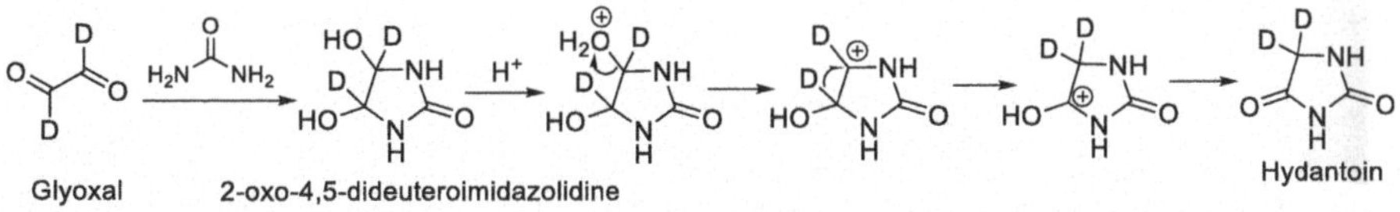

Fig. 4.35 Deuterated glyoxal reacts with urea in prebiotic conditions, leading to 5,5′-dideuterium-hydantoin and suggesting a pinacol-pinacolone rearrangement in the prebiotic synthesis of hydantoin. The named compounds are formed and identified in prebiotic model reactions

Assuming the significant formation of hydantoins in organic-rich prebiotic environments, as the experimental models suggests, what is their potential as a precursor of nucleobases? It seems logical to think that the nitrogen invested in the formation of hydantoins is saved from other less reactive forms (e.g., tholins) and could be recycled in the formation of other prebiotically relevant molecules. In our laboratory, we tested whether hydantoin could be a precursor of nucleobases in a plausible prebiotic environment (Menor-Salván et al. 2018). We observed that the ultraviolet irradiation of a solution of urea and hydantoin, which has been concentrated by evaporation or by formation of a water-urea eutectic in ice, could lead to the formation of pyrimidines and purines (Fig. 4.40b). The generated molecular diversity included triazines, tetra-substituted pyrimidines, triamino pyrimidines, uric acid, guanine, and xanthine, but no evidence of pteridines (see Sect. 4.6), indicating that decomposition in the experimental conditions does not lead to glyoxal. Instead, we found that malonic and amino malonic acid are formed, which could be prebiotic precursors of pyrimidines after condensation with urea (Fig. 4.40a).

The prebiotic formation of malonic acid and its derivatives could constitute another interesting previous step in the origin of purines and pyrimidines, although it does not totally explain the chemical space generated. Both acids are easily synthesized in prebiotic model reactions based on methane/acetylene irradiation or energization, or by cyanide hydrolysis. In fact, the formation of amino malonic acid by HCN polymerization is a basic feature of the cyanide chemistry (Johnson and

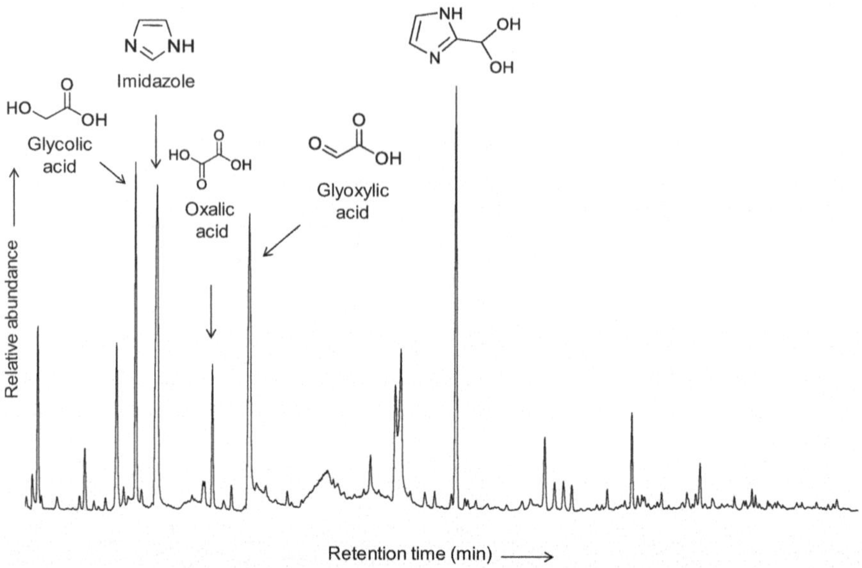

Fig. 4.36 The ultraviolet irradiation of a nitrogen-based, acetylene-containing atmosphere over an icy ammonia solution, leading to imidazole. The probable precursor of imidazole is glyoxal. If the solution contained urea, glyoxylate would react to form hydantoin derivatives (adapted and updated from Menor-Salván and Marín-Yaseli 2013)

Nicolet 1914; Fig. 4.12). The mechanism of its formation by irradiation of hydantoin-rich urea eutectic, however, is not straightforward and needs to be elucidated.

After creating a viscous solution of urea/guanidine by evaporation at moderate temperatures (a drying pool model), we have shown that the addition of malonic acid and repeated drying-rewetting cycles during several days will lead to the formation of a pyrimidine assemblage, including barbituric acid and 2,4,5- and 2,4,6-triamino pyrimidines (Fig. 4.41). The resulting composition of this experiment strikingly fits with the composition found in hydantoin-urea prebiotic model reactions, demonstrating that the malonic-urea-hydantoin system leads to a pyrimidine and triazine assemblage that is perfect for further steps in chemical evolution: nucleosidation and self-assembly. It is still necessary to explain some mechanistic details and elucidate reactions (particularly regarding the formation of 2,4,5-triaminopyrimidine, which is formed at a significant yield). However, it seems to be clear that the initial steps include urea decomposition with the release of ammonia and isomerization to cyanate.

Urea decomposition could provide the reactants necessary for the formation of monoamido malonic acid and malondiamide, which condenses with urea or guanidine to form 2,4,6-substituted pyrimidines (Fig. 4.42). Also, malondiamide could self-condense or react with malonic acid to form 2-substituted 4,6-dihydroxypyrimidines

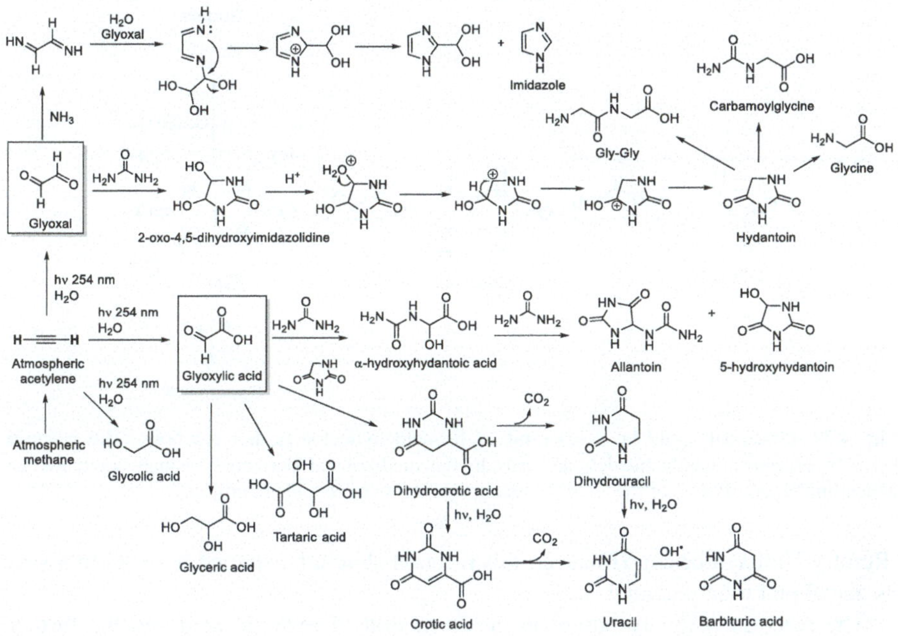

Fig. 4.37 Prebiotic chemistry pathways involving hydantoins (excluding purine photodegradation reactions). If we assume the formation of urea, glyoxylic acid, and glyoxal (which would not necessarily involve acetylene, a very efficient precursor of glyoxylate and glyoxal), hydantoins and its derivatives are expected in the prebiotic chemical space. The formation of hydantoin explains the preponderance of glycine in prebiotic model reactions, which could serve as nitrogen reservoir and could lead to the formation of nucleobases

Fig. 4.38 Glycouril crystals formed after leaving a solution containing urea and glyoxal in water at room temperature for a few days. The solution also contained hydantoin and glycine as the main components (Menor-Salván, previously unpublished results)

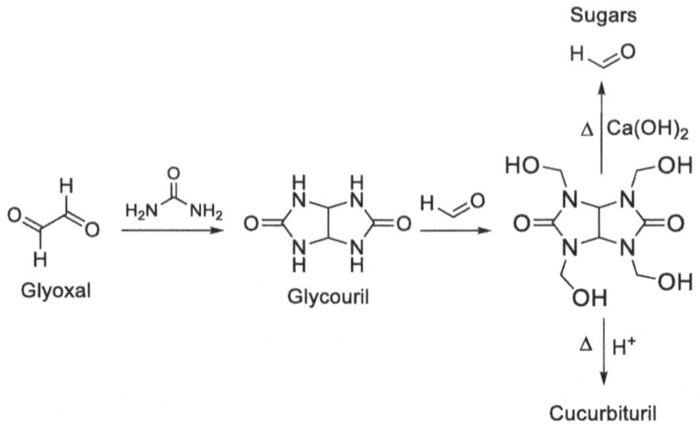

Fig. 4.39 Glycouril could be a reservoir of formaldehyde for further reactions. The reaction between glycouril and formaldehyde leads to the methyolated derivative, which could evolve depending on conditions of the formose reaction or the macrocycle cucurbituril

(Remfry-Hull synthesis). However, this reaction should be studied in detail to assess its actual prebiotic potential.

The reaction pathway based on the formation of malonic acid and the further synthesis of malondiamide and monoamido malonic acid, boosted by urea decomposition and/or ammonia present in the system, is consistent with both the high yield of 2-amino-4,6-dhihydroxy pyrimidine found in malonic acid/urea eutectic drying-wetting experiments and the composition found in the prebiotic model reaction of ultraviolet irradiation of hydantoin-urea solutions. However, the synthesis of amino malonic acid could be the source of the purines uric acid and guanine, through formation of tetrasubstituted pyrimidines (Fig. 4.43). Uramil could condense with urea or cyanate to form pseudouric and uric acids (Schmidt 1950), and the 4,5-diamino pyrimidines could undergo an easy Traube reaction to form other purines.

At this point, the reader probably agrees with the idea that a plausible origin of urea and hydantoin in prebiotic environments is one of the possible previous steps in the origin of nucleobases, linked with other parallel chemistries (cyanide, formamide) in a sort of prebiotic "proto-metabolism." Whether the formation of viscous "little warm ponds" rich in organic solutes (especially urea), which are clear points of accumulation of nucleobases, are relevant for the further steps in the origin of nucleic acids is one of the current open questions.

4.7 The Chemical Predestination of Purines and Pterins

It is not possible to think about the prebiotic chemistry of purines without considering the pteridines (derivatives of pyrazino-[2,3-*d*]-pyrimidine). After the discovery of the pteridine structure in biological compounds—xanthopterin, leucopterin,

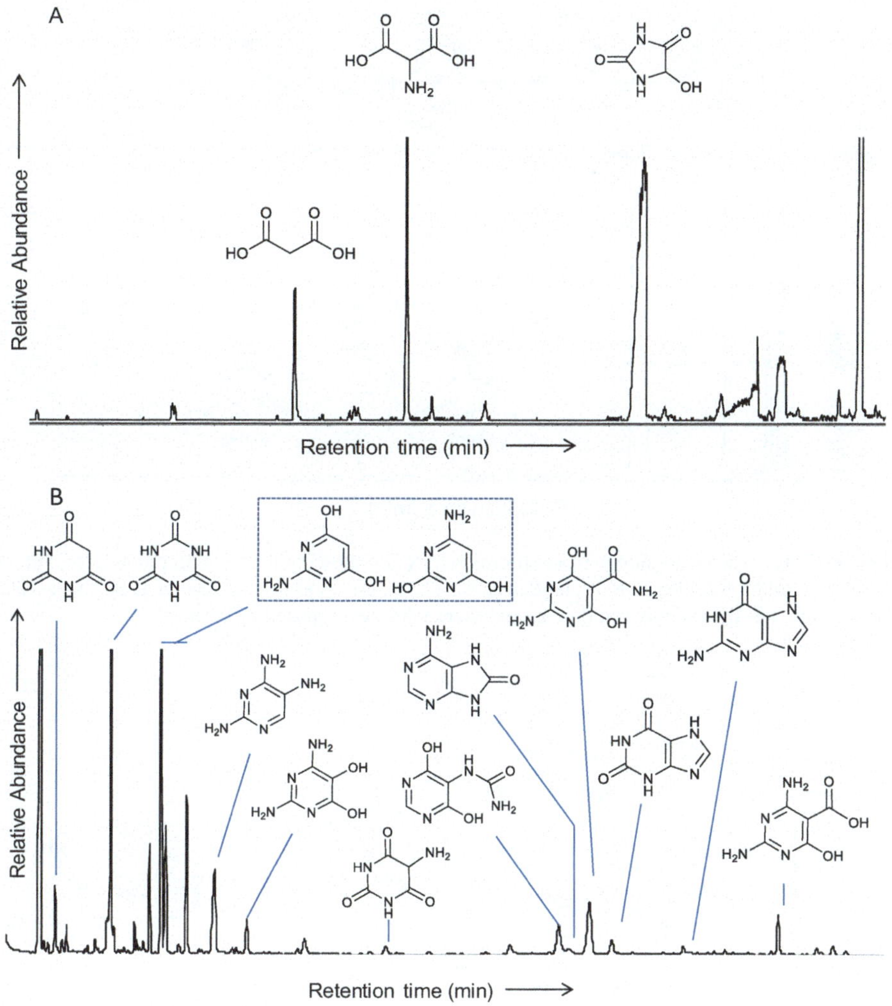

Fig. 4.40 The main products of ultraviolet irradiation (254 nm) of an hydantoin-urea solution dried to form a viscous solution under inert atmosphere. (**a**) Lower retention time section of a gas chromatography–mass spectrometry chromatogram of the product, showing the malonic and amino malonic acids. (**b**) Main bases formed after irradiation, which includes triazines, pyrimidines, and purines. In lower proportions, 2,4,6-triamino pyrimidine and its derivatives, as well as 2,4-diamino purine, were also identified. When using freezing to form a urea eutectic solution and performing the experiment at a lower temperature, uric acid is strongly increased (here it is found in very low proportions) and tetraaminopyrimidine could be identified

and isoxanthopterin from the pigment of butterfly wings (Rembold and Gyure 1972)—the ubiquity and biochemical connection of purines and pteridines have been very well established. The most important biological pteridines are the pterins, derivatives of 2-amino-4-keto-pteridine, which is the core of essential cofactors

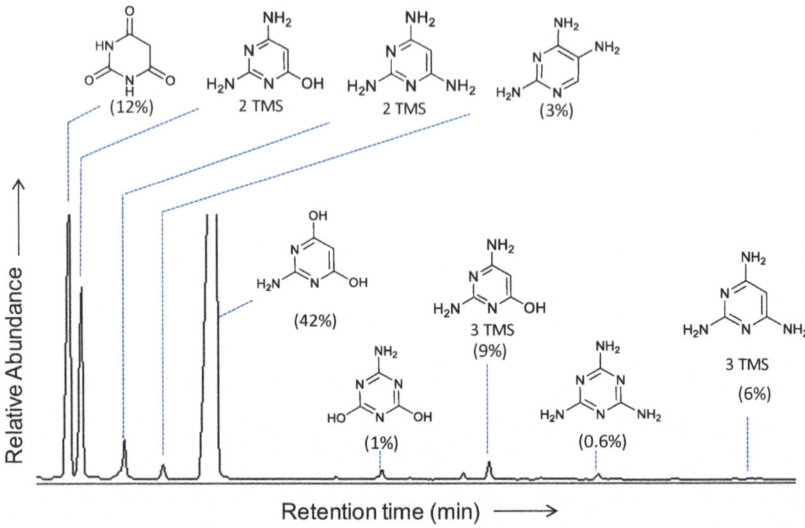

Fig. 4.41 Gas chromatography–mass spectrometry trace showing the reaction products of a urea/guanidine solution subjected to drying/rewetting cycles, after addition of malonic acid. The excess of 2-amino-4,6-dihydroxypyrimidine is consistent with the proposed mechanism

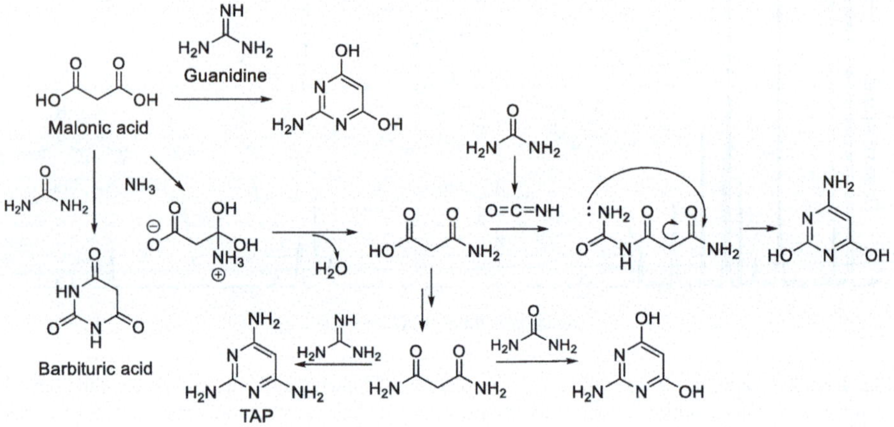

Fig. 4.42 Possible reactions involving malonic acid, urea, and guanidine during drying-wetting cycles of a solution containing these reactants

such as dihydrobiopterin, molybdopterin, and folic acid. The biosynthesis of pterins starts with the nucleoside guanosine, in a pathway similar to the plausible prebiotic origin of pteridine compounds. The close relationship between guanine and pterin, the identity between the biosynthetic pathway and its prebiotic chemistry, and the coevolution of both purines and pterins as biochemical cofactors allow us to consider it as the best example of chemical determinism. The biochemical advantage of the pterins as cofactors is their ability to form reduced (dihydro and

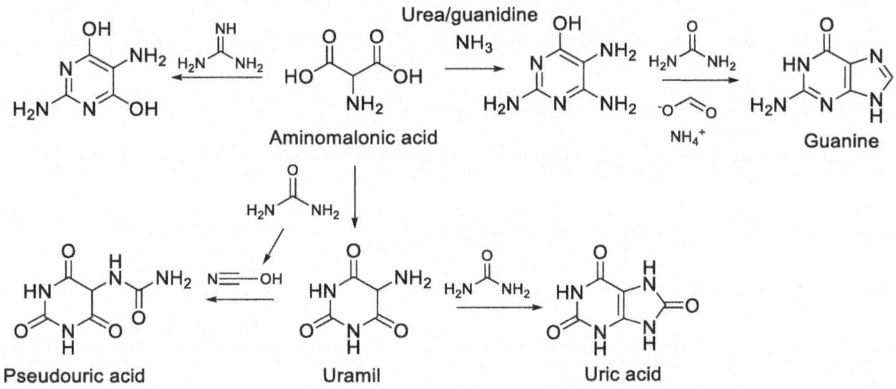

Fig. 4.43 Prebiotic synthesis of purines from amino malonic acid. All the depicted compounds were found in prebiotic model reactions involving urea/hydantoins (Menor-Salván et al. 2018)

tetrahydro) derivatives, allowing them to be the essential agents in several biological redox reactions for the synthesis and maintenance of DNA. Pterins have purine-like chemical and physico-chemical properties, so it is easy to imagine that, in the first steps of chemical evolution, both structures were available to be incorporated in pre-RNA structures and both purine and pterins began their roles as cofactors together. Hence, we could speak about the chemical predestination of both heterocycles, to the point that purine-based cofactors and nucleic acid components imply that pterins will be present in the system. Thus, the biochemistry of pterins and the common prebiotic origin may be linked and the consequence may be chemically determined. When it comes to prebiotic chemistry, purines and pterins seem destined to be together and share the same fate.

There is a strong connection between the prebiotic chemistry of sugars and other aldehydes and the formation of pteridine derivatives. In our postulated "warm little pond" rich in urea—or any other system where 5,6-diaminopyrimidine and/or purine and their nucleosides are formed—the synthesis of pteridines should be expected. In fact, during laboratory practice, some prebiotic reactions involving purines that lead to colorful solutions or products (different tones of bright yellow, orange, red, pink, or purple) usually involve the formation of a pteridine ring.

The most straightforward route to pteridine is the condensation between 5,6-pyrimidinediamines and glyoxal, glyoxylic, or pyruvic acids (a Gabriel-Isay reaction), which could take place in prebiotic conditions if the 5,6-pyrimidinediamine moiety is formed. In the case of 4-hydroxy-2,5,6-diaminopyrimidine, the resulting products of a reaction with glyoxal and glyoxylic acid are pterin and a mixture of xanthopterin and isoxanthopterin, respectively (Fig. 4.44). In prebiotic mild conditions at slightly acidic, neutral, or basic conditions (range $6 < pH < 9$), there is a significantly higher yield of isoxanthopterin; in acidic conditions, however, the synthesis leads to almost pure xanthopterin. This pH dependence is easily explained by the well-known difference in nucleophilicity and acidity of 5-amino and 6-amino groups (Pfleiderer 1964). In neutral, basic, or

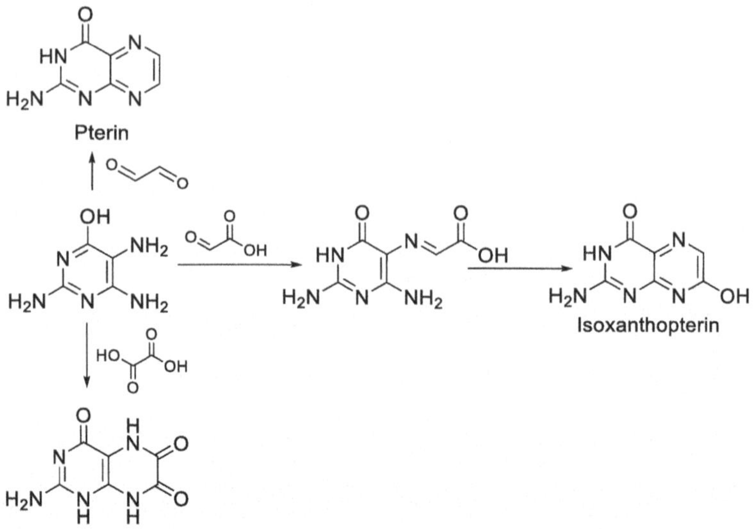

Fig. 4.44 Prebiotic synthesis of pterins. At a pH range from moderate acidic to basic, 7-pterins are expected to dominate the mixture

mildly acid solutions, aldehydes tend to react preferentially with the 5-amino group, which have exalted nucleophilicity in mild conditions; in stronger acid solution, the preference is the 6-amino group due to the preferential protonation of the 5-amino group. This difference has important consequences regarding the prebiotic synthesis of pterins.

It is possible to obtain 5,6-diaminopyrimines by irradiation of urea/hydantoin solutions, after formation of a concentrated solution by evaporation or freezing-melting cycles, or by hydrolysis of cyanide solutions, for which we confirmed the hypothesis of prebiotic pteridine formation by Gabriel-Isay condensation as suggested by Marin-Yaseli and coworkers (Menor-Salván et al. 2018; Marín-Yaseli et al. 2015). Thus, we tested whether nucleosides as guanosine could be synthesized in the same environment and in plausible prebiotic conditions, without intervention or isolation of intermediates, given that the components present in the prebiotic milieu (e.g., urea) will not magically disappear from the system. Hence, we performed a one-pot formylation of 4-hydroxy-2,5,6-triaminopyrimidine in diluted urea/ammonium formate/water eutectic (a solvent described in Burcar et al. 2016) and Traube's cyclization to obtain guanine in the presence of ribose. As expected, a complex mixture was formed, dominated by the presence of pterins, particularly isoneopterin and neopterin (Fig. 4.45). The formation of neopterins involves two possible pathways: (1) the ribosylation of the 5-amino group in the 5,6-diaminopyrimidine moiety, followed by Amadori rearrangement of the adduct and subsequent cyclization to form the 7-pterin (Tschesche et al. 1962); and (2) the reaction of guanine with 1,2-dicarbonylic compounds or the hydrolytic ring opening of guanosine, to yield pterin, xanthopterins or 6-neopterin (Albert 1957). This

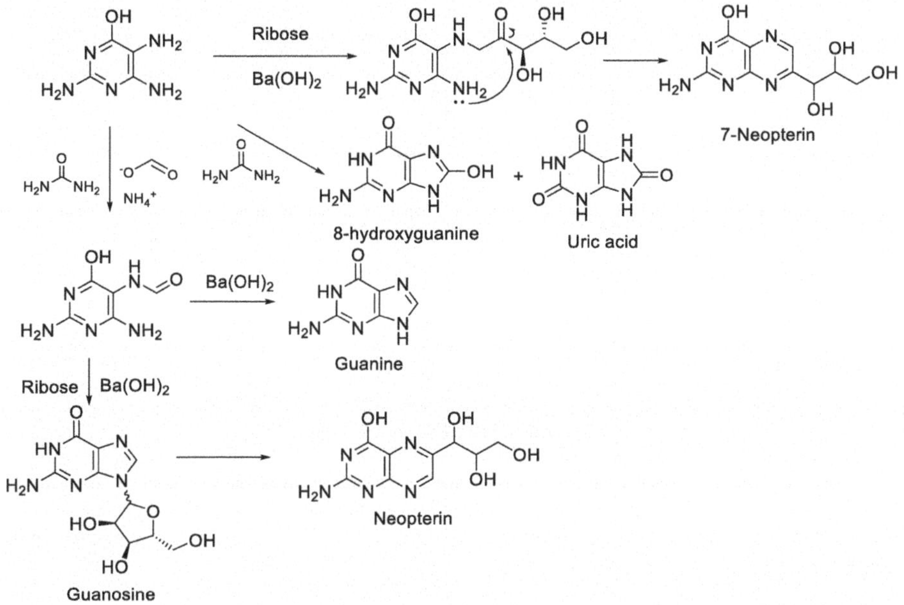

Fig. 4.45 One-pot prebiotic synthesis of guanosine and neopterins, starting with the sulfate salt of 2,5,6-triamino-4-hydroxypyrimidine. Making the pH basic with barium hydroxide both eliminates the sulfate and allows the cyclization of the purine or pteridine ring (Menor-Salván et al. 2017, 2018)

reaction, involving guanosine or guanine, should be considered in prebiotic chemistry systems; if guanosine could be formed, it is likely that pterin derivatives will be formed to some extent. In addition, the method used to form pterins, very readily in presence of carbonylic reactants such as glyoxylic acid, affects the stability of purine nucleosides in the prebiotic milieu.

The potential prebiotic formation of sugars by a formose reaction is another potential source of pterins that will affect the final nucleobase-nucleoside composition in prebiotic systems. Degradation and browning reactions of monosaccharides, glyceraldehyde, and 1,3-dihydroxyacetona are possible sources of glyoxylic acid (Novotný et al. 2008). Glycine (likely abundant in the prebiotic world) undergoes Strecker degradation by a reaction with glyoxal, releasing glyoxylic acid and formaldehyde. Thus, potential prebiotic carbohydrates could interfere with the prebiotic chemistry of pyrimidine and purine nucleobases, favoring the formation of pteridines. In recent experiments (manuscript in preparation), we showed that 2,4,6-triaminopyrimidine nucleosides could be synthesized directly starting with the formose reaction (Fig. 4.46). Interestingly, the experiments showed a high yield formation of isoxanthopterin. The mechanism of the formation of pterin remains unexplained, but it could involve the associated generation of glyoxylate and its reaction with 2,6-diaminopurine, which, in turn, we observed could be formed by reaction between 2,4,6-triaminopurine and urea. The problem with this pathway is

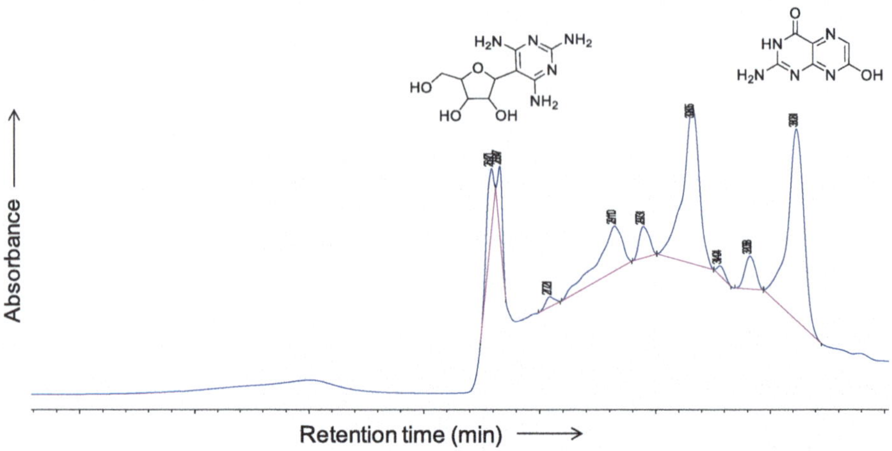

Fig. 4.46 High-performance liquid chromatography chromatogram of the nucleobase containing-fraction of the synthesis of nucleosides by a formose reaction and 2,4,6-triaminopyrimidine. The mechanism of formation of isoxanthopterin is not yet explained. (Data adapted from a manuscript in preparation)

that neither urea nor diaminopurine have been observed in the experiments. Hence, this significant formation of isoxanthopterin has been unexplained to date.

The easy prebiotic formation of 7-pterins, which are analogs of guanine, and the suggestion that 7-pterins almost invariably will accompany purines in prebiotic systems, lead to question of whether pterins could have been ancient components of proto-nucleic acids. Isoxanthopterin deoxyribonucleotides could be incorporated in DNA sequences with minimal structural and base pairing alterations (Lehbauer and Pfleiderer 2001). The incorporation of pterin nucleotides into DNA convert the pteridine ring in a very useful probe, given its high fluorescence (Datta et al. 2012). From a prebiotic point of view, it is very likely that the pterins and purines are both available and that is not possible to have the synthesis of purines or their nucleotides without the concomitant formation of pteridines, especially pterins. The pterins have a similar difficulty as the purines in forming nucleosides because neither of the two react directly with sugars to yield the corresponding osides (aside from the reaction with exocyclic amines). Moreover, attempts to get prebiotic nucleosides of pterins lead to the formation of neopterin or similar molecules. However, neopterin is an interesting molecule. It is an analog of guanine; both can form stable Watson-Crick base pairs with cytidine because the pK_a of neopterins is in the range of 8.0–9.9 (depending on the source), which is comparable to the pK_a of guanosine (Krishnamurthy 2012). Also, neopterin contains a glycerol moiety, which is easily phosphorylated in prebiotic conditions (Burcar et al. 2016). The pseudo-nucleotide

7-neopterin phosphate could be inserted in an oligonucleotide chain. A molecular dynamics simulation indicates that it could substitute a guanine nucleotide and still be able to base pair with complementary cytosine, forming a relatively stable structure (Fig. 4.47).

Given the reasonable formation of pterins in prebiotic model reactions involving urea or cyanide, their properties, and their biochemical functions, it would be worthwhile for future studies to investigate the role of the pteridine ring in chemical evolution, including the possibility of pteridine derivatives as ancient pre-RNA components.

4.8 Concluding Remarks

Model prebiotic reactions increasingly point to urea and cyanide as key precursors of chemical evolution, as well as to "little warm ponds" enriched with urea and other organics as the most appropriate place for the processes immediately preceding life as we know it. In adddition, there are beautiful consistency and robustness in prebiotic chemistry (which I like to consider as the small molecule-geochemistry connected subfield of chemical evolution). Related intermediates, such as urea, formamide, and cyanide, came from even simpler molecules (methane, nitrogen, carbon oxides) and can lead to a very predictable chemical space. This is also perfectly consistent with the structure and composition of extant biopolymers.

This chapter showed that urea, which has been formed in prebiotic model reactions since Miller's experiment of 1953, could be both a precursor or part of a scenario where the components of nucleic acid could be formed and assembled. Aside from the chemical reactions in which urea is involved, its physical-chemical properties could have allowed the formation of viscous, highly concentrated brines that maintained fluidity, protected, and promoted several prebiotic processes. If we add urea chemistry and physics to molecules that likely could have been formed during the same period (e.g., other nitrogen molecules such as hydantoin, cyanide, formamide, and ammonia; carbonyl compounds such as glyoxal, glyoxylic, oxalic, acetic, and formic acids), we could show a very accurate picture of the first steps in chemical evolution and realize that this picture is predictable and robust.

It is thus easy to think that the composition of organic life—or the chemical inventory of the environments where life could have emerged—is chemically determined and not a product of random chemical processes. This assumption piqued our imagination: Probably, we will not need to think about exotic extraterrestrial molecular biology; instead, when this momentous event comes, we may easily recognize and interact with extraterrestrial life because it will be composed of recognizable biopolymers and biochemical mechanisms that we will able to study. We hope that space exploration projects will soon test the predictive power of prebiotic chemistry. If we are correct, these explorations will show the universality of the chemical space that preceded the emergence of life on Earth. For example, the (hopefully) future mission Dragonfly to Titan (Turtle et al. 2018) has prebiotic

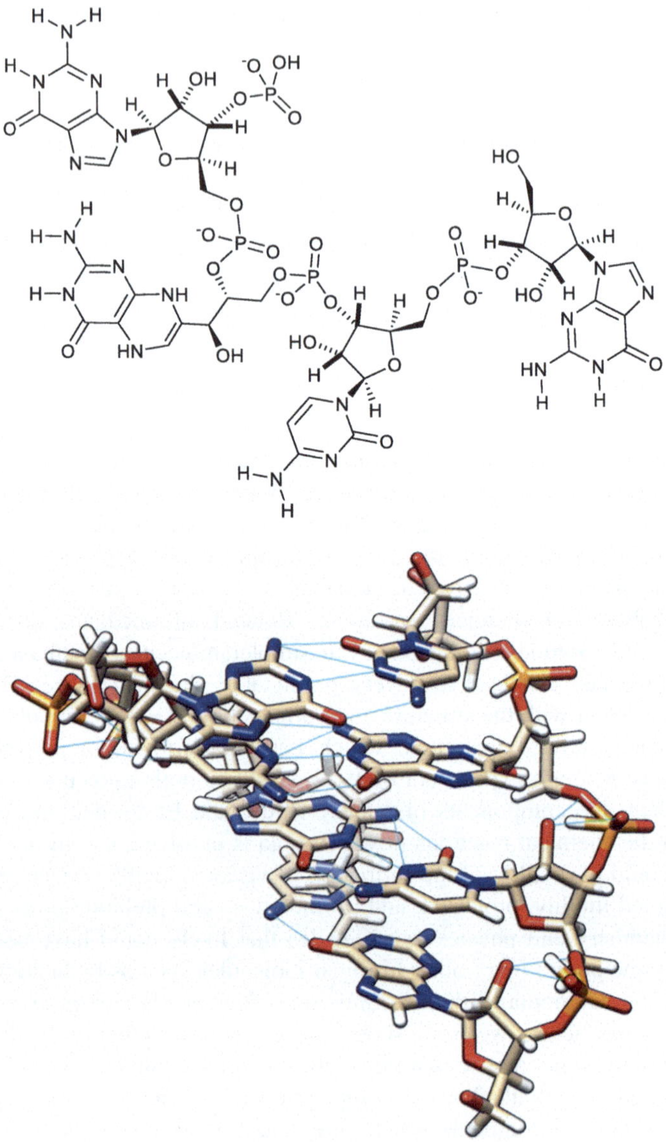

Fig. 4.47 A hypothetical tetranucleotide with 7-neopterin (up) and the structure of a tetranucleotide chain with a 7-neopterin insertion, hybridized with a complementary GCGC tetranucleotide in an energy-minimized A-form helix structure (structure simulated with Chimera software). After insertion of neopterin pseudonucleotide, the pterin ring still form H-bonds with cytosine

chemistry as one of its main goals. Titan (and other icy worlds with subsurface liquid water) could show us a detailed picture of the first steps of chemical evolution, "frozen" under the low temperatures of Saturn's moon. If missions such as

Dragonfly become reality, they will be a big test for all of the laboratory work that chemists have done since the nineteenth century.

Prebiotic chemistry is dynamic, and its laboratory approaches require a high grade of speculation. Some authors consider prebiotic chemistry as syntheses that merely use plausible prebiotic precursors and mild conditions. However, reactions that work in the laboratory using clean multistep syntheses or pure reagents may not work as beautifully when ingredients that should be present in the natural system are added, or when the constraints of mineralogy or geochemistry are considered. We have a big limitation in our ability to experimentally uncover the mechanisms of the origin and chemical evolution of nucleic acids and life: It is not possible to simulate the capacity of nature to carry out an experiment at a planetary scale with infinite time (on a human scale). Thus, we gradually assembled small pieces and realized that the rules of chemistry should follow well-determined pathways.

This realization is not a small achievement. This chapter showed that, if methane and molecular nitrogen are present in a prebiotic atmosphere, cyanide and acetylene will also be present, along with urea, hydantoins, glyoxal, glyoxylate, and other molecules. Atmospheric methane predestines the presence of a diversity of nucleobases and probably nucleotides (and peptide-like molecules), as well as the possibility of the formation of small water ponds enriched with organics. Because of the colligative properties of urea, these ponds can form relatively stable, viscous, and fluid environments with large temperature ranges for chemical evolution.

Thus, in my opinion, space exploration is a necessary test for the chemistry we study in our laboratories. That exploration should include probes or rovers to explore places where active or "frozen" prebiotic chemistry is present in the solar system, as well as future discoveries in the exoplanet quest. In this quest, finding actual life is a big goal and would be an inflexion point in human history. However (and we should make an outreach effort to explain this), finding places where chemical evolution or prebiotic chemistry is occurring is an equally important goal. It is not a problem if Titan, Europa, and Earth-like exoplanets do not harbor life. However, if they harbor chemical processes like those we presumptuously call *prebiotic chemistry*—including those explained in this book, keeping in mind what Charles Darwin wrote in his letter exposing his theories to the botanist Asa Gray in 1857: "This sketch is most imperfect; but in so short a space I cannot make it better. Your imagination must fill up many wide blanks. Without some reflexion it will appear all rubbish; perhaps it will appear so after reflexion"—this would be a major finding with important implications in Science and Society. It should not be considered the completion of this subfield within chemical evolution, though. Several details, mechanisms, and reactions still remain to be explained or explored, although, in my opinion, we will have reached a very significant goal in human knowledge.

Acknowledgements This chapter was supported by the National Science Foundation (NSF) and the National Aeronautics and Space Administration Astrobiology Program, under the NSF Center for Chemical Evolution at Georgia Institute of Technology (CHE-1004570, CHE-1504217). I wish to acknowledge the fruitful discussions with Nicholas V. Hud and his constant and unconditional support. This chapter and the whole book has been possible thanks to my friends, colleagues, and staff at Georgia Tech, to whom I will be always indebted for their support.

References

Adcock CT, Hausrath EM, Forster PM (2013) Readily available phosphate from minerals in early aqueous environments on Mars. Nat Geosci 6(10):824–827

Airapetian VS, Glocer A, Gronoff G, Hébrard E, Danchi W (2016) Prebiotic chemistry and atmospheric warming of early Earth by an active young Sun. Nat Geosci 9(6):452–455

Albert A (1957) The transformation of purines into pteridines. Biochem J 65(1):124

Bada JL (2004) How life began on Earth: a status report. Earth Planet Sci Lett 226:1–15

Baeyer A (1864) Mittheilungen aus dem organischen Laboratorium des Gewerbeinstitutes in Berlin: Untersuchungen über die Harnsäuregruppe. Justus Liebigs Ann Chem 130(2):129–175

Becker S, Thoma I, Deutsch A, Gehrke T, Mayer P, Zipse H, Carell T (2016) A high-yielding, strictly regioselective prebiotic purine nucleoside formation pathway. Science 352 (6287):833–836

Bendich A, Getler H, Brown GB (1949) A synthesis of isotopic cytosine and a study of its metabolism in the rat. J Biol Chem 177(2):565–570

Bendich A, Furst SS, Brown GB (1950) On the role of 2,6-diaminopurine in the biosynthesis of nucleic acid guanine. J Biol Chem 185(1):423–433

Ben-Ishai D, Altman J, Bernstein Z (1977) The reactions of ureas with glyoxylic acid and methyl glyoxylate. Tetrahedron 33(10):1191–1196

Benner SA, Kim HJ, Carrigan MA (2012) Asphalt, water, and the prebiotic synthesis of ribose, ribonucleosides, and RNA. Acc Chem Res 45(12):2025–2034

Bernstein MP, Dworkin JP, Sandford SA, Cooper GW, Allamandola LJ (2002) Racemic amino acids from the ultraviolet photolysis of interstellar ice analogues. Nature 416(6879):401

Bredlow LA (2010) Diogenes Laertius, Vidas y Opiniones de los Filósofos Ilustres (Life and opinions of eminent philosophers). Translation to Spanish and Comments. Lucina, Zamora (Spain)

Brooke TY, Tokunaga AT, Weaver HA, Crovisier J, Bockelée-Morvan D, Crisp D (1996) Detection of acetylene in the infrared spectrum of comet Hyakutake. Nature 383(6601):606

Burcar B, Pasek M, Gull M, Cafferty BJ, Velasco F, Hud NV, Menor-Salván C (2016) Darwin's warm little pond: a one-pot reaction for prebiotic phosphorylation and the mobilization of phosphate from minerals in a urea-based solvent. Angew Chem Int Ed 55(42):13249–13253

Burton AS, Stern JC, Elsila JE, Glavin DP, Dworkin JP (2012) Understanding prebiotic chemistry through the analysis of extraterrestrial amino acids and nucleobases in meteorites. Chem Soc Rev 41(16):5459–5472

Butch C, Cope ED, Pollet P, Gelbaum L, Krishnamurthy R, Liotta CL (2013) Production of tartrates by cyanide-mediated dimerization of glyoxylate: a potential abiotic pathway to the citric acid cycle. J Am Chem Soc 135(36):13440–13445

Cafferty BJ, Hud NV (2014) Abiotic synthesis of RNA in water: a common goal of prebiotic chemistry and bottom-up synthetic biology. Curr Opin Chem Biol 22:146–157

Cafferty BJ, Fialho DM, Khanam J, Krishnamurthy R, Hud NV (2016) Spontaneous formation and base pairing of plausible prebiotic nucleotides in water. Nat Commun 7:11328

Callahan MP, Smith KE, Cleaves HJ, Ruzicka J, Stern JC, Glavin DP et al (2011) Carbonaceous meteorites contain a wide range of extraterrestrial nucleobases. Proc Natl Acad Sci U S A 108 (34):13995–13998

Carracido JR (1920) Filogenia química de la molécula albuminoide. Revista de la Real Academia de Ciencias Físicas, Exactas y Naturales, vol 20

Carter MK (1951) The story of barbituric acid. J Chem Educ 28(10):524

Cavalieri LF, Brown GB (1949) The exchange between a formamido group and formamide, studied with C13,1. J Am Chem Soc 71(6):2246–2247

Chawla M, Oliva R, Bujnicki JM, Cavallo L (2015) An atlas of RNA base pairs involving modified nucleobases with optimal geometries and accurate energies. Nucleic Acids Res 43 (14):6714–6729

Chen MC, Cafferty BJ, Mamajanov I, Gállego I, Khanam J, Krishnamurthy R, Hud NV (2013) Spontaneous prebiotic formation of a β-ribofuranoside that self-assembles with a complementary heterocycle. J Am Chem Soc 136(15):5640–5646

Cleaves HJ (2008) The prebiotic geochemistry of formaldehyde. Precambrian Res 164:111–118

Cleaves HJ, Nelson KE, Miller SL (2006) The prebiotic synthesis of pyrimidines in frozen solution. Naturwissenschaften 93(5):228–231

Cleaves HJ, Scott AM, Hill FC, Leszczynski J, Sahai N, Hazen R (2012) Mineral–organic interfacial processes: potential roles in the origins of life. Chem Soc Rev 41(16):5502–5525

Crick FHC (1968) The origin of the genetic code. J Mol Biol 38:367–379

Curd P (2007) Anaxagoras of Clazomenae: fragments. Text and translation with notes and essays. University of Toronto Press, Toronto

Damer B (2016) A field trip to the Archaean in search of Darwin's warm little pond. Life 6(2):21

Datta K, Johnson NP, Villani G, Marcus AH, von Hippel PH (2012) Characterization of the 6-methyl isoxanthopterin (6-MI) base analog dimer, a spectroscopic probe for monitoring guanine base conformations at specific sites in nucleic acids. Nucleic Acids Res 40 (3):1191–1202

Davis TL (1921) The action of ammonia water on dicyandiamide. J Am Chem Soc 43 (10):2230–2233

Day A, Arnold AP, Blanch RJ, Snushall B (2001) Controlling factors in the synthesis of cucurbituril and its homologues. J Org Chem 66(24):8094–8100

De Marcellus P, Bertrand M, Nuevo M, Westall F, Le Sergeant d'Hendecourt L (2011) Prebiotic significance of extraterrestrial ice photochemistry: detection of hydantoin in organic residues. Astrobiology 11(9):847–854

de Marcellus P, Meinert C, Myrgorodska I, Nahon L, Buhse T, d'Hendecourt LLS, Meierhenrich UJ (2015) Aldehydes and sugars from evolved precometary ice analogs: importance of ices in astrochemical and prebiotic evolution. Proc Natl Acad Sci U S A 112(4):965–970

Delidovich IV, Taran OP, Simonov AN, Matvienko LG, Parmon VN (2011) Photoinduced catalytic synthesis of biologically important metabolites from formaldehyde and ammonia under plausible "prebiotic" conditions. Adv Space Res 48(3):441–449

Des Marais DJ, Nuth JA III, Allamandola LJ, Boss AP, Farmer JD, Hoehler TM, Spormann AM (2008) The NASA astrobiology roadmap. Astrobiology 8(4):715–730

Ellington A, Szostak J (1990) In vitro selection of RNA molecules that bind specific ligands. Nature 346:818–822

Eschenmoser A (2007) The search for the chemistry of life's origin. Tetrahedron 63 (52):12821–12844

Fahrenbach AC, Giurgiu C, Tam CP, Li L, Hongo Y, Aono M, Szostak JW (2017) Common and potentially prebiotic origin for precursors of nucleotide synthesis and activation. J Am Chem Soc 139(26):8780–8783

Fenton HJH (1882) XXXVIII.—Transformation of urea into cyanamide. J Chem Soc Trans 41:262–263

Ferris JP, Hagan WH (1984) HCN and chemical evolution: the possible role of cyano compounds in prebiotic synthesis. Tetrahedron 40(7):1093–1120

Ferris JP, Orgel LE (1966) An unusual photochemical rearrangement in the synthesis of adenine from hydrogen cyanide. J Am Chem Soc 88(5):1074–1074

Ferris JP, Sanchez RA, Orgel LE (1968) Studies in prebiotic synthesis of pyrimidines. J Mol Biol 33:693–704

Fialho DM, Clarke KC, Moore MK, Schuster GB, Krishnamurthy R, Hud NV (2018) Glycosylation of a model proto-RNA nucleobase with non-ribose sugars: implications for the prebiotic synthesis of nucleosides. Org Biomol Chem 16(8):1263–1271

Forsythe JG, Yu SS, Mamajanov I, Grover MA, Krishnamurthy R, Fernández FM, Hud NV (2015) Ester-mediated amide bond formation driven by wet–dry cycles: a possible path to polypeptides on the prebiotic earth. Angew Chem Int Ed 54(34):9871–9875

Francis BR (2015) The hypothesis that the genetic code originated in coupled synthesis of proteins and the evolutionary predecessors of nucleic acids in primitive cells. Life 5:467–505

Gautier A (1884) Nouvelle methode de synthese de composes organiques azotes; synthese de la xanthine et de la methylxanthine. Bulletin de la Societe Chimique de France 42:141–146

Goesmann F, Rosenbauer H, Bredehöft JH, Cabane M, Ehrenfreund P, Gautier T, McKenna-Lawlor S (2015) Organic compounds on comet 67P/Churyumov-Gerasimenko revealed by COSAC mass spectrometry. Science 349(6247):aab0689

Grdadolnik J, Maréchal Y (2002) Urea and urea–water solutions—an infrared study. J Mol Struct 615(1–3):177–189

Grimaux E (1879) Synthese des dérivés uriques de la série de l'alloxane. Bull Soc Chim Fr 31:146

Grover M, He C, Hsieh M-C, Yu S-S (2015) A chemical engineering perspective on the origins of life. PRO 3(2):309–338

Guerrier-Takada C, Gardiner K, Marsh T, Pace N, Altman S (1983) The RNA moiety of ribonuclease P is the catalytic subunit of the enzyme. Cell 35(3):849–857

Guilbert W (1986) The RNA world. Nature 319:618

Gull M, Pasek MA (2013) Was struvite a prebiotic mineral? Life 3:321–330

Handschuh GJ, Orgel LE (1973) Struvite and prebiotic phosphorylation. Science 179 (4072):483–484

Hayashi Y, Katsumoto Y, Omori S, Kishii N, Yasuda A (2007) Liquid structure of the urea−water system studied by dielectric spectroscopy. J Phys Chem B 111(5):1076–1080

Hayem G (1888) Revue des Sciences Médicales en France et a l'etranger. G. Masson ed. Paris

He C, Gállego I, Laughlin B, Grover MA, Hud NV (2017) A viscous solvent enables information transfer from gene-length nucleic acids in a model prebiotic replication cycle. Nat Chem 9 (4):318–324

Hilbert GE, Johnson TB (1930) Researches on pyrimidines. CXVII. A method for the synthesis of nucleosides. J Am Chem Soc 52(11):4489–4494

Holm RH (2003) Electron transfer: iron–sulfur clusters. In: McCleverty JA, Meyer TJ (eds) Comprehensive coordination chemistry II, vol 8: Bio-coordination chemistry. Pergamon, Oxford, pp 61–90

Hud NV, Cafferty BJ, Krishnamurthy R, Williams LD (2013) The origin of RNA and "my grandfather's axe". Chem Biol 20(4):466–474

Hurst DT (1979) An introduction to the chemistry and biochemistry of pyrimidines, purines and pteridines. Wiley, New York

Hysell M, Siegel JA, Yitzhak T (2005) Synthesis and stability of exocyclic triazine nucleosides. Org Biomol Chem 3:2946–2952

Jeilani YA, Orlando TM, Pope A, Pirim C, Nguyen MT (2014) Prebiotic synthesis of triazines from urea: a theoretical study of free radical routes to melamine, ammeline, ammelide and cyanuric acid. RSC Adv 4(61):32375

Johnson TB (1914) The origin of purines in plants. J Am Chem Soc 36(2):337–345

Johnson TB, Nicolet BH (1914) Researches on pyrimidines. LXVI. The formation of pyrimidines from diethyl aminomalonate and aminomalonic nitrile. J Am Chem Soc 36(2):345–355

Johnson AP, Cleaves HJ, Dworkin JP, Glavin DP, Lazcano A, Bada JL (2008) The Miller volcanic spark discharge experiment. Science 322(5900):404

Kameda Y, Sasaki M, Hino S, Amo Y, Usuki T (2006) Hydration structure around the carbonyl group of a urea molecule in concentrated aqueous solutions studied by neutron diffraction with 12C/13C isotopic substitution. Bull Chem Soc Jpn 79(9):1367–1371

Kaur S, Sharma P, Wetmore SD (2017) Structural and electronic properties of barbituric acid and melamine-containing ribonucleosides as plausible components of prebiotic RNA: implications for prebiotic self-assembly. Phys Chem Chem Phys 19(45):30762–30771

Kenner GW, Taylor CW, Todd AR (1949) 348. Experiments on the synthesis of purine nucleosides. Part XXIII. A new synthesis of adenosine. J Chem Soc (Resumed), 1620–1624

Kim HJ, Furukawa Y, Kakegawa T, Bita A, Scorei R, Benner SA (2016) Evaporite borate-containing mineral ensembles make phosphate available and regiospecifically phosphorylate ribonucleosides: borate as a multifaceted problem solver in prebiotic chemistry. Angew Chem Int Ed 15816–15820

Kinoshita H (1953) Synthesis of melamine from urea. Rev Phys Chem Jpn 23:1–9

Kočandrle R, Kleisner K (2013) Evolution born of moisture: analogies and parallels between Anaximander's ideas on origin of life and man and later pre-Darwinian and Darwinian evolutionary concepts. J Hist Biol 46(1):103–124

Krishnamurthy R (2012) Role of pKa of nucleobases in the origins of chemical evolution. Acc Chem Res 45(12):2035–2044

Kruger K, Grabowski PJ, Zaug AJ, Sands J, Gottschling DE, Cech TR (1982) Self-splicing RNA: autoexcision and autocyclization of the ribosomal RNA intervening sequence of Tetrahymena. Cell 31(1):147–157

Kurzer F (1956) Biuret and related compounds. Chem Rev 56(1):95–197

Kusama K, Prescott DM, Fröholm LO, Cohn WE (1966) The formation and metabolic interchanges of pseudouridine in *Tetrahymena pyriformis*. J Biol Chem 241(17):4086–4091

Lazcano A (2009) Complexity, self-organization and the origin of life: the happy liaison? Orig Life Self-Organ Biol Evol 13–22

Lehbauer J, Pfleiderer W (2001) Nucleotides, Part LXIX, Synthesis of phosphoramidite building blocks of isoxanthopterin N8-(2′-deoxy-β-D-ribonucleosides): new fluorescence markers for oligonucleotide synthesis. Helv Chim Acta 84(8):2330–2342

Li X, Chin DN, Whitesides GM (1996) Synthesis and evaluation of thioether-based tris-melamines as components of self-assembled aggregates based on the CA.M lattice. J Org Chem 61:1779–1786

Löb W (1913) Über das Verhalten des Formamids unter der Wirkung der stillen Entladung Ein Beitrag zur Frage der Stickstoff-Assimilation. Eur J Inorg Chem 46(1):684–697

Lowe CU, Rees MW, Markham RFRS (1963) Synthesis of complex organic compounds from simple precursors: formation of amino-acids, amino-acid polymers, fatty acids and purines from ammonium cyanide. Nature 199(4890):219–222

Marín-Yaseli MR, Mompeán C, Ruiz-Bermejo M (2015) A prebiotic synthesis of pterins. Chem Eur J 21(39):13531–13534

Marín-Yaseli MR, González-Toril E, Mompeán C, Ruiz-Bermejo M (2016) The role of aqueous aerosols in the "Glyoxylate Scenario": an experimental approach. Chem Eur J 22 (36):12785–12799

Martin W, Russell MJ (2007) On the origin of biochemistry at an alkaline hydrothermal vent. Philos Trans R Soc 362:1887–1925

Martins S, Martins SIFS, Jongen WMF (2001) A review of Maillard reaction in food and implications to kinetic modelling. Trends Food Sci Technol 11:364–373

Meinert C, Myrgorodska I, De Marcellus P, Buhse T, Nahon L, Hoffmann SV, Meierhenrich UJ (2016) Ribose and related sugars from ultraviolet irradiation of interstellar ice analogs. Science 352(6282):208–212

Menor-Salván C, Marín-Yaseli MR (2012) Prebiotic chemistry in eutectic solutions at the water–ice matrix. Chem Soc Rev 41(16):5404–5415

Menor-Salván C, Marín-Yaseli MR (2013) A new route for the prebiotic synthesis of nucleobases and hydantoins in water/ice solutions involving the photochemistry of acetylene. Chem Eur J 19 (20):6488–6497

Menor-Salván C, Ruiz-Bermejo D, Guzmán MI, Osuna-Esteban S, Veintemillas-Verdaguer S (2009) Synthesis of pyrimidines and triazines in ice: implications for the prebiotic chemistry of nucleobases. Chem Eur J 15(17):4411–4418

Menor-Salván C, Fialho D, Hud NV (2017) Urea: a key prebiotic reagent in the origin of nucleic acids. In: Astrobiology science conference 2017 (AbSciCon 2017), Mesa, AZ, USA
Menor-Salván C, Fialho D, Hud NV (2018) Prebiotic origin of nucleobases, pterins and purine nucleosides in urea and hydantoin rich scenario. Implication of the Traube synthesis. Chem Eur J
Miller SL (1953) A production of amino acids under possible primitive earth conditions. Science 117(3046):528–529
Miller SL (1957) The formation of organic compounds on the primitive Earth. Ann N Y Acad Sci 69(2):260–275
Mitchell HK, Nyc JF (1947) Intermediates in the synthesis of orotic acid from oxalacetic ester and urea. J Am Chem Soc 69(3):674–677
Nelson KE, Robertson MP, Levy M, Miller SL (2001) Concentration by evaporation and the prebiotic synthesis of cytosine. Orig Life Evol Biosph 31(3):221–229
Noller HF, Hoffart V, Zimniak L (1992) Unusual resistance of peptidyl transferase to protein extraction procedures. Science 256:1416–1419
Novotný O, Cejpek K, Velíšek J (2008) Formation of carboxylic acids during degradation of monosaccharides. Czech J Food Sci 26(2):117–131
Nuevo M, Bredehöft JH, Meierhenrich UJ, d'Hendecourt L, Thiemann WH (2010) Urea, glycolic acid, and glycerol in an organic residue produced by ultraviolet irradiation of interstellar/pre-cometary ice analogs. Astrobiology 20:245–256
Orgel LE (1968) Evolution of the genetic apparatus. J Mol Biol 38:381–393
Orgel LE (2004a) Prebiotic chemistry and the origin of the RNA world. Crit Rev Biochem Mol Biol 39(2):99–123
Orgel LE (2004b) Prebiotic adenine revisited: eutectics and photochemistry. Orig Life Evol Biosph 34(4):361–369
Oró J (1960) Synthesis of adenine from ammonium cyanide. Biochem Biophys Res Commun 2 (6):407–412
Oró J, Kimball AP (1961) Synthesis of purines under possible primitive earth conditions. I. Adenine from hydrogen cyanide. Arch Biochem Biophys 94(2):217–227
Oró J, Kimball AP (1962) Synthesis of purines under possible primitive earth conditions: II. Purine intermediates from hydrogen cyanide. Arch Biochem Biophys 96(2):293–313
Parnica J, Antalik M (2014) Urea and guanidine salts as novel components for deep eutectic solvents. J Mol Liq 197:23–26. https://doi.org/10.1016/j.molliq.2014.04.016
Pascal R, Taillades J, Commeyras A (1980) Systèmes de strecker et apparentes—XII: Catalyse par les aldehydes de l'hydratation intramoleculaire des α-aminonitriles. Tetrahedron 36 (20–21):2999–3008
Patel BH, Percivalle C, Ritson DJ, Duffy CD, Sutherland JD (2015) Common origins of RNA, protein and lipid precursors in a cyanosulfidic protometabolism. Nat Chem 7(4):301–307
Peretó J, Bada JL, Lazcano A (2009) Charles Darwin and the origin of life. Orig Life Evol Biosph 39(5):395–406
Perri MJ, Seitzinger S, Turpin BJ (2009) Secondary organic aerosol production from aqueous photooxidation of glycolaldehyde: laboratory experiments. Atmos Environ 43(8):1487–1497
Pfleiderer W (1964) Recent developments in the chemistry of pteridines. Angew Chem Int Ed 3 (2):114–132
Powner MW, Gerland B, Sutherland JD (2009) Synthesis of activated pyrimidine ribonucleotides in prebiotically plausible conditions. Nature 459(7244):239–242
Powrie WD, Wu CH, Molund VP (1986) Browning reaction systems as sources of mutagens and antimutagens. Environ Health Perspect 67:47
Ramberg PJ (2000) The death of vitalism and the birth of organic chemistry: Wohler's urea synthesis and the disciplinary identity of organic chemistry. Ambix 47(3):170–195
Rembold H, Gyure WL (1972) Biochemistry of the pteridines. Angew Chem Int Ed 11 (12):1061–1072
Ruiz-Bermejo M, Menor-Salván C, Osuna-Esteban S, Veintemillas-Verdaguer S (2006) Prebiotic microreactors: a synthesis of purines and dihydroxy compounds in aqueous aerosol. Orig Life Evol Biosph 37:123–142

Ruiz-Bermejo M, Menor-Salván C, Osuna-Esteban S, Veintemillas-Verdaguer S (2007) Prebiotic microreactors: a synthesis of purines and dihydroxy compounds in aqueous aerosol. Orig Life Evol Biosph 37(2):123–142

Ruiz-Bermejo M, Menor-Salván C, de la Fuente JL, Mateo-Martí E, Osuna-Esteban S, Martín-Gago JÁ, Veintemillas-Verdaguer S (2009a) CH4/N2/H2-spark hydrophobic tholins: a systematic approach to the characterisation of tholins. Part II. Icarus 204(2):672–680

Ruiz-Bermejo M, Rogero C, Menor-Salván C, Osuna-Esteban S, Angel Martin-Gago J, Veintemillas-Verdaguer S (2009b) Thermal wet decomposition of Prussian Blue: implications for prebiotic chemistry. Chem Biodivers 6:1309–1322

Russell MJ, Martin W (2004) The rocky roots of the acetyl-CoA pathway. Trends Biochem Sci 29:358–363

Saladino R, Crestini C, Ciciriello F, Pino S, Costanzo G, Di Mauro E (2009) From formamide to RNA: the roles of formamide and water in the evolution of chemical information. Res Microbiol 160(7):441–448

Saladino R, Botta G, Pino S, Costanzo G, Di Mauro E (2012) From the one-carbon amide formamide to RNA all the steps are prebiotically possible. Biochimie 94(7):1451–1456

Sanchez RA, Orgel LE (1970) Studies in prebiotic synthesis: V. Synthesis and photoanomerization of pyrimidine nucleosides. J Mol Biol 47(3):531–543

Sarker PK, Takahashi JI, Obayashi Y, Kaneko T, Kobayashi K (2013) Photo-alteration of hydantoins against UV light and its relevance to prebiotic chemistry. Adv Space Res 51 (12):2235–2240

Schmidt G (1950) Nucleic acids, purines, and pyrimidines. Annu Rev Biochem 19(1):149–186

Schwartz AW (2006) Phosphorus in prebiotic chemistry. Philos Trans R Soc Lond Ser B Biol Sci 361(1474):1743–1749; discussion 1749

Shaw E (1950) A new synthesis of the purines adenine, hypoxanthine, xanthine and isoguanine. J Biol Chem 185:439–447

Smith EL, Abbott AP, Ryder KS (2014) Deep Eutectic Solvents (DESs) and their applications. Chem Rev 114(21):11060–11082

Smolin EM, Rapoport L (1959) s-Triazines and derivatives. In: Weissberger A (ed) The chemistry of heterocyclic compounds. A series of monographs. Interscience, New York

Soper AK, Castner EW Jr, Luzar A (2003) Impact of urea on water structure: a clue to its properties as a denaturant? Biophys Chem 105(2–3):649–666

Stokes RH (1965) Tracer diffusion in binary solutions subject to a dimerization equilibrium. J Phys Chem 69(11):4012–4017

ten Cate MGJ, Omerović M, Oshovsky GV, Crego-Calama M, Rei houdt DN (2005) Self-assembly and stability of double rosette nanostructures with biological functionalities. Org Biomol Chem 3(20):3727–3733

Tolleson W, Diachenko G, Heller D (2009) Background paper on the chemistry of melamine alone and in combination with related compounds. World Health Organization, Geneva

Traube W (1900) Ueber eine neue Synthese des Guanins und Xanthins. Ber Dtsch Chem Ges 33 (1):1371–1383

Traube W (1904) Der aufbau der xanthinbasen aus der cyanessigsäure. Synthese des hypoxanthins und adenins. Eur J Org Chem 331(1):64–88

Ts'o POP, Melvin IS, Olson AC (1963) Interaction and association of bases and nucleosides in aqueous solutions. J Am Chem Soc 85:1289–1296

Tschesche R, Hess B, Ziegler I, Machleidt H (1962) Über pteridine, XVII. Trennung von synthetischem Biopterin und Isobiopterin. Eur J Org Chem 658(1):193–201

Tsipis CA, Karipidis PA (2003) Mechanism of a chemical classic: quantum chemical investigation of the autocatalyzed reaction of the serendipitous Wöhler synthesis of urea. J Am Chem Soc 125:2307–2318

Turtle EP, Barnes JW, Trainer MG, Lorenz RD, Hibbard KE, Adams DS, Freissinet C (2018, March) Dragonfly: in situ exploration of Titan's organic chemistry and habitability. In: Lunar and planetary science conference, vol 49

Viqueira FD (2010) El evolucionismo de Rodríguez Carracido. Nuevas consideraciones. Anales Real Academia Nacional de Farmacia 76(4):479–491
Wächtershäuser G (1988) Before enzymes and templates: theory of surface metabolism. Microbiol Rev 52:452–484
Wagner AJ, Blackmond DG (2016) The future of prebiotic chemistry. ACS Cent Sci 2 (11):775–777. https://doi.org/10.1021/acscentsci.6b00336
Ware E (1950) The chemistry of the hydantoins. Chem Rev 46(3):403–470
Weber AL (2001) The sugar model: catalysis by amines and amino acid products. Orig Life Evol Biosph 31(1–2):71–86
Woese CR (1967) The genetic code: the molecular basis for genetic expression. Harper and Row, New York
Wöhler F (1828) Ueber künstliche bildung des harnstoffs. Ann Phys 88(2):253–256
Yamagata Y, Sasaki K, Takaoka O, Sano S, Inomata K, Kanemitsu K, Matsumoto I (1990) Prebiotic synthesis of orotic acid parallel to the biosynthetic pathway. Orig Life Evol Biosph 20(5):389–399
Zaia D a M, Zaia CTBV, De Santana H (2008) Which amino acids should be used in prebiotic chemistry studies? Orig Life Evol Biosph 38(6):469–488

Chapter 5
Searching for Possible Ancestors of RNA: The Self-Assembly Hypothesis for the Origin of Proto-RNA

Brian J. Cafferty, David M. Fialho, and Nicholas V. Hud

Abstract There are currently two main schools of thought regarding the origins of RNA. In one school, RNA is considered to be a product of nonenzymatic, prebiotic reactions. In the other, RNA is considered to be a product of chemical and/or biological evolution. The numerous challenges to demonstrating a plausible prebiotic synthesis of RNA support the hypothesis that life started with an ancestral RNA-like polymer, or proto-RNA. If RNA is an "invention" of early life, then it is logical to assume that identifying the chemical structure of proto-RNA, and intermediate pre-RNAs, would require exploration of a seemingly insurmountable number of possible proto-RNA building blocks and prebiotic reactions. Here we report progress toward finding a proto-RNA that is the product of molecular self-assembly. Results obtained thus far demonstrate that seemingly minor changes to the structure of the extant building blocks of RNA (e.g., the substitution of uracil by barbituric acid) alleviate several long-standing problems associated with finding a prebiotic synthesis for RNA.

5.1 Introduction

In May of 1953, Stanley Miller reported the production of amino acids in his earliest spark-discharge experiment (Miller 1953). In addition to the fundamental importance of this study to origins of life research, the Miller-Urey experiment inspired other researchers to explore various model prebiotic reactions for the possible abiotic production of biological building blocks. Of particular relevance to this volume is the study reported by Oró in 1960. Because Miller had determined that amino acids were produced via the Strecker synthesis in his spark-discharge experiment (Miller 1957; Strecker 1850), with HCN as an intermediate, Oró decided to explore

B. J. Cafferty · D. M. Fialho · N. V. Hud (✉)
School of Chemistry and Biochemistry, NSF-NASA Center for Chemical Evolution, Georgia Institute of Technology, Atlanta, GA, USA
e-mail: nick.hud@chemistry.gatech.edu

© Springer International Publishing AG, part of Springer Nature 2018
C. Menor-Salván (ed.), *Prebiotic Chemistry and Chemical Evolution of Nucleic Acids*, Nucleic Acids and Molecular Biology 35,
https://doi.org/10.1007/978-3-319-93584-3_5

reactions that take place in concentrated solutions of ammonium cyanide (Oró 2002). Oró's investigations resulted in his serendipitous discovery of an abiotic route to adenine (Oró 1960).

Coincidentally, also in May of 1953, James Watson and Francis Crick published their double helix model for DNA (Watson and Crick 1953). While elucidation of the structure of DNA would prove key to understanding the functionality of DNA in living organisms (Olby 2003), the structure of nucleic acids (namely, RNA) is also part of many hypotheses regarding the origins of life. The simplicity and elegance of nature's mechanism for information transfer via Watson-Crick base pairs has caused many origins researchers to assume that these hydrogen-bonded structures have been utilized from the earliest stages of life. Indeed, it is often assumed that base pairing by the canonical RNA nucleobases was sufficient for information transfer before the emergence of nucleic acid polymerases, an assumption that we believe to be questionable, at best (Hud and Anet 2000; Hud et al. 2007; Engelhart and Hud 2010) (and see below).

In the 1960s, Leslie Orgel, Francis Crick, Alex Rich, and Carl Woese each independently proposed that RNA could have been the original polymer for information storage (like DNA today) and for chemical catalysis (currently the domain of proteins) (Crick 1968; Orgel 1968; Woese 1967; Rich 1962). In the early 1980s, the discovery of catalytic RNA (Kruger et al. 1982; Guerrier-Takada et al. 1983) greatly increased interest in this hypothesis, which became known as the "RNA world" hypothesis (Gesteland et al. 2006). Despite its popularity and numerous arguments made in support of this hypothesis (Gesteland et al. 2006), there are reasons to seriously question the validity of an early RNA world (Bowman et al. 2015; Hud et al. 2013; Bernhardt 2012; Kurland 2010). The lack of a plausible prebiotic route to RNA polymers, despite decades of effort by many very talented chemists, represents a specific and persistent challenge to the RNA world hypothesis. At present, we do not have a plausible prebiotic synthesis for RNA polymers or even for each of the extant RNA mononucleosides (a long-standing problem for prebiotic chemists that has become known as *the Nucleoside Problem*). Progress made toward demonstrating a prebiotic route to RNA has been summarized in previous reviews (Engelhart and Hud 2010; Hud et al. 2013; Monnard 2016; Cafferty and Hud 2014; Schwartz 2006, 1997; Orgel 2004), and this pursuit continues in several laboratories around the globe (Saladino et al. 2018; Dorr et al. 2012; Benner et al. 2012; Powner et al. 2011).

The numerous difficulties associated with the prebiotic synthesis of RNA have inspired some chemists to entertain the hypothesis that RNA is actually a product of chemical and/or biological evolution (Hud et al. 2013; Joyce et al. 1987). Orgel referred to this possible reality as a "gloomy prospect" (Orgel 1998), because if RNA was not a product of prebiotic chemistry, then uncovering the chemical origins of life may require an exhaustive search of an immense "chemical space" for molecules that could have functioned as a proto-RNA. Despite his seemingly pessimistic tone, Orgel came to accept that RNA is a product of evolution and encouraged origins researchers to consider, and examine, alternative polymers that may have preceded RNA (Orgel 2004).

In this chapter, we discuss the progress made, particularly in our laboratory, in the search for an ancestral nucleic acid that could have self-assembled from plausible prebiotic molecules on the early Earth. Like many other researchers, one author (N.V.H.) was originally enamored with the idea that RNA was the "first molecule of life." However, years of (mostly negative) experiments (Hud 2017) eventually led to the conclusion that the time had come to face Orgel's gloomy prospect and to begin searching for possible candidates for, and synthetic routes to, a plausible proto-RNA. Fortunately, as discussed in this chapter, this exploration has produced some attractive candidate molecules that, in some cases, provide solutions to more than one challenge facing the prebiotic synthesis of a polymer that may have served as an early ancestor of RNA.

5.2 Guiding Hypotheses of Our Approach to Finding Candidates for Proto-RNA

With the perspective provided by six decades of research into prebiotic chemistry and extensive biophysical studies of DNA and RNA over the same time period, we believe that the failure to find a simple and robust prebiotic synthesis of RNA is a strong indication of the need to reconsider what the Miller-Urey experiment and Watson-Crick base pairs tell us about the origins of RNA. We do not wish to diminish the importance of the contributions that these two advances have provided to origins of life research. However, we believe it is time to consider the possibility that these and related discoveries have more subtle, yet equally important, lessons to teach us about the origins of life.

Along with several proteinogenic α-amino acids (i.e., glycine, alanine, and aspartic acid), Miller found that even his earliest spark-discharge reactions produced non-proteinogenic amino acids, including α-amino-*n*-butyric acid and β-alanine, as well as other compounds that he could not identify at that time (Miller 1953). A reanalysis of Miller's samples with more advanced analytical techniques has confirmed that there are many non-biological molecules produced in spark-discharge reactions (Johnson et al. 2008). This observation supports a hypothesis that has inspired much of the research described in this chapter: *There were molecules present on the prebiotic Earth that were essential for the emergence of life, but their roles in the origins of life have thus far been overlooked because these molecules are now absent from, or only play a minor role in, contemporary life.* A recent example of the potential validity and usefulness of this hypothesis is the discovery of a plausible prebiotic route to polypeptides. Briefly, without chemical activation, amino acids dried at temperatures up to around 100 °C form peptide bonds in only low yields (e.g., 1–2%) and are typically restricted to dipeptides (Rode 1999). However, if hydroxy acids—which were also produced in the spark-discharge experiment (Miller and Urey 1959)—are dried along with amino acids, then peptide bonds are formed in good yields at temperatures as low as 65 °C (Forsythe et al. 2015).

Turning to the duplex structure of nucleic acids, the centrality of Watson-Crick base pairs to information storage and transfer in extant organisms, and the ease of abiotic nucleobase synthesis (Oró 1960; Menor-Salván et al. 2009; Saladino et al. 2004), it is easy to believe that life has always used the four canonical bases of RNA (i.e., A, G, C, and U). Moreover, many models for the prebiotic origins of RNA have assumed that mononucleotides came before RNA polymers on the prebiotic Earth and that de novo synthesis of RNA polymers was the result of the polymerization of chemically activated nucleotides (Rajamani et al. 2008; Ferris et al. 1996). We have, however, long questioned the feasibility of such scenarios. The mononucleotides of RNA do not form Watson-Crick base pairs in water; instead, they form vertical stacks with their hydrogen-bonding edges exposed to solvent (Pongs and Ts'o 1971). Even when concentrated to the point of crystallization, the nucleobases tend to favor Hoogsteen base pairing over Watson-Crick base pairing (Hoogsteen 1959). It is only when multiple nucleotides are connected together in an oligonucleotide that the nucleobases form Watson-Crick base pairs with a complementary oligonucleotide as part of a double helix. These facts beg the question, *Why would prebiotic chemistry have connected the canonical nucleobases along a common backbone so that they would form base pairs if the monomers were not forming base pairs prior to polymerization?* We have called this conundrum "the Paradox of Base Pairing" (Hud et al. 2007). In short, if there were not a preselection of pairing bases (i.e., heterocycles that form Watson-Crick or Watson-Crick-like base pairs), then the abiotic polymerization of nucleobases (and similar heterocycles) on the prebiotic Earth would *not* have led to polymers with the ability to form duplexes. Instead, it can be assumed that these polymers would have contained many other heterocycles present in comparable abundances—due to random incorporation of pairing and non-pairing bases—that do not form stable Watson-Crick base pairs.

Finally, in our search for molecules and reactions that could have given rise to the first RNA, we seek robust molecules and robust reactions. That is, far from being content with simply showing that arguably plausible prebiotic molecules or reaction can give rise to a molecule with a structure similar to RNA, we are looking for proto-RNA building blocks that have a stability that is sufficient for these molecules to have accumulated to an appreciable steady-state concentration on the prebiotic Earth. Moreover, we seek reactions that exhibit some level of selectivity, so that key building blocks are not "wasted" by unproductive reactions, while also restricting ourselves to reactions that do not require chemical activation by high-energy compounds. We believe that the probability that life emerged from such molecules and reactions is, by definition, much higher than proposals in which life emerged from molecules that would have been rare on the prebiotic Earth and/or from reactions that need to be carefully orchestrated in order to favor one particular product. Additionally, holding to these principles will help all of us working on the origins of life to guard against prematurely ending the search for the origins of biopolymers by "settling" for a less-than-inspiring (and historically flawed) solution.

5.3 RNA Evolution and *My Grandfather's Axe*

Our assertion that the nucleobases of proto-RNA (or proto-nucleobases) must have paired as monomers before their connection along a common backbone is compatible with the wider hypothesis that RNA evolved from an earlier polymer. Through the eyes of a chemist, the structure of RNA appears modular. That is, RNA is comprised of three chemically distinct molecular subunits: the nucleobases, which serve as the recognition units (**RU**) of RNA and appear to be the products of prebiotic reactions involving HCN or related molecules (e.g., formamide, urea); ribose, which serves as a trifunctional connector (**TC**) and could have originally been produced by prebiotic reactions involving formaldehyde or related molecules (e.g., glycolaldehyde, glyoxylate); and phosphate, which serves as an ionized linker (**IL**) and is an inorganic molecule that is typically sequestered in minerals. There are numerous problems facing the de novo incorporation of each of these functional groups into RNA, several of which we will discuss in this chapter; a more comprehensive assessment of these problems has been presented in our previously published reviews (Engelhart and Hud 2010; Hud et al. 2013; Cafferty and Hud 2014, 2015).

We have hypothesized that each of these three components of RNA was changed during the early stages of life as the chemical structure of RNA underwent refinement (Hud et al. 2013). A conceptual illustration of how we imagine RNA evolution to have taken place is shown in Fig. 5.1. This hypothetical path begins with proto-RNA, the first and oldest ancestor of RNA, in which all molecular subunits are different from those of extant RNA. Between proto-RNA and RNA, we show two polymers, labeled pre-RNA$_i$ and pre-RNA$_j$, which represent two intermediates along a path of evolution from proto-RNA to RNA. The original subunits of proto-RNA

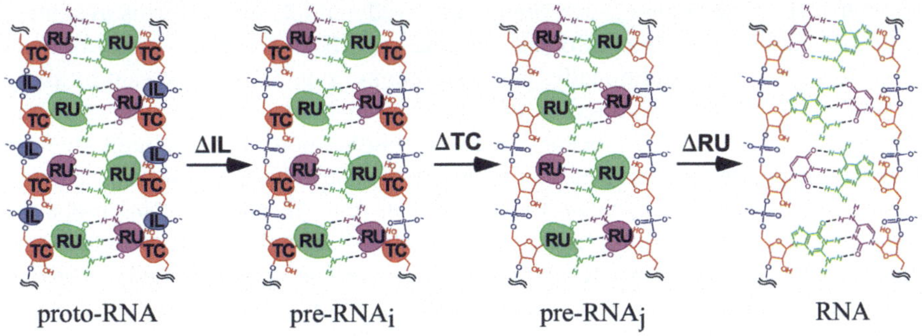

Fig. 5.1 A schematic illustration of RNA evolution, from proto-RNA through multiple intermediate pre-RNAs. The subunits of proto-RNA and the pre-RNAs are labeled **RU** for recognition unit (A, C, G, U in extant RNA), **TC** for trifunctional connector (ribose in extant RNA), and **IL** for ionized linker (phosphate in extant RNA). The subunits of proto-RNA and pre-RNAs are shown as nondescript molecular entities because their identities are unknown. Specific functional groups of RNA are retrained in this hypothetical extrapolation back to proto-RNA to emphasize that some features of RNA may have been retained since the time of proto-RNA due to the constraints imposed by the principle of continuity (see text)

are depicted as nondescript shapes because their identities are not known. Nevertheless, for each of these subunits, we have retained some atoms over the course of extrapolation from RNA back to proto-RNA. These atoms indicate the chemical features that we hypothesize to have remained the same in terms of atomic species, charge, and position since the time of the earliest nucleic acids. We retain these elements in accordance with the "principle of continuity," proposed five decades ago by Orgel (1968), which states that sequential molecular species along an evolutionary path must have features in common that allow them to function similarly. In Fig. 5.1, we have retained the groups involved in hydrogen-bonding between the recognition units (labeled as **RU**) based on the expectation that older pre-RNA molecules were "forward compatible" for information transfer with each successive pre-RNA (up to RNA); otherwise, it would not have been possible for sequence information to be maintained as the chemical structure of the pre-RNAs changed over time. We have retained the oxygen atoms between the ionized linkers (labeled **IL**) and the trifunctional connectors (labeled **TC**) to indicate our expectation that the nature of the linkages between these subunits has not changed dramatically, being phosphodiester linkages as they are today, or perhaps ester linkages in proto-RNA and earlier pre-RNAs (if phosphate entered at a later stage). The OH group that is now at the 2′ position of ribose is also indicated as an ancient feature because this group is essential for stabilizing complex RNA tertiary structures, including those that can function as catalysts [e.g., the ribosome in its most ancestral form (Petrov et al. 2014)]. Finally, each **IL** is shown as having a negative charge, as the polyanionic nature of the RNA backbone is arguably, based on the results of experiments with non-ionized structural isomers of RNA, necessary for RNA to function as an informational polymer (Benner et al. 2004).

RNA evolution is depicted in Fig. 5.1 as occurring by the sequential and single-step replacement of each subunit, but this is merely an illustration. It is entirely possible that optimization of each subunit involved multiple replacements over time, with there being more than two intermediate pre-RNAs along the pathway from proto-RNA to RNA. Additionally, the illustration in Fig. 5.1 is not meant to imply that subunit evolution occurred in a particular order, as multiple changes to the different subunits may have been interleaved. Furthermore, some subunit changes may have happened independently of others, while other changes may have required a concerted replacement of two different subunits (e.g., **IL** and **TC** changing at the same time to retain the overall length of the backbone). Finally, it may have been that the backbone of proto- or pre-RNA was heterogeneous and was eventually refined to contain only (deoxy)ribose phosphate (Cafferty and Hud 2014; Gavette et al. 2016; Krishnamurthy 2015). In any case, as stated by the principle of continuity, each pre-RNA would have had to have been compatible with at least its immediate predecessor and its immediate successor (Yang et al. 2007).

The proposal that RNA evolved from a proto-RNA raises an intriguing philosophical question: *Should we consider proto-RNA and RNA to be the same molecule?* From a chemical perspective, proto-RNA and extant RNA are not the same molecule; their chemical structures are different. However, proto-RNA and any one of the pre-RNAs along the path to RNA would have served the same function at different times during the emergence and early evolution of life. Thus, from a functional perspective, proto-RNA, the pre-RNAs, and RNA are all the same molecule, much like the human hand is the same biological entity as the pectoral fins of our ancestral fish (Onimaru et al. 2015; Shubin 2008). The question of whether or not proto-RNA and RNA are the same molecule is reminiscent of a philosophical puzzle known as *My Grandfather's Axe* (a modern version of *The Ship of Theseus*, which was debated by the great philosophers of ancient Greece). Briefly, *My Grandfather's Axe* can be told as: *A man claims that his axe once belonged to his grandfather. He goes on to admit, "Well, of course, my father replaced the handle and I replaced the head."* Is the man's axe truly his grandfather's axe? The answer, of course, depends on whether the physical material of which the axe is made defines the identity of the axe or if its identity is defined by a continuity of use. We favor the latter criterion for RNA (Hud et al. 2013).

5.4 Candidates for the Proto-nucleobases: Resolving *the Paradox of Base Pairing*

The chemical space containing possible subunits of proto-RNA and pre-RNAs is potentially vast, but it may be constrained by some reasonable assumptions. Based on the abovementioned principle of continuity, Miller, Orgel, and Crick each assumed that the earlier nucleobases would have been structurally similar to the extant nucleobases (Crick 1968; Orgel 1968; Kolb et al. 1994). It might also be assumed that the proto-nucleobases were synthesized by the same, or similar, abiotic reactions that first produced the extant nucleobases. If so, then RNA chemical evolution could have transpired by the incremental replacement of subunits with structurally and chemically similar molecules that were already present in the local environment, rather than requiring the emergence of a ribozyme or enzyme for the de novo synthesis of each new subunit. These assumptions, arguably, restrict the chemical space of possible proto-/pre-RNA nucleobases to that defined by the purines, pyrimidines, and closely related heterocycles with exocyclic groups of -H, =O, or $-NH_2$. The chemical space of 91 chemical structures defined by these criteria is shown in Fig. 5.2. The 1,3,5-triazines are included along with the purines and pyrimidines because of their close structural similarity to pyrimidines and because these compounds are produced in model prebiotic reactions and have been found in carbonaceous meteorites along with the canonical pyrimidine and purine nucleobases of RNA (Menor-Salván et al. 2009; Hayatsu et al. 1975; Fripiat and Cruzcump 1974; Harvey et al. 1971).

Orgel appears to be the first to propose that RNA has evolved from its original form. Crick reported in 1968 that Orgel had shared this idea with him (Crick 1968). In particular, Orgel suggested that the first nucleic acids might have only contained adenine and hypoxanthine as nucleobases (Fig. 5.2, **10** and **19**, respectively) (hypoxanthine being the nucleobase of the nucleoside inosine). As Crick discussed, the

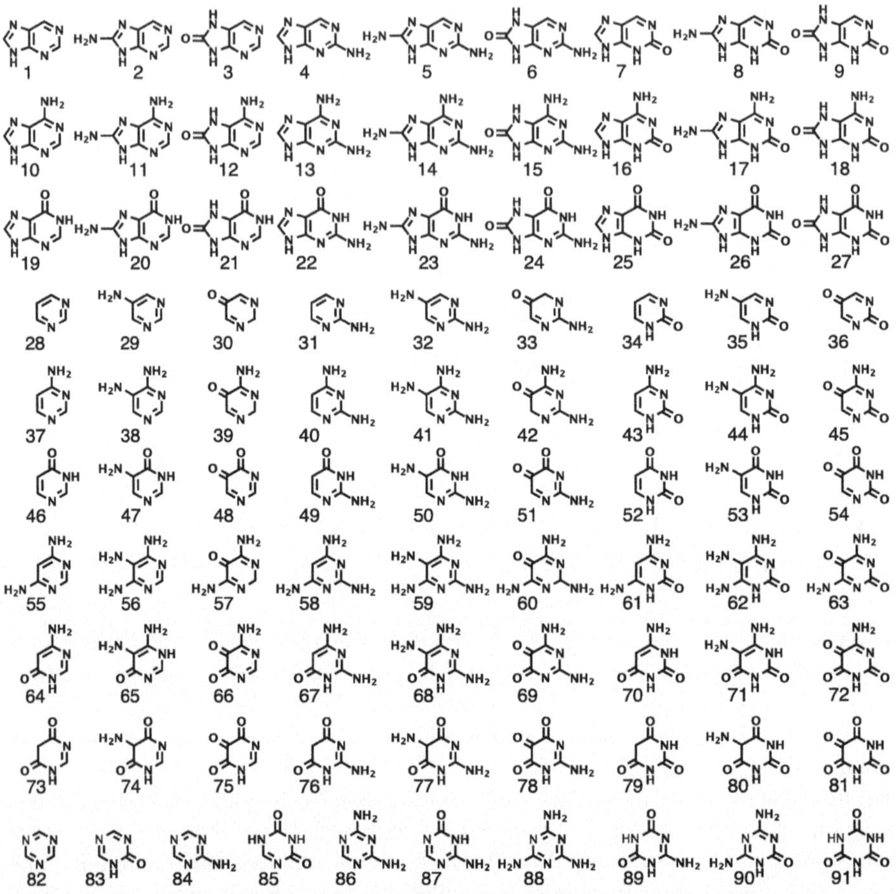

Fig. 5.2 The chemical space considered by the authors as a reasonable starting point to explore for candidate proto-nucleobases of ancestral RNAs, including proto-RNA and intermediate pre-RNAs. The extant nucleobases of RNA, adenine (**10**), guanine (**22**), cytosine (**43**), and uracil (**52**), were used to define most of this chemical space; molecules **1–27** are all possible purines, and molecules **28–81** are all possible pyrimidines for which the exocyclic groups can be –H, =O, or $-NH_2$. The bottom row, molecules **82–91**, are the 1,3,5-triazines with the same three possible exocyclic groups. For some heterocycles different oxidation states or different tautomers may be available, but these have been omitted for simplicity

hydrogen bond donor and acceptor groups of adenine and hypoxanthine are such that adenine and hypoxanthine could form Watson-Crick-like base pairs (Fig. 5.3). Subsequent experiments have, however, revealed that adenine-hypoxanthine base pairs are not particularly stable within a double helix. Moreover, homo-adenosine and homo-inosine polymers actually favor formation of a triple helix that is held together by base triplets containing one adenine base that is hydrogen bonded to two hypoxanthine bases, even in solutions containing a 1:1 molar ratio of these polymers (Howard and Miles 1977; Arnott and Bond 1973).

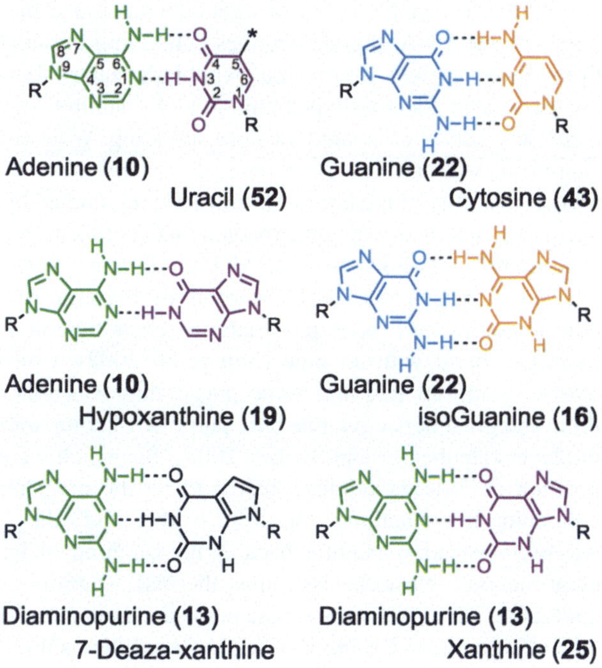

Fig. 5.3 Hydrogen-bonded structures of the purine-purine base pairs mentioned in the text, with the Watson-Crick base pairs shown for comparison. Color code indicates H-bonding configuration of the Watson-Crick faces: Green is for A and A-like purines; purple is for U and U-like purines; blue is for G and G-like purines; orange is for C and C-like purines. The asterisk on U indicate the methyl group on the thymine (T) base of DNA

In theory, guanine and isoguanine can form a Watson-Crick-like purine-purine base pair (Fig. 5.3). However, duplex formation by oligonucleotides containing only these two purines is also challenging. Switzer et al. demonstrated that oligonucleotides containing only guanine and isoguanine nucleobases prefer to form four-stranded structures (Roberts et al. 1997). These very stable quadruplex assemblies are held together by guanine tetrads and/or isoguanine tetrads (known as G-tetrads and isoG-tetrads). The formation of G-tetrads from guanosine monomers has been known for decades (Yu et al. 2008; Gellert et al. 1962). While G-tetrads are of interest in contemporary biology (Rhodes and Lipps 2015) and for the development of novel materials (Davis and Spada 2007), the relevance of these structures to the origins of life is unclear. Some researchers have suggested that G-tetrads could have been important for the emergence of nucleic acids because G-quadruplexes formed by RNA and DNA oligonucleotides can be very stable, and G-tetrads are the only structures that will assemble in water from any of the four extant nucleobase monomers (Cassidy et al. 2014). It is not obvious, however, how the G-tetrad could have been used as a motif for storage and transfer of information given that it only contains one species of nucleobase. It is worth noting that quadruplexes containing guanine and isoguanine tetrads might not have been as favorable (or problematic) for proto-RNA or for pre-RNAs with an earlier non-RNA backbone, as duplex formation is apparently more favorable than quadruplex formation for some synthetic oligonucleotides that contain a backbone sugar that is different from ribofuranose and 2′-deoxyribofuranose (i.e., the sugars that are used in extant RNA and DNA, respectively) (Groebke et al. 1998; Krishnamurthy et al. 1996).

In 2007, Battersby and co-workers demonstrated the formation of duplexes with homo-purine DNA oligonucleotides containing the nucleobases adenine, hypoxanthine, guanine, and isoguanine. Although stable above room temperature, these duplexes with adenine•hypoxanthine and guanine•isoguanine base pairs (Fig. 5.3) were less stable compared to corresponding Watson-Crick DNA duplexes with canonical A•T and G•C base pairs (Battersby et al. 2007). We speculate that the reduced stability of the homo-purine duplexes studied by Battersby might be, in part, due to the lack of an exocyclic oxygen on C2 of the hypoxanthine base (Fig. 5.3). We propose this possibility because McLaughlin and co-workers have reported that the exocyclic oxygen on the C2 carbon of thymine (5-methyl-U) is an important contributor to A•T base pair stability even though it does not participate in a hydrogen bond with adenine (Sun et al. 2002). Consistent with this possibility, Switzer et al. reported that homo-purine duplexes containing adenine•7-deaza-xanthine and guanine•isoguanine base pairs are of comparable stability to Watson-Crick duplexes (Heuberger and Switzer 2008), for which the exocyclic oxygen on the C2 position of 7-deaza-xanthine seems to be the main difference from the duplexes containing hypoxanthine (Fig. 5.3). In this study the 7-deaza-xanthine nucleobase was used instead of xanthine because the xanthine nucleoside is particularly prone to depurination. Nevertheless, the thermal stability of homo-purine duplexes containing adenine•xanthine base pairs also indicates that the presence of an oxygen at the C2 position of xanthine (7-deaza-xanthine) is important for stabilization of this purine-purine base pair (Buckley et al. 2011).

To explore the possibility that RNA started with only purine recognition units—which was attractive because of the propensity of purines over pyrimidines to assemble at the monomer level (e.g., G-tetrads)—our laboratory tested the ability for short, chemically activated homo-purine oligonucleotides to assemble and polymerize in water. It was observed that tetranucleotides containing only purine bases, with sequences that could form mini-duplexes with purine-purine base pairs, polymerize (i.e., ligate) at least 200-fold more efficiently than mini-duplexes with Watson-Crick base pairs, presumably due to a greater tendency for end-to-end stacking of purine-purine base pairs as compared to purine-pyrimidine base pairs (Kuruvilla et al. 2013). However, our laboratory eventually abandoned exploration of the homo-purine start to RNA because we never found conditions under which two complementary purine *mono*nucleotides formed assemblies—a property that would have facilitated de novo formation of oligonucleotides (see below).

More recently, our group began exploring the chemical space of the pyrimidine and triazine heterocycles shown in Fig. 5.2 for a possible solution to the *Paradox of Base Pairing*. These studies have primarily focused on two pyrimidines, 2,4,6-traiminopyrimidine (**TAP**) and barbituric acid (**BA**) (Fig. 5.2, **58** and **79**), and their triazine analogs, melamine (**MA**) and cyanuric acid (**CA**) (Fig. 5.2, **89** and **92**). These molecules have been explored since the 1990s as recognition units for the assembly of synthetic molecules in nonpolar solvents (e.g., $CDCl_3$) (Seto and Whitesides 1990; Lehn et al. 1990), but relatively few investigations have been reported that examine the ability of **MA** and **CA**, or their derivatives (Ma and Bong 2011), to assemble in water because these heterocycles were known to coprecipitate in a 1:1 complex from water (Seto and Whitesides 1990). A number of supramolecular assemblies have been designed around the concept that **MA** or **TAP** and **BA** or **CA** can form hexad "rosettes" (Seto and Whitesides 1990; Lehn et al. 1990; Rakotondradany et al. 2005; Prins et al. 2001). We hypothesized that these pairing partners might form soluble hexad assemblies if at least one of the two pairing partners was modified with a chemical appendage to block co-crystallization/ precipitation and enhance water solubility (e.g., by the addition of a moiety with an electrostatic charge). Because hexads have such large hydrophobic surfaces, we also hypothesized that, once hexads are formed, the unfavorable exposure of their surfaces to water would lead to the formation of linear assemblies of stacked hexads. The predicted assemblies are shown schematically in Fig. 5.4a.

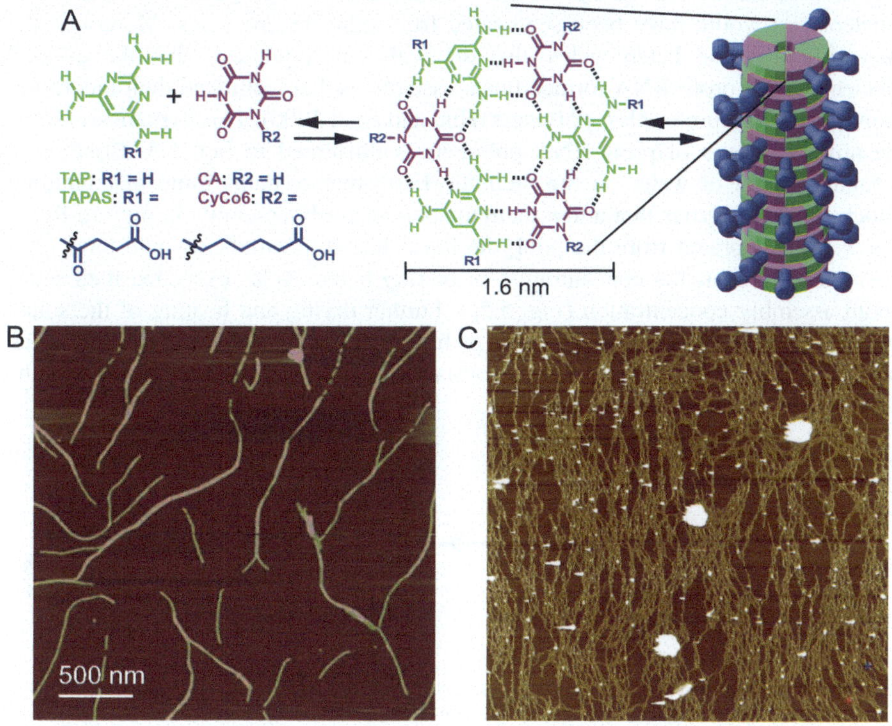

Fig. 5.4 (**a**) Chemical structures of 2,4,6-triaminopyrimidine (**TAP**), a succinic acid conjugate of **TAP** (**TAPAS**), cyanuric acid (**CA**), and a hexanoic acid conjugate of **CA** (**CyCo6**). Also shown are the hexad structures that can be formed by **TAP**, **CA**, and their conjugates, as well as with melamine and barbituric acid. The hexad is shown to be in exchange with a linear supramolecular assembly of stacked hexads. (**b**) Topographical AFM images of assemblies formed by **TAPAS** and **CA**. Reproduced with permission from reference (Cafferty et al. 2013). (**c**) Topographical AFM images of assemblies formed by **TAP** and **CyCo6**. Reproduced with permission from reference (Cafferty et al. 2014). Images shown in (**b**) and (**c**) were obtained with freshly cleaved mica as a substrate. (**b**) and (**c**) are shown at the same scale. Height measurements indicate that the fibers in (**b**) and (**c**) have a diameter of 2 nm, which is consistent with the predicted assembly shown in (**a**)

This hypothesis for hexad assembly was first tested with a modified form of **TAP**, named **TAPAS**, which is an amide-linked conjugate of **TAP** and succinic acid (Fig. 5.4a). When mixed together, **TAPAS** and **CA** formed extremely long fibers in water with thicknesses that correspond to that predicted by the diameter of a **TAP-CA** hexad (Fig. 5.4b). The association of **TAPAS** and **CA** is surprisingly efficient, with a minimal assembly concentration (i.e., the temperature-dependent concentration required by both compounds to form assemblies, which is abbreviated MAC) of 3.5 mM at room temperature (Cafferty et al. 2013). Similarly, **CA** modified by conjugation with a hexanoic acid tail, named **CyCo6**, also forms fibers when mixed with **TAP** (Fig. 5.4c).

The spontaneous assembly of **TAPAS** and **CA** suggested that there could be a pyrimidine and/or triazine proto-RNA base pair that assembled at the monomer level. Such a base pair could have circumvented the problems expressed as the *Paradox of Base Pairing* by providing a mechanism by which the first proto-nucleobases would have been segregated from other heterocycles. Moreover, the assembly of these heterocycles also fits with our hypothesis that the original nucleobases of proto-RNA formed linear assemblies that facilitated their incorporation into the first proto-RNA polymers (Hud and Anet 2000). This hypothesis for the de novo synthesis of proto-RNA polymers is illustrated in Fig. 5.5. Briefly, we imagine a pool of water on the prebiotic Earth that contains molecular building blocks with properties that allow them to spontaneously assemble into proto-RNA. As water evaporated from the pool, the linear hexad assemblies would have been favored to form as the concentration of pairing heterocycles exceeded their minimum assembly concentration (Fig. 5.5a). Further drying and heating of the solute molecules results in the condensation of backbone-forming subunits on the hexad assemblies, which would result in the formation of six proto-RNA polymers with the

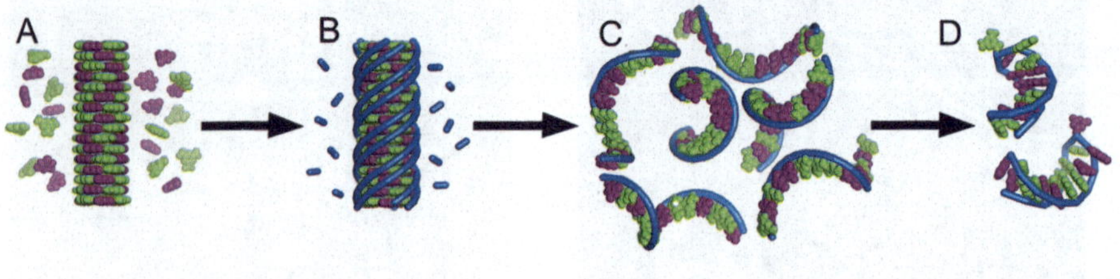

Fig. 5.5 Model for the de novo synthesis of proto-RNA. (**a**) The evaporation of water from an aqueous solution containing heterocyclic molecules causes the formation of a hexad stack from proto-nucleobases, such as 2,4,6-triaminopyrimidine (**TAP**) and barbituric acid (**BA**), or molecules with a similar propensity for self-assembly. (**b**) Further evaporation and increased temperature drive the formation of backbones from other molecules in the pool that connect the pre-assembled nucleobases into proto-RNA polymers. (**c**) Rehydration of the pool, and perhaps changes in solution conditions (i.e., pH, ionic strength), results in the separation of proto-RNA molecules from the hexad assembly. (**d**) While this mode of de novo proto-RNA synthesis would have resulted (ideally) in the production of polymers with random nucleobase sequences, a subset of these sequences would have formed stable secondary and tertiary structures. Such structures could have protected particular sequences against hydrolysis and served as starting points for the selection of sequences with favorable properties (e.g., ligand binding, catalytic activity)

heterocycles as side chains (Fig. 5.5b). Stacked hexads formed from *only* monomers (i.e., formed without a template strand) that underwent polymerization would have produced covalent polymers of random sequence. Because each hexad requires three of each of the two paring heterocycles, the condensation of hexads in a particular stack would result in the production of three proto-RNA strands of a particular (random) sequence and three proto-RNA strands that are complementary in sequence to the other three. In this model for de novo synthesis of proto-RNA, the strands synthesized as part of a hexad assembly would be released from the hexameric assembly when the dried pool is rehydrated (Fig. 5.5c), perhaps with strand separation from the hexad assembly being facilitated by additional changes in solution conditions [e.g., pH, which can be very affective for destabilizing hexad assemblies (Cafferty et al. 2014)]. Depending upon the exact sequence of these random proto-RNA polymers, some could fold into stable secondary and tertiary structures (Fig. 5.5d). It is in this state that we propose that proto-ribozymes (polymers with catalytic activity) would be expected to have first appeared. A related model for proto-RNA replication is discussed below.

5.5 Candidates for the Proto-nucleobases: Addressing *the Nucleoside Problem*

An abiotic mechanism for the segregation and spatial organization of pairing heterocycles is, of course, only one of the multiple steps necessary for proto-RNA synthesis. As illustrated by the second step of our model for de novo proto-RNA synthesis (Fig. 5.5), we expect that the non-covalent assembly of the proto-nucleobases would have been followed by the covalent coupling of these molecules along a common backbone. In extant RNA polymers, the covalent bond between the nucleobases and the backbone is the glycosidic bond between the nucleobases and ribose. As mentioned above, the lack of a plausible prebiotic route to extant nucleobase ribosylation (i.e., nucleoside formation) is a long-standing problem in origins of life research, so much so that it is referred to as *the Nucleoside Problem*.

While Orgel and co-workers only observed the formation of adenosine in 1–5% yields when adenine was dried and heated with ribose, these researchers observed that in the same reactions the exocyclic nitrogen of adenine was glycosylated by ribose in 50–70% yield (Fuller et al. 1972a). Given that the exocyclic amine of adenine is analogous to the exocyclic amines at the 4 and 6 positions of **TAP** (structures **10** and **58** in Fig. 5.2), we decided to explore the possibility that the exocyclic amines of **TAP** would also react with ribose. This was found to be the case, with exocyclic *N*-ribonucleosides being formed with **TAP** in solution and at temperatures as low as 5 °C (Chen et al. 2014). Unexpectedly, we discovered that the C5 carbon of **TAP** also reacts with ribose to form *C*-nucleosides (Fig. 5.6) (Chen et al. 2014). Overall, the combination of *N*- and *C*-nucleosides represented a combined yield of greater than 60%. A molecule that we have named **TARC**, the

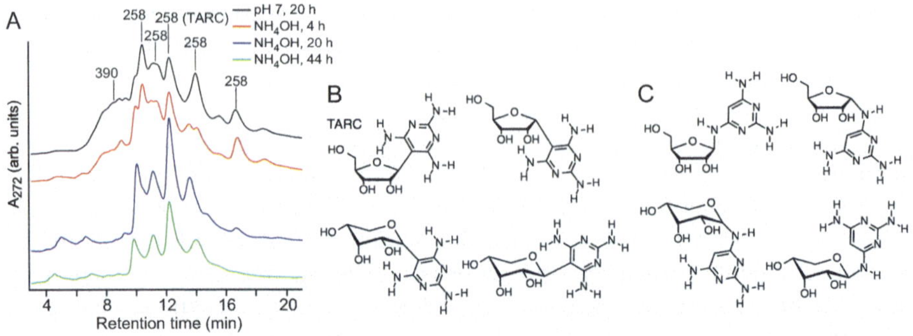

Fig. 5.6 Analysis of molecules formed during the repeated drying of ribose with 2,4,6-triaminoprymidine (**TAP**). (**a**) Top, black curve is UV-monitored HPLC chromatogram of products produced when **TAP** was dried and ribose at 35 °C for 10 days, with rehydration each day at room temperature. Red, blue, and green curves show the selective hydrolysis of products after exposure to 10 M ammonium hydroxide at 65 °C for 4, 20, and 44 h, respectively. Peak labels are m/z values based on simultaneous monitoring of products upon elution from column by mass spectrometry. Values of 258 m/z are consistent with the formation of a glyosidic bond between **TAP** and one ribose molecule (i.e., single ribosylation). The products lost with increased time of incubation with ammonium hydroxide are believed to be **TAP** ribosylated on an exocyclic amine (i.e., *N*-nucleosides), which are expected to hydrolyze rapidly under basic conditions. The products that are stable to base are believed to be *C*-nucleosides (i.e., **TAP** glycosylated by the formation of a covalent bond with a ring carbon). (**b**) The four *N*-ribosides that are possible by the formation of a glycosidic bond between an exocyclic amine of **TAP** and ribose, two with ribose in the furanose form and two in the pyranose form, and two with **TAP** in the α-configuration and two with **TAP** in β-configuration. (**c**) The four *C*-nucleosides that are possible upon the reaction of ribose with **TAP**. Note that four dominant peaks that remain in the chromatogram for a sample subjected to basic conditions for 44 h. are consistent with the four possible C5 nucleosides (i.e., α-*C*-furanoside, β-*C*-furanoside (named **TARC**), α-*C*-pyranoside, β-*C*-pyranoside) being most stable to hydrolysis. Reprinted with permission from the supporting information of reference (Chen et al. 2014)

C-nucleoside of ribose and **TAP** with ribose in the furanose ring configuration and **TAP** in the β-orientation at the anomeric carbon of ribose (like that of extant nucleosides), turned out to be the nucleoside formed in the highest yield and to be of greater stability than the *N*-nucleosides.

Benner and co-workers have also reported nucleoside formation by a noncanonical pyrimidine nucleobase and ribose in model prebiotic reactions (Kim and Benner 2015). These investigators demonstrated that 6-aminouracil (Fig. 5.2, **70**) forms *C*-nucleosides in greater than 50% yield, but exocyclic *N*-nucleosides were not detected. In contrast to studies involving **TAP**, Benner and co-workers found that the ribopyranosides of 6-aminouracil were produced in greater yields (ca. 3:1) compared to the ribofuranosides.

More recently, we explored glycosylation of the pyrimidine **BA** (Fig. 5.2, **79**) and the triazine **MA** (Fig. 5.2, **88**) by ribose in model prebiotic reactions (Cafferty et al. 2016). These two heterocycles gave some of the most promising results thus far for prebiotic nucleoside formation. It was found that **MA** formed glycosides with ribose-5-phosphate (**R5P**) in yields of over 50%, and the yields of **BA** glycosylation

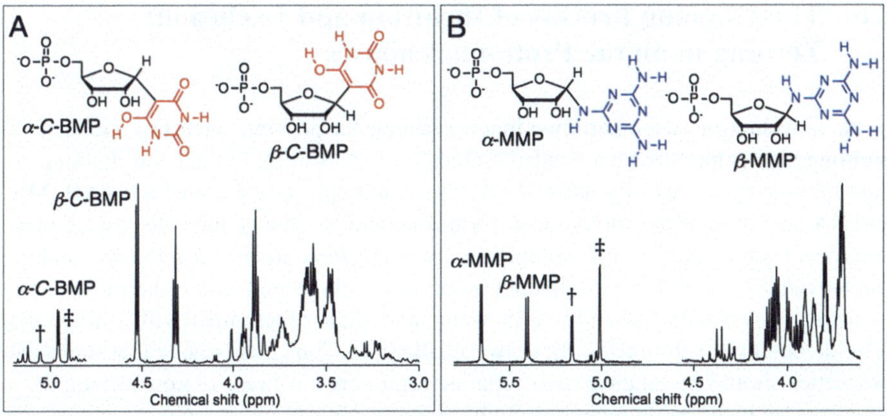

Fig. 5.7 (**a**) Chemical structures of the two *C*-nucleotide anomers of **BA**-ribosyl-monophosphate (***C*-BMP**) and ^{1}H NMR spectrum of a **BA-R5P** crude reaction mixture revealing the formation of α-***C*-BMP** and β-***C*-BMP**. (**b**) Chemical structures of the two anomers of **MA**-ribosyl-monophosphate (**MMP**) and the ^{1}H NMR spectrum of a **MA-R5P** crude reaction mixture revealing the formation of α-**MMP** and β-**MMP**. The anomeric proton resonances for each nucleotide are labeled, and those of **R5P** are marked with dagger indicating α-**R5P** and double dagger indicating β-**R5P**. Relative integrated intensities of the nucleotide anomeric resonances show that, for these two reactions, the total ***C*-BMP** yield was 82%, and the total **MMP** yield was 55%. The HOD peaks have been removed from the NMR spectra for clarity. Adapted from reference (Cafferty et al. 2016)

by ribose-5-phosphate exceeded 80%. Good yields were also observed for the reaction of these two bases with (unphosphorylated) ribose, but the restriction of **R5P** in the furanose conformation made the products of the reaction of **R5P** with **BA** and **MA** particularly “clean” and easy to analyze. For example, the ^{1}H NMR spectra shown in Fig. 5.7 of crude reaction products revealed that the reaction of **R5P** with **BA** in a 1:1 molar ratio produces almost exclusively the two possible *C*-nucleosides, while a 1:1 reaction of **R5P** with **MA** produces the two expected exocyclic *N*-nucleosides. In this same work, it was shown that, like the modified forms of **TAP** and **CA** discussed above, these **R5P** glycosides of **BA** and **MA** form supramolecular assemblies when mixed in water, either in purified forms or when mixed as components of their crude reaction products (Cafferty et al. 2016). A key demonstration of the ability of these supramolecular assemblies to select proto-nucleotides that could have been compatible with backbone formation was the finding that assemblies formed within mixtures containing **BA** and approximately equal amounts of the α- and β-anomer of the **MA**-ribosyl-monophosphate (**MMP**), which are produced together upon the reaction of **MA** with **R5P** (Fig. 5.7c), preferentially incorporated the β-anomer over the α-form. Unexpectedly, it was also observed that a mixture of the α- and β-forms of **MMP** increases in the proportion of the β-anomer to the α-forms when assemblies are formed with **BA** (Cafferty et al. 2016). Thus, it appears that the supramolecular assemblies, formed by the stacking of **MA-BA** hexads, can shift the equilibrium between the α- and β-anomers of **MA** nucleotides (which undergo exchange in solution) by preferentially incorporating and stabilizing the β-anomer.

5.6 The Ongoing Process of Inclusion and Exclusion: Zeroing in on the Proto-nucleobases

In the introduction, we mentioned that the search for possible ancestral nucleic acid building blocks has already revealed molecules that solve more than one challenge to the prebiotic synthesis of a proto-RNA. The results presented above for **TAP**, **MA**, and **BA** are some of our motivations for this statement. As we have discussed, these heterocycles provide possible solutions to both the *Paradox of Base Pairing* and the *Nucleoside Problem*. It is two distinct properties of these molecules, their propensity to form non-covalent assemblies in water and their reactivities with ribose, that provide potential solutions to these two challenges. Thus, we believe that these two properties should be taken as two separate arguments in favor of considering these heterocycles as possible recognition units of proto-RNA.

We are compiling a list of criteria by which to include and exclude specific heterocycles as possible proto-nucleobases. The application of these criteria to the heterocycles in Fig. 5.2, which is our working model of the chemical space of potential proto-nucleobases, is shown graphically in Fig. 5.8. In addition to the abovementioned requirements that a heterocycle be able to form Watson-Crick-like base pairs (i.e., with at least two hydrogen bonds), assembles as monomers in water, and form nucleosides with ribose, we also consider their likelihood to have been present on the prebiotic Earth and the chemical stability of these molecules.

Beginning with Oró's synthesis of adenine from concentrated ammonium cyanide, all four nucleobases of extant RNA have been produced in model prebiotic reactions (Oró 1960; Saladino et al. 2004; Robertson and Miller 1995; Sanchez et al. 1968; Ferris et al. 1968). Three out of four (adenine, guanine, uracil) have also been isolated from carbonaceous chondrites (Callahan et al. 2011; Martins et al. 2008; Stoks and Schwartz 1979). We consider either a report of the production of a particular heterocycle in a model prebiotic reaction or its isolation from a meteorite to be sufficient reason (for now) to mark a heterocycle as having a valid prebiotic synthesis. Among the molecules that meet this particular criterion are **TAP**, **MA**, **BA**, and **CA**. In the case of **MA** and **CA**, like the extant nucleobases, these molecules have been reported to be both produced in model prebiotic reactions and to be found in some carbonaceous chondrites (Hayatsu et al. 1968). Moreover, **CA**, **BA**, and **MA** have been reported to form in the same model prebiotic reaction where urea serves as a primary feedstock (Menor-Salván et al. 2009).

Chemical stability is widely considered an important property of molecules that participated in the origins of life. Stability can be defined as resistance to chemical degradation and/or chemical modification. For example, the susceptibility of cytosine to deamination (i.e., conversion to uracil) has long been cited as a possible reason to eliminate cytosine from the list of prebiotic nucleobases (Shapiro 1999). The reactions leading to degradation or irreversible modification can vary between molecules of interest and can depend strongly on environmental conditions. In the case of cytosine, Levy and Miller reported a half-life of 340 years at 25 °C and only 19 h at 100 °C (Levy and Miller 1998). Moreover, in concentrated solutions of urea,

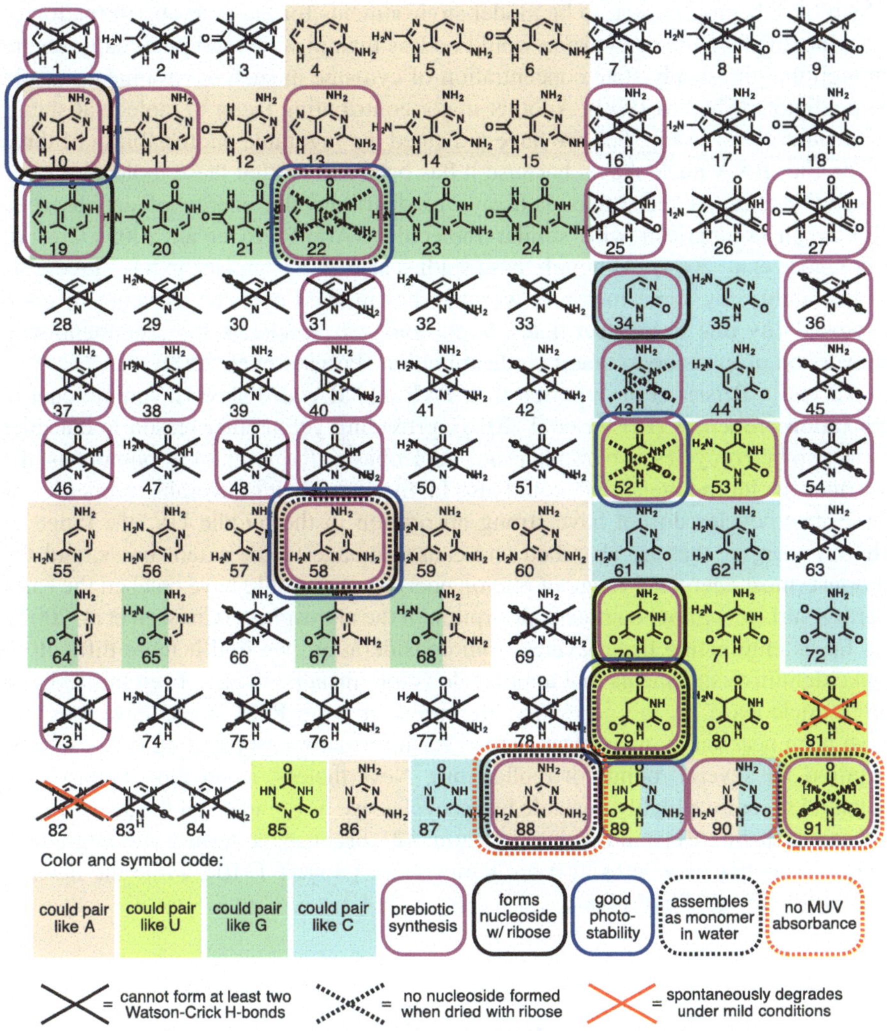

Fig. 5.8 Evaluation of proto-nucleobase candidacy of the chemical space defined in Fig. 5.2. Heterocycles that, when incorporated as a nucleoside, are not expected to form a Watson-Crick-like base pair with at least two H-bonds with an extant nucleobase (i.e., not forward compatible for information transfer) are crossed out. Heterocycles that are expected to be forward compatible are shown on colored backgrounds, yellow for pairing with adenine, cyan for pairing with guanine, orange for pairing with uracil, and green for pairing with cytosine. Heterocycles known to self-assemble as monomers in water are surrounded by dashed boxes (Cafferty et al. 2013; Davis 2004; Bohanon et al. 1995). Heterocycles shown to form nucleosides with sugars are boxed in black (Fuller et al. 1972a, b; Chen et al. 2014; Kim and Benner 2015; Cafferty et al. 2016; Sheng et al. 2009; Wulff and Clarkson 1994). The three extant nucleobases that have not been demonstrated to form nucleosides with preformed sugars are crossed out with dashed lines. Heterocycles identified in meteorites (Callahan et al. 2011; Botta and Bada 2002) or in model prebiotic reactions (Menor-Salván et al. 2009; Menor-Salván and Marin-Yaseli 2013; Nuevo et al. 2012; Barks et al. 2010; Cleaves et al. 2006; Borquez et al. 2005; Saladino et al. 2001; Ferris and Hagan 1984) are boxed in magenta

which have been proposed to be model environments for the prebiotic formation of cytosine, the half-life of cytosine would be less than in water and potentially severe enough that the steady-state concentration of cytosine in such environments would be negligible (Shapiro 1999). Another major contributing factor to molecular stability is photolysis. In Fig. 5.8, we have indicated the favorable photostability of three of the four RNA nucleobases because it has been shown that these molecules have unusually efficient relaxation pathways to distribute the energy of an absorbed photon, in comparison with similar molecules (Abo-Riziq et al. 2005; Crespo-Hernandez et al. 2004). Although these studies also show cytosine to have favorable photostability, by some measure, we have not marked cytosine as being of good photostability due to the fact that UV photons can accelerate the aforementioned conversion of cytosine to uracil by deamination (Powner et al. 2009).

Recently, Brister et al. reported that **TAP** and **BA** also show favorable energy relaxation properties (Brister et al. 2016), providing yet another reason to consider these two heterocycles as possible proto-nucleobases. Regarding UV photolysis, the triazines, including melamine, could also be listed as having favorable properties as these heterocycles do not have strong absorption in the middle UV, the range of photon energies that are absorbed and cause damage to the extant bases, and the shortest (most energetic) range of photon energies that would have reached the early Earth's surface without complete absorption in the atmosphere (Cnossen et al. 2007). As far as eliminating heterocycles from consideration, we find it more difficult to make definitive statements that a molecule is too unstable to have been utilized as a proto-nucleobase, as the necessary lifetime of a proto-RNA nucleobase may not have been necessarily more than needed for information propagation (i.e., the time required for several rounds of replication). Nevertheless, as an illustration of the potential to remove highly unstable heterocycles, we have indicated the instability of alloxan (structure **81**) and triazine (structure **82**), because the former will degrade to CO_2 in a matter of months at room temperature (Bogert 1910), while the latter is susceptible to rapid hydrolysis.

5.7 Considering Alternatives to Ribose in the Backbone of Ancestral RNAs

We now address the possibility that the RNA backbone has evolved since its earliest form (Fig. 5.1). There are several reasons to expect that the trifunctional connector (**TC**) of proto-RNA was not ribose, and perhaps not even a sugar. First of all, the prebiotic synthesis of ribose remains an unsolved problem. The reaction most commonly cited for prebiotic sugar formation, the formose reaction, is inherently nonselective; it produces a complex mixture of distinct chemical species (Sagi et al. 2012; Shapiro 1988). Linear and branched sugars alike are formed (Decker et al. 1982). Ribose is formed in very small amounts (ca. 1%) (Shapiro 1988) and is relatively unstable compared to many other sugars (Larralde et al. 1995) and much less stable than possible non-sugar **TC**s (Hud et al. 2013). There have been several work-arounds proposed to address the problem of prebiotic ribose synthesis and degradation, the two most widely discussed proposals coming from the Benner and the Sutherland laboratories.

Benner and co-workers have demonstrated that ribose, in the presence of borate, is preferentially stabilized against degradation over other sugars (Ricardo et al. 2004; Kim et al. 2011). Based on this observation, these researchers have argued that borate minerals on the prebiotic Earth could have facilitated the synthesis and buildup of ribose. Some geologists have, however, questioned the presence of borate minerals on the prebiotic Earth (Grew et al. 2011). We think that it may be a "chemical coincidence" that the sugar found to be preferentially stabilized by borate is also the sugar chosen by life for the backbone of RNA. The 2′,3′-*cis*-diol of ribose in the furanose ring configuration provides a favorable site for borate coordination, as the ribose-borate complex holds ribose in a cyclic hemiacetal conformation, which protects against degradation. On the other hand, as discussed below, we speculate that the 2′,3′-*cis*-diol is also a reason that ribose was selected for RNA because the placement of the 2′ oxygen of ribose near the 3′ oxygen facilitates the formation of complex RNA structures and also provides a "built-in" mechanism for RNA strand cleavage—the latter property may have become advantageous as pre-RNA took on the role of an ephemeral molecule that transmitted genetic information in the cell, rather than providing a medium for long-term information storage.

A second potential resolution to the problem of prebiotic ribose synthesis and stability comes along with a proposal to ameliorate the inability of the extant pyrimidine nucleobases to form nucleosides with ribose in model prebiotic reactions. Following the lead of Orgel and co-workers (Sanchez and Orgel 1970), Sutherland and co-workers developed a synthesis for cyclic cytidine monophosphate that involves the sequential addition and reaction of small molecules, a synthetic pathway that results in the simultaneous synthesis of the ribose sugar and the cytosine nucleobase (Powner et al. 2009). This approach seems to necessitate acceptance of ribose as the first **TC** of genetic polymers, as the proposed nucleoside synthesis would not be compatible with any sugar other than ribose.

We interpret the difficulty in finding a selective synthesis for ribose (or a conjoined synthesis of ribose as part of a prebiotic route to guanosine and adenosine), as well as the exquisite match of ribose with the structural and chemical features of RNA, as signs that the backbone of RNA has been optimized by evolution. This optimization could have included the selection of ribose during the time that RNA (or its immediate pre-RNAs) was taking on increasingly complex roles in nascent biology (e.g., coded protein synthesis).

As further support for this position, synthetic organic chemists have demonstrated that nucleic acid oligomers prepared with a variety of non-ribose sugars as the **TC** can still form stable duplexes, including nucleic acids with tetrose, pentose, and hexoses sugars, as well as alternative conformations of ribose (e.g., β-pyranosyl). A few examples of such nonnatural nucleic acids are shown in Fig. 5.9, along with RNA for comparison. Reviews of these investigations, including comparisons of

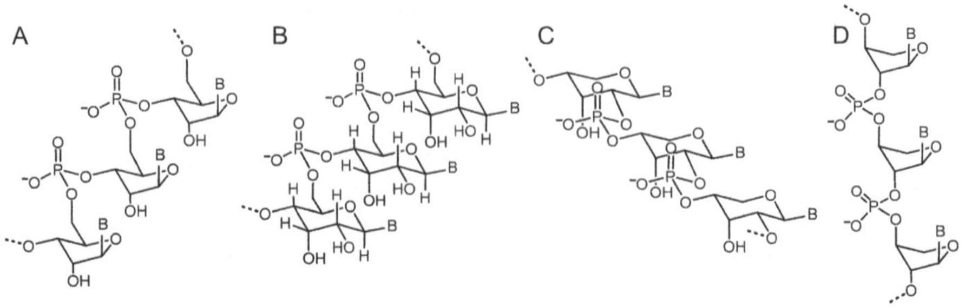

Fig. 5.9 Structures of RNA and selected analogs that maintain the phosphodiester backbone linkage, but with different sugars acting as the trifunctional connectors, all of which are known to support duplex formation (but not necessarily Watson-Crick base pairs). (**a**) Extant RNA, with β-ribofuranose; (**b**) β-allopyranosyl nucleic acid (Eschenmoser 1999); (**c**) pRNA, β-pyranosyl-RNA, with ribose in its pyranose ring form (Bolli et al. 1997); and (**d**) TNA, α-threofuranosyl nucleic acid, with a four-carbon sugar (Schöning et al. 2000). For all structures shown, B represents one of the canonical nucleobases

properties of these nonnatural nucleic acids, have already been published (Eschenmoser 1999, 2004, 2007; Egli et al. 2006; Benner 2004; Herdewijn 2001; Schöning et al. 2000; Bolli et al. 1997; Pitsch et al. 1993). In brief, these studies have shown that some sugars do not support Watson-Crick duplex formation, or the duplexes that are formed exhibit substantial sequence-dependent stability. In contrast, some of these alternative nucleic acids actually form duplexes that are more stable than those of RNA (Eschenmoser 1999). Such results caused Eschenmoser, an authority on this topic, to conclude that ribose was not selected by nature because of its unique or superior ability to stabilize Watson-Crick base pairs, but rather because ribose provides an optimal balance between duplex stability and the ability to adopt the less regular structures, such as those exemplified by the variety of RNA structures necessary to construct the ribosome (Ban et al. 2000; Wimberly et al. 2000). Eschenmoser's conclusion is consistent with the hypothesis that ribose was selected in the latter stages of nucleic acid evolution for its functional properties, rather than for prebiotic abundance or favorable prebiotic reactivity.

Previously, we have discussed some of the challenges facing the prebiotic synthesis of polymers with the extant backbone of RNA, and we have discussed possible ancestral units for both the **TC** and the **IL** (Engelhart and Hud 2010; Hud et al. 2013; Cafferty and Hud 2014, 2015). We will therefore focus the present discussion on recent results regarding the possibility that the earliest sugar(s) to act as a **TC**(s) in the nucleic acid backbone was not ribose. As discussed above, we have discovered that **TAP**, a plausible proto-nucleobase, reacts readily with ribose to form nucleosides (i.e., β-ribofuranosyl glycosides of **TAP**). Because nucleoside formation was potentially an important criterion for the selection of the first sugar to act as a **TC** in nucleic acids, we investigated the potential for **TAP** to be glycosylated by a variety of sugars. In Fig. 5.10, we show the structures of 17 sugars and modified sugars that were tested for their ability to react with **TAP**. We chose **TAP** for this investigation, instead of **MA** or **BA**, because we had found that this heterocycle

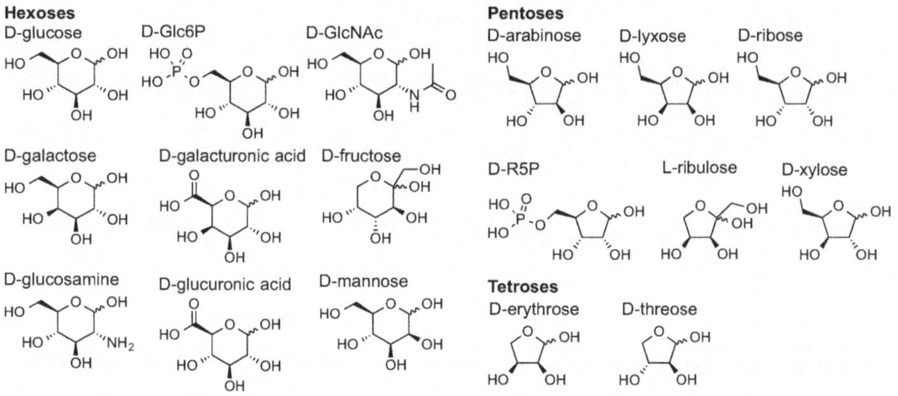

Fig. 5.10 Sugars investigated for their reactivity with TAP. Although the hexoses are shown as pyranoses and pentoses as furanose, most sugars exist in equilibria between several cyclic hemiacetal/hemiketal forms and linear aldehyde/ketone forms. Reprinted with permission from Fialho et al. (2018)

forms both *C*-nucleosides and exocyclic *N*-nucleosides. For these experiments, **TAP** was incubated with various sugars in water at 85 °C for 24 h to assess whether or not glycosides (formally only referred to as "nucleosides" where the sugar is ribose) could be formed (Fialho et al. 2018). The sugars surveyed differed in their chemical structures and properties (ketoses and aldoses; tetroses, pentoses, and hexoses; neutral, anionic, and cationic). Although the reaction conditions chosen for this study were optimized for the reaction of **TAP** with glucose and its derivatives, all sugars tested successfully reacted with **TAP** to form glycosides (Fialho et al. 2018). Interestingly, aldosides were detected in the reactions of **TAP** with ribulose and fructose (both ketoses), suggesting that carbonyl migration had occurred to form aldoses (either ribose or arabinose in the case of ribulose and either glucose or mannose in the case of fructose), which then reacted with **TAP** to form aldosides. Ketosides were not explicitly detected, but their existence could not be ruled out.

It was previously found that the reaction of **TAP** with ribose forms at least eight ribosides, with the β-*C*-ribofuranoside produced in the greatest yield (Fialho et al. 2018). To gain more information on the chemical selectivity of the reaction of **TAP** with sugars, products from reactions of **TAP** with glucose, *N*-acetylglucosamine (**GlcNAc**), and glucose-6-phosphate (**Glc6P**) were isolated and structurally characterized. By a combination of NMR spectroscopy techniques, it was ultimately found that the reaction of **TAP** with glucose produces two β-*N*-pyranosides and one β-*C*-pyranoside, the reaction of **TAP** with **Glc6P** produces two β-*N*-pyranosides and one β-*C*-pyranoside, and the reaction of **TAP** with **GlcNAc** produces only two β-*N*-pyranosides (Fig. 5.11). That is, in the reaction of **TAP** with glucose and its derivatives, the product is always a β-pyranoside. It seems that the inherent stability of one chair form of glucopyranose (and its derivatives) gives greater selectivity and fewer reaction products than, for example, ribose, which gives many possible glycoside products.

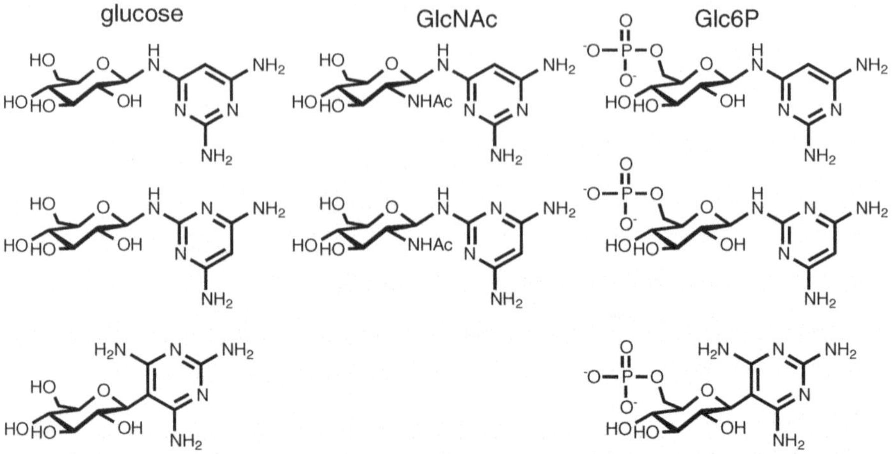

Fig. 5.11 Chemical structures of the products formed from the reaction of **TAP** with glucose, **GlcNAc**, or **Glc6P**. All glycosides identified are β-pyranosides. No *C*-glycosides were detected in the reactions of **TAP** with **GlcNAc**. Reprinted with permission from reference (Fialho et al. 2018)

Overall, our study of **TAP** glycosylation by various sugars suggests that β-ribofuranosides would not have been the primary, and certainly not the only, potential proto-nucleoside present on the prebiotic Earth if **TAP** and other heterocycles with similar reactivity coexisted with a complex mixture of sugars. As noted above, Benner and co-workers have proposed borate coordination could have provided a mechanism for the more selective production of ribose in the formose reaction (Ricardo et al. 2004; Kim et al. 2011). Our observation that **TAP** produces aldose glycosides when reacted with ketose sugars (i.e., ribulose and fructose), presumably by catalyzing carbonyl migration prior to glycoside formation, presents a different possible scenario for the formation of glycosides that would also circumvent the problems associated with aldose sugar instability by utilizing ketose sugars as a feedstock (Larralde et al. 1995). This scenario looks quite plausible from a prebiotic chemistry perspective given that Krishnamurthy and co-workers have shown that ketopentoses, including ribulose, are formed selectively in good yields from glyceraldehyde and dihydroxyfumarate in a model prebiotic reaction (Sagi et al. 2012). Similarly, Weber has shown that ketohexoses, including fructose, are formed selectively and in good yields from glyceraldehyde in a model prebiotic reaction (Weber 1992). Thus, ketose sugars may have been more abundant than aldose sugars on the prebiotic Earth, but heterocycles such as **TAP** could have still selected aldoses—due to their reactivity—within mixtures of carbohydrates for aldoside production.

5.8 Comments on the Merits and Challenges of Phosphate in the Backbone of Proto-RNA

In previous communications, we have discussed some of the challenges associated with phosphate being the original **IL** of proto-RNA (Engelhart and Hud 2010; Hud et al. 2013). In short, the prebiotic challenges of phosphate incorporation into early nucleic acids are mainly issues of accessibility and the thermodynamics of phosphodiester bond formation. Possible solutions to these challenges are discussed in Chap. 6 by Pasek. We will therefore limit our discussion of phosphate as an early **IL**. Just as ribose looks optimal for its function as the **TC** in extant RNA, phosphate is, arguably, even more uniquely suited to function as the **IL**. In his thoughtful paper "Why Nature Chose Phosphates" (Westheimer 1987), Frank H. Westheimer presented many reasons why phosphate possesses chemical properties that make it ideal for its many functions in biochemistry, including as the **IL** of RNA and DNA. In short, phospho(di)ester linkages are sufficiently labile that enzymes can hydrolyze nucleic acids for proofreading or to recycle monomers, while the negative charge provides protection of the phosphorus center of these polymers against nucleophilic attack. Westheimer, Benner, and others have also concluded that the electrostatic charge of the phosphate group provides RNA and DNA with characteristics that appear necessary for molecules to be able to remain soluble and maintain a consistent double helix structure for essentially all possible nucleotide sequences (Benner et al. 2004; Westheimer 1987). Moreover, a wide variety of synthetic nucleic acid analogs have been prepared with non-furanose ribose sugars as the **TC**s, but still with phosphate as the **IL**, that still function well as informational pairing systems (Eschenmoser 2011). A few examples are shown above in Fig. 5.9. These synthetic examples illustrate the possibility that phosphate could have linked together a different sugar before evolution refined RNA to use β-D-ribofuranosyl nucleosides.

5.9 A Proposed Mechanism for the Prebiotic Replication and Early Sequence Evolution of Proto-RNA

Our hypothesis that the self-assembly of stacked, H-bonded monomers with Watson-Crick-like pairing allowed for the de novo synthesis of proto-RNA polymers has included, from its inception, our proposal that the same mechanism would have allowed for the replication of existing proto-RNA polymers (i.e., those previously synthesized by the de novo method) (Hud and Anet 2000). In Fig. 5.12, we provide a schematic representation of our hypothesis for proto-RNA replication, which can be

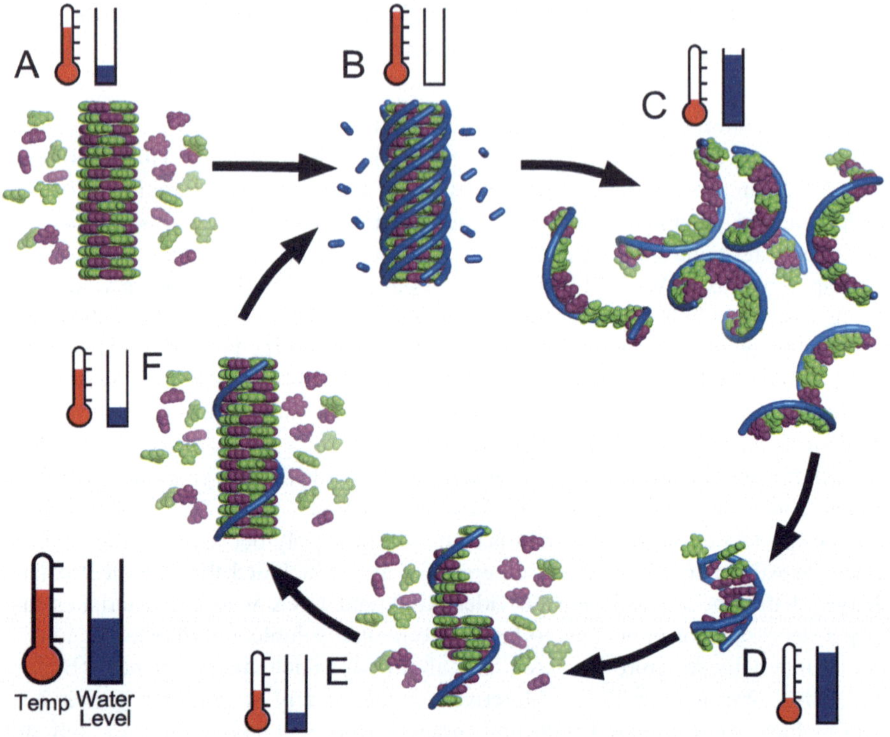

Fig. 5.12 A hypothetical cycle for the de novo synthesis, replication, and evolution of proto-RNA polymers via hexad stack formation that is driven by oscillations in temperature and level of hydration. (**a**) Formation of a hexad stack by heterocycles (e.g., melamine and barbituric acid), which is driven by the increase in heterocycle concentration upon the evaporation of water. (**b**) Further evaporation of water leads to the increased local concentration of trifunctional connectors (**TC**) and ionized linkers (**IL**). An increase in temperature leads to the formation of polymers that connect the heterocycles assembled in the hexad stack. (**c**) Rehydration results in sufficient dilution and change in solution conditions (e.g., change in ionic strength, pH) that the newly formed proto-RNA polymers are released from their hexameric assembly. (**d**) A subset of polymers formed with random sequences, i.e., those formed de novo from stacks shown in **a**, will have sequences that fold into stable structures that protect the polymers against hydrolysis. (**e** and **f**) As water is removed by evaporation, those polymers that survived the fully hydrated state will begin to unfold as the concentration of monomers increases and hexad formation becomes more energetically favored over intramolecular folding. This cycle is complete when backbones are formed along stacks that were nucleated by pre-existing proto-RNA polymers (F→B)

thought of as an enzyme-free polymerase chain reaction (PCR) that is driven by environmental oscillations in temperature and levels of hydration [the original inspiration for this hypothesis (Hud 2017)].

In our model, proto-RNA polymers produced by de novo proto-RNA synthesis provide polymers with random sequences of nucleobases that initially serve as templates for the replication cycle. As discussed above, the production of these polymers is driven by water evaporation and elevated temperature (Fig. 5.12a, b).

Rehydration, potentially resulting from rain falling on a dried pond, causes separation of the strands from the hexad stack (Fig. 5.12c). A subset of the (initially) random sequences form stable folded structures that provide resistance against backbone and glycosidic bond hydrolysis during the hydrated phase (Fig. 5.12d). These sequences that resist hydrolysis survive long enough to act as templates for the formation of hexad stacks during the next dry phase (Fig. 5.12e, f). Unlike the hexad stacks formed without a pre-existing polymer (e.g., Fig. 5.12a), the hexad stacks that are nucleated by a pre-existing polymer will have a registration of hexads within the stack that is dictated by the sequences of bases along the nucleating polymer (Fig. 5.12f). When backbones are formed along the free bases that have entered into the hexad stack with the pre-existing polymer (Fig. 5.12b), three of the resulting polymers will have sequences that are complementary (i.e., able to form base pairs) to the pre-existing polymer, while the other two will be exact copies. Because the strands being replicated will be those that preferentially survived the wet phases of the cycle, this process will both increase the number of polymers in a given environment as well as enrich the sequences of these polymers with those that fold into defined structures. Thus, this hypothetical cycle could provide a mechanism by which proto-RNA sequences evolve toward more stable structures, which would also increase the likelihood that sequences with catalytic activity emerge, as enzymes and ribozymes are drawn from sequences that form well-defined three-dimensional structures—a subset of all possible sequences.

Thus far, our description of the hypothetical cycle of proto-RNA formation and replication shown in Fig. 5.12 has only elicited changes in water levels as a means to increase the concentration of monomeric heterocycles and backbone building blocks, in the case of evaporation, and as a means to facilitate the separation of proto-RNA polymers from a hexad stack, in the case of rehydration. However, water evaporation would cause other changes in the properties of the solution that could have also assisted the replication cycle. For example, strand separation from the hexad stack could also be facilitated by the reduced concentration of metal ions (which may be necessary for stack formation) or changes in pH [which can greatly alter the stability of hexad pairing (Cafferty et al. 2014)]. The increase in viscosity that is associated with evaporation of water from aqueous solutions could have helped keep pre-existing proto-RNA polymers from reannealing into hexad stacks long enough for each to act as a template for the organization of monomeric heterocycles (Fig. 5.12e, f), an essential component of replication that can be blocked by the reannealing of the polymers, a long-appreciated problem known as the *Product Inhibition Problem* or the *Strand Inhibition Problem* (Grossmann et al. 2008; Fernando et al. 2007). We have recently published a proof-of-principle demonstration of how viscosity could have circumvented the *Strand Inhibition Problem* (He et al. 2017), and a detailed description and discussion of how a viscous solvent could have facilitated information transfer is provided in Chap. 7 by He and Gállego.

5.10 Conclusions and the Path Forward

In this chapter, we have reviewed our progress toward finding candidates for proto-RNA. While Orgel understandably labeled the possibility that RNA evolved from another molecule as a "gloomy prospect" because of the many molecules we might have to search through to find the ancestor of RNA, the results summarized here provide reason to be optimistic about finding RNA-like molecules that will spontaneously assemble from plausible prebiotic molecular building blocks. At this point, we feel that 2,4,6-triaminopyrimidine, melamine, and barbituric acid represent excellent candidates for the nucleobases of an early, if not the earliest, form of pre-RNA. This opinion is based on the various properties of these molecules that provide solutions to multiple challenges facing the prebiotic formation of RNA, as well as our conclusion that few other molecules within the chemical space around the extant bases can exhibit the same propensity for assembly in water, glycoside formation with ribose (and other sugars), formation in the same model prebiotic reaction, and H-bonding with extant nucleobases (i.e., forward compatibility) (Li et al. 2016).

The next important challenge to address is finding a plausible chemical structure for the proto-RNA backbone. As mentioned above, synthetic chemists have shown that many changes can be made to the nucleic acid backbone without disrupting its ability to form a Watson-Crick duplex [often cited as a minimal criteria for accepting a backbone as a potential ancestor of the RNA backbone (Yang et al. 2007)], from changes as subtle as changing the connectivity of the phosphate-ribose backbone from the 3′–5′ linkage to the 2′–5′ linkage (Sheng et al. 2014; Horowitz et al. 2009) to the complete replacement of the extant backbone with a *N*-(2-aminoethyl)-glycine polymer (known as the peptide nucleic acid backbone) (Nielsen 2007). The results reviewed in this chapter provide some support for the possibility that an earlier form of RNA utilized a different sugar than ribose, as our model proto-nucleobases are glycosylated by a wide range of sugars. Nevertheless, we believe that it is important to keep considering potential candidates for the backbone for proto-RNA until one is found that allows the spontaneous formation of RNA-like polymers with lengths sufficient to exhibit catalytic activity (e.g., 25–50 nucleotides), as well as being formed in good yield. Finally, we have proposed that the initial backbone may have been a polymer that was formed independent of the nucleobases but which had a structure that was a perfect match with the spatial organization of heterocycles in a hexad stack (Hud et al. 2013). In this "polymer fusion" model proto-RNA is hypothesized to be the merger of two polymers, one being the hexad stack, a non-covalent polymer, and the backbone, a covalent polymer, such as a polypeptide with side chains that can react with the bases that are interspersed with residues that provide the charge necessary for solubility (Mittapalli et al. 2007). Time will tell which model for the prebiotic emergence of nucleic acids can be demonstrated in the laboratory. Until then, we encourage researches to remain open-minded as well as scientifically critical about all models currently being considered for the origins of nucleic acids.

Acknowledgments We thank Profs. Frank A. L. Anet, Ram Krishnamurthy, Gary B. Schuster, and Loren Dean Williams, who have been valuable collaborators and consultants on much of the work described in this chapter from our laboratory. This work was supported by the NSF and the NASA Astrobiology Program, under the NSF Center for Chemical Evolution (CHE-1504217) and the NASA Exobiology Program NNX13AI02G.

References

Abo-Riziq A, Grace L, Nir E, Kabelac M, Hobza P, de Vries MS (2005) Photochemical selectivity in guanine-cytosine base-pair structures. Proc Natl Acad Sci U S A 102:20–23

Arnott S, Bond PJ (1973) Triple-stranded polynucleotide helix containing only purine bases. Science 181:68

Ban N, Nissen P, Hansen J, Moore PB, Steitz TA (2000) The complete atomic structure of the large ribosomal subunit at 2.4 Ångstrom resolution. Science 289:905–920

Barks HL, Buckley R, Grieves GA, Di Mauro E, Hud NV, Orlando TM (2010) Guanine, adenine, and hypoxanthine production in UV-irradiated formamide solutions: relaxation of the requirements for prebiotic purine nucleobase formation. Chembiochem 11:1240–1243

Battersby TR, Albalos M, Friesenhahn MJ (2007) An unusual mode of DNA duplex association: Watson-Crick interaction of all-purine deoxyribonucleic acids. Chem Biol 14:525–531

Benner SA (2004) Understanding nucleic acids using synthetic chemistry. Acc Chem Res 37:784–797

Benner SA, Ricardo A, Carrigan MA (2004) Is there a common chemical model for life in the universe? Curr Opin Chem Biol 8:672–689

Benner SA, Kim H-J, Carrigan MA (2012) Asphalt, water, and the prebiotic synthesis of ribose, ribonucleosides, and RNA. Acc Chem Res 45:2025–2034

Bernhardt HS (2012) The RNA world hypothesis: the worst theory of the early evolution of life (except for all the others)(a). Biol Direct 7:23–23

Bogert MT (1910) The instability of alloxan. J Am Chem Soc 32:809–810

Bohanon TM, Denzinger S, Fink R, Paulus W, Ringsdorf H, Weck M (1995) Barbituric-acid 2,4,6-triaminopyrimidine aggregates in water and their competitive interaction with a monolayer of barbituric-acid lipids at the gas-water interface. Angew Chem Int Ed Engl 34:58–60

Bolli M, Micura R, Eschenmoser A (1997) Pyranosyl-RNA: chiroselective self-assembly of base sequences by ligative oligomerization of tetra nucleotide-2′,3′-cyclophosphates (with a commentary concerning the origin of biomolecular homochirality). Chem Biol 4:309–320

Borquez E, Cleaves HJ, Lazcano A, Miller SL (2005) An investigation of prebiotic purine synthesis from the hydrolysis of HCN polymers. Orig Life Evol Biosph 35:79–90

Botta O, Bada JL (2002) Extraterrestrial organic compounds in meteorites. Surv Geophys 23:411–467

Bowman J, Hud N, Williams L (2015) The ribosome challenge to the RNA world. J Mol Evol 80:143–161

Brister MM, Pollum M, Crespo-Hernandez CE (2016) Photochemical etiology of promising ancestors of the RNA nucleobases. Phys Chem Chem Phys 18:20097–20103

Buckley R, Enekwa CD, Williams LD, Hud NV (2011) Molecular recognition of Watson-Crick-like purine-purine base pairs. Chembiochem 12:2155–2158

Cafferty BJ, Hud NV (2014) Abiotic synthesis of RNA in water: a common goal of prebiotic chemistry and bottom-up synthetic biology. Curr Opin Chem Biol 22:146–157

Cafferty BJ, Hud NV (2015) Was a pyrimidine-pyrimidine base pair the ancestor of Watson-Crick base pairs? Insights from a systematic approach to the origin of RNA. Isr J Chem 55:891–905

Cafferty BJ, Gallego I, Chen MC, Farley KI, Eritja R, Hud NV (2013) Efficient self-assembly in water of long noncovalent polymers by nucleobase analogues. J Am Chem Soc 135:2447–2450

Cafferty BJ, Avirah RR, Schuster GB, Hud NV (2014) Ultra-sensitive pH control of supramolecular polymers and hydrogels: pK(a) matching of biomimetic monomers. Chem Sci 5:4681–4686
Cafferty BJ, Fialho DM, Khanam J, Krishnamurthy R, Hud NV (2016) Spontaneous formation and base pairing of plausible prebiotic nucleotides in water. Nat Commun 7:11328
Callahan MP, Smith KE, Cleaves HJ, Ruzicka J, Stern JC, Glavin DP, House CH, Dworkin JP (2011) Carbonaceous meteorites contain a wide range of extraterrestrial nucleobases. Proc Natl Acad Sci U S A 108:13995–13998
Cassidy LM, Burcar BT, Stevens W, Moriarty EM, McGown LB (2014) Guanine-centric self-assembly of nucleotides in water: an important consideration in prebiotic chemistry. Astrobiology 14:876–886
Chen MC, Cafferty BJ, Mamajanov I, Gállego I, Khanam J, Krishnamurthy R, Hud NV (2014) Spontaneous prebiotic formation of a β-ribofuranoside that self-assembles with a complementary heterocycle. J Am Chem Soc 136:5640–5646
Cleaves HJ, Nelson KE, Miller SL (2006) The prebiotic synthesis of pyrimidines in frozen solution. Naturwissenschaften 93:228–231
Cnossen I, Sanz-Forcada J, Favata F, Witasse O, Zegers T, Arnold NF (2007) Habitat of early life: solar X-ray and UV radiation at Earth's surface 4-3.5 billion years ago. J Geophys Res Planets 112:10
Crespo-Hernandez CE, Cohen B, Hare PM, Kohler B (2004) Ultrafast excited-state dynamics in nucleic acids. Chem Rev 104:1977–2019
Crick FHC (1968) The origin of the genetic code. J Mol Biol 38:367–379
Davis JT (2004) G-quartets 40 years later: from 5'-GMP to molecular biology and supramolecular chemistry. Angew Chem Int Ed Engl 43:668–698
Davis JT, Spada GP (2007) Supramolecular architectures generated by self-assembly of guanosine derivatives. Chem Soc Rev 36:296–313
Decker P, Schweer P, Pohlmann R (1982) Identification of formose sugars, presumable prebiotic metabolites, using capillary gas chromatography/gas chromatography-mass spectroscopy of n-butoxime trifluoroacetates on OV-225. J Chromatogr 244:281–291
Dorr M, Loffler PMG, Monnard PA (2012) Non-enzymatic polymerization of nucleic acids from monomers: monomer self-condensation and template-directed reactions. Curr Org Synth 9:735–763
Egli M, Pallan PS, Pattanayek R, Wilds CJ, Lubini P, Minasov G, Dobler M, Leumann CJ, Eschenmoser A (2006) Crystal structure of homo-DNA and Nature's choice of pentose over hexose in the genetic system. J Am Chem Soc 128:10847–10856
Engelhart AE, Hud NV (2010) Primitive genetic polymers. Cold Spring Harb Perspect Biol 2: a002196. https://doi.org/10.1101/cshperspect.a002196
Eschenmoser A (1999) Chemical etiology of nucleic acid structure. Science 284:2118–2124
Eschenmoser A (2004) The TNA-family of nucleic acid systems: properties and prospects. Orig Life Evol Biosph 34:277–306
Eschenmoser A (2007) The search for the chemistry of life's origin. Tetrahedron 63:12821–12844
Eschenmoser A (2011) Etiology of potentially primordial biomolecular structures: from vitamin B12 to the nucleic acids and an inquiry into the chemistry of life's origin: A retrospective. Angew Chem Int Ed Engl 50:12412–12472
Fernando C, Von Kiedrowski G, Szathmary E (2007) A stochastic model of nonenzymatic nucleic acid replication: "Elongators" sequester replicators. J Mol Evol 64:572–585
Ferris JP, Hagan WJ (1984) HCN and chemical evolution - the possible role of cyano compounds in prebiotic synthesis. Tetrahedron 40:1093–1120
Ferris JP, Sanchez RA, Orgel LE (1968) Synthesis of pyrimidines from cyanoacetylene and cyanate. J Mol Biol 33:693–704
Ferris JP, Hill AR, Liu RH, Orgel LE (1996) Synthesis of long prebiotic oligomers on mineral surfaces. Nature 381:59–61

Fialho DM, Clarke KC, Moore MK, Schuster GB, Krishnamurthy R, Hud NV (2018) Glycosylation of a model proto-RNA nucleobase with non-ribose sugars: implications for the prebiotic synthesis of nucleosides. Org Biomol Chem 16:1263–1271

Forsythe JG, Yu SS, Mamajanov I, Grover MA, Krishnamurthy R, Fernandez FM, Hud NV (2015) Ester-mediated amide bond formation driven by wet-dry cycles: a possible path to polypeptides on the prebiotic Earth. Angew Chem Int Ed Engl 54:9871–9875

Fripiat JJ, Cruzcump M (1974) Clays as catalysts for natural processes. Annu Rev Earth Planet Sci 2:239–256

Fuller WD, Sanchez RA, Orgel LE (1972a) Studies in prebiotic synthesis. VI. Synthesis of purine nucleosides. J Mol Biol 67:25–33

Fuller WD, Sanchez RA, Orgel LE (1972b) Studies in prebiotic synthesis: VII. Solid-state synthesis of purine nucleosides. J Mol Evol 1:249–257

Gavette JV, Stoop M, Hud NV, Krishnamurthy R (2016) RNA-DNA chimeras in the context of an RNA world transition to an RNA/DNA world. Angew Chem Int Ed Engl 55:13204–13209

Gellert M, Lipsett MN, Davies DR (1962) Helix formation by guanylic acid. Proc Natl Acad Sci U S A 48:2013–2018

Gesteland RF, Cech TR, Atkins JF (2006) The RNA world, 3rd edn. Cold Spring Harbor Laboratory Press, Cold Spring Harbor, NY

Grew ES, Bada JL, Hazen RM (2011) Borate minerals and origin of the RNA world. Origins Life Evol Biosph 41:307–316

Groebke K, Hunziker J, Fraser W, Peng L, Diederichsen U, Zimmermann K, Holzner A, Leumann C, Eschenmoser A (1998) Why pentose- and not hexose-nucleic acids? Purine-purine pairing in homo-DNA: guanine, isoguanine, 2,6-diaminopurine, and xanthine. Helv Chim Acta 81:375–474

Grossmann TN, Strohbach A, Seitz O (2008) Achieving turnover in DNA-templated reactions. Chembiochem 9:2185–2192

Guerrier-Takada C, Gardiner K, Marsh T, Pace N, Altman S (1983) The RNA moiety of ribonuclease P is the catalytic subunit of the enzyme. Cell 35:849–857

Harvey GR, Degens ET, Mopper K (1971) Synthesis of nitrogen heterocycles on kaolinite from CO_2 and NH_3. Naturwissenschaften 58:624–625

Hayatsu R, Studier MH, Oda A, Fuse K, Anders E (1968) Origin of organic matter in early solar system 2. Nitrogen compounds. Geochim Cosmochim Acta 32:175–190

Hayatsu R, Studier MH, Moore LP, Anders E (1975) Purines and triazines in murchison meteorite. Geochim Cosmochim Acta 39:471–488

He C, Gallego I, Laughlin B, Grover MA, Hud NV (2017) A viscous solvent enables information transfer from gene-length nucleic acids in a model prebiotic replication cycle. Nat Chem 9:318–324

Herdewijn P (2001) TNA as a potential alternative to natural nucleic acids. Angew Chem Int Ed Eng 40:2249–2251

Heuberger BD, Switzer C (2008) An alternative nucleobase code: characterization of purine-purine DNA double helices bearing guanine-isoguanine and diaminopurine-7-deaza-xanthine base pairs. Chembiochem 9:2779–2783

Hoogsteen K (1959) The structure of crystals containing a hydrogen-bonded complex of 1-methylthymine and 9-methyladenine. Acta Crystallogr 12:822–823

Horowitz ED, Lilavivat S, Holladay BW, Germann MW, Hud NV (2009) Solution structure and thermodynamics of 2′,5′ RNA intercalation. J Am Chem Soc 131:5831–5838

Howard FB, Miles HT (1977) Interaction of poly(A) and poly(I), a reinvestigation. Biochemistry 16:4647–4650

Hud NV (2017) Our odyssey to find a plausible prebiotic path to RNA: the first twenty years. Synlett 28:36–55

Hud NV, Anet FAL (2000) Intercalation-mediated synthesis and replication: a new approach to the origin of life. J Theor Biol 205:543–562

Hud NV, Jain SS, Li X, Lynn DG (2007) Addressing the problems of base pairing and strand cyclization in template-directed synthesis. Chem Biodivers 4:768–783

Hud NV, Cafferty BJ, Krishnamurthy R, Williams LD (2013) The origin of RNA and 'My grandfather's axe'. Chem Biol 20:466–474

Johnson AP, Cleaves HJ, Dworkin JP, Glavin DP, Lazcano A, Bada JL (2008) The Miller volcanic spark discharge experiment. Science 322:404–404

Joyce GF, Schwartz AW, Miller SL, Orgel LE (1987) The case for an ancestral genetic system involving simple analogs of the nucleotides. Proc Natl Acad Sci U S A 84:4398–4402

Kim HJ, Benner SA (2015) Prebiotic glycosylation of uracil with electron-donating substituents. Astrobiology 15:301–306

Kim HJ, Ricardo A, Illangkoon HI, Kim MJ, Carrigan MA, Frye F, Benner SA (2011) Synthesis of carbohydrates in mineral-guided prebiotic cycles. J Am Chem Soc 133:9457–9468

Kolb VM, Dworkin JP, Miller SL (1994) Alternative bases in the RNA world: the prebiotic synthesis of urazole and its ribosides. J Mol Evol 38:549–557

Krishnamurthy R (2015) On the emergence of RNA. Isr J Chem 55:837–850

Krishnamurthy R, Pitsch S, Minton M, Miculka C, Windhab N, Eschenmoser A (1996) Pyranosyl-RNA: base pairing between homochiral oligonucleotide strands of opposite sense of chirality. Angew Chem Int Ed Engl 35:1537–1541

Kruger K, Grabowski PJ, Zaug AJ, Sands J, Gottschling DE, Cech TR (1982) Self-splicing RNA: autoexcision and autocyclization of the ribosomal RNA intervening sequence of *Tetrahymena*. Cell 31:147–157

Kurland CG (2010) The RNA dreamtime. BioEssays 32:866–871

Kuruvilla E, Schuster GB, Hud NV (2013) Enhanced non-enzymatic ligation of homo-purine miniduplexes: Support for greater base stacking in a pre-RNA world. Chembiochem 14 (1):45–48. https://doi.org/10.1002/cbic.201200601

Larralde R, Robertson MP, Miller SL (1995) Rates of decomposition of ribose and other sugars: implications for chemical evolution. Proc Natl Acad Sci U S A 92:8158–8160

Lehn J-M, Mascal M, Decian A, Fischer J (1990) Molecular recognition directed self-assembly of ordered supramolecular strands by cocrystallization of complementary molecular components. J Chem Soc Chem Commun:479–481

Levy M, Miller SL (1998) The stability of the RNA bases: implications for the origin of life. Proc Natl Acad Sci U S A 95:7933–7938

Li C, Cafferty BJ, Karunakaran SC, Schuster GB, Hud NV (2016) Formation of supramolecular assemblies and liquid crystals by purine nucleobases and cyanuric acid in water: Implications for the possible origins of RNA. Phys Chem Chem Phys 18:20091–20096

Ma M, Bong D (2011) Determinants of cyanuric acid and melamine assembly in water. Langmuir 27:8841–8853

Martins Z, Botta O, Fogel ML, Sephton MA, Glavin DP, Watson JS, Dworkin JP, Schwartz AW, Ehrenfreund P (2008) Extraterrestrial nucleobases in the Murchison meteorite. Earth Planet Sci Lett 270:130–136

Menor-Salván C, Marin-Yaseli MR (2013) A new route for the prebiotic synthesis of nucleobases and hydantoins in water/ice solutions involving the photochemistry of acetylene. Chem Eur J 19:6488–6497

Menor-Salván C, Ruiz-Bermejo DM, Guzman MI, Osuna-Esteban S, Veintemillas-Verdaguer S (2009) Synthesis of pyrimidines and triazines in ice: implications for the prebiotic chemistry of nucleobases. Chem Eur J 15:4411–4418

Miller SL (1953) A production of amino acids under possible primitive Earth conditions. Science 117:528–529

Miller SL (1957) The mechanism of synthesis of amino acids by electric discharges. Biochim Biophys Acta 23:480–489

Miller SL, Urey HC (1959) Organic compound synthesis on the primitive Earth. Science 130:245–251

Mittapalli GK, Osornio YM, Guerrero MA, Reddy KR, Krishnamurthy R, Eschenmoser A (2007) Mapping the landscape of potentially primordial informational oligomers: oligodipeptides tagged with 2,4-disubstituted 5-aminopyrimidines as recognition elements. Angew Chem Int Ed Engl 46:2478–2484

Monnard P-A (2016) Taming prebiotic chemistry: the role of heterogeneous and interfacial catalysis in the emergence of a prebiotic catalytic/information polymer system. Life 6:E40

Nielsen PE (2007) Peptide nucleic acids and the origin of life. Chem Biodivers 4:1996–2002

Nuevo M, Milam SN, Sandford SA (2012) Nucleobases and prebiotic molecules in organic residues produced from the ultraviolet photo-irradiation of pyrimidine in NH3 and H2O+NH3 ices. Astrobiology 12:295–314

Olby R (2003) Quiet debut for the double helix. Nature 421:402–405

Onimaru K, Kuraku S, Takagi W, Hyodo S, Sharpe J, Tanaka M (2015) A shift in anterior–posterior positional information underlies the fin-to-limb evolution. elife 4:e07048

Orgel LE (1968) Evolution of the genetic apparatus. J Mol Biol 38:381–393

Orgel LE (1998) The origin of life – a review of facts and speculations. TIBS 23:491–495

Orgel LE (2004) Prebiotic chemistry and the origin of the RNA world. Crit Rev Biochem Mol Biol 39:99–123

Oró J (1960) Synthesis of adenine from ammonium cyanide. Biochem Biophys Res Commun 2:407–412

Oró J (2002) In: Schopf W (ed) Life's origin: the beginnings of biological evolution. University of California, Berkeley, pp 7–45

Petrov AS, Bernier CR, Hsiao CL, Norris AM, Kovacs NA, Waterbury CC, Stepanov VG, Harvey SC, Fox GE, Wartell RM et al (2014) Evolution of the ribosome at atomic resolution. Proc Natl Acad Sci U S A 111:10251–10256

Pitsch S, Wendeborn S, Jaun B, Eschenmoser A (1993) Why pentose- and not hexose-nucleic acids? Part VII. Pyranosyl-RNA('p-RNA'). Helv Chim Acta 76:2161–2183

Pongs O, Ts'o POP (1971) Polymerization of unprotected 2′-deoxyribonucleoside 5′-phosphates at elevated temperature. J Am Chem Soc 93:5241–5250

Powner MW, Gerland B, Sutherland JD (2009) Synthesis of activated pyrimidine ribonucleotides in prebiotically plausible conditions. Nature 459:239–242

Powner MW, Sutherland JD, Szostak JW (2011) The origins of nucleotides. Synlett 14:1956–1964

Prins LJ, Reinhoudt DN, Timmerman P (2001) Noncovalent synthesis using hydrogen bonding. Angew Chem Int Ed 40:2382–2426

Rajamani S, Vlassov A, Benner S, Coombs A, Olasagasti F, Deamer D (2008) Lipid-assisted synthesis of RNA-like polymers from mononucleotides. Orig Life Evol Biosph 38:57–74

Rakotondradany F, Palmer A, Toader V, Chen BZ, Whitehead MA, Sleiman HF (2005) Hydrogen-bond self-assembly of DNA-analogues into hexameric rosettes. Chem Comm:5441–5443

Rhodes D, Lipps HJ (2015) G-quadruplexes and their regulatory roles in biology. Nucleic Acids Res 43:8627–8637

Ricardo A, Carrigan MA, Olcott AN, Benner SA (2004) Borate minerals stabilize ribose. Science 303:196–196

Rich A (1962) In: Kasha M, Pullmann B (eds) Horizons in biochemistry. Academic Press, New York, pp 103–126

Roberts C, Chaput JC, Switzer C (1997) Beyond guanine quartets: cation-induced formation of homogenous and chimeric DNA tetraplexes incorporating iso-guanine and guanine. Chem Biol 4:899–908

Robertson MP, Miller SL (1995) An efficient prebiotic synthesis of cytosine and uracil (vol 375, pg 772, 1995). Nature 377:257–257

Rode BM (1999) Peptide and the origin of life. Peptides 20:773–786

Sagi VN, Punna V, Hu F, Meher G, Krishnamurthy R (2012) Exploratory experiments on the chemistry of the "glyoxylate scenario": Formation of ketosugars from dihydroxyfumarate. J Am Chem Soc 134:3577–3589

Saladino R, Crestini C, Costanzo G, Negri R, DiMauro E (2001) A possible prebiotic synthesis of purine, adenine, cytosine, and 4(3H)-pyrimidinone from formamide: implications for the origin of life. Bioorgan Med Chem 9:1249–1253

Saladino R, Crestini C, Costanzo G, DiMauro E (2004) Advances in the prebiotic synthesis of nucleic acids bases: implications for the origin of life. Curr Org Chem 8:1425–1443

Saladino R, Šponer JE, Šponer J, Di Mauro E (2018) Rewarming the primordial soup: revisitations and rediscoveries in prebiotic chemistry. Chembiochem 19(1):22–25

Sanchez RA, Orgel LE (1970) Studies in prebiotic synthesis. V. Synthesis and photoanomerization of pyrimidine nucleosides. J Mol Biol 47:531–543

Sanchez RA, Ferris JP, Orgel LE (1968) Studies in prebiotic synthesis. 4. Conversion of 4-aminoimidazole-5-carbonitrile derivatives to purines. J Mol Biol 38:121–128

Schöning KU, Scholz P, Guntha S, Wu X, Krishnamurthy R, Eschenmoser A (2000) Chemical etiology of nucleic acid structure: the alpha-threofuranosyl-(3′->2′) oligonucleotide system. Science 290:1347–1351

Schwartz AW (1997) Speculation on the RNA precursor problem. J Theor Biol 187:523–527

Schwartz AW (2006) Phosphorus in prebiotic chemistry. Philos Trans R Soc Lond B:1743–1749

Seto CT, Whitesides GM (1990) Self-assembly based on the cyanuric acid melamine lattice. J Am Chem Soc 112:6409–6411

Shapiro R (1988) Prebiotic ribose synthesis: a critical analysis. Orig Life Evol Biosph 18:71–85

Shapiro R (1999) Prebiotic cytosine synthesis: a critical analysis and implications for the origin of life. Proc Natl Acad Sci U S A 96:4396–4401

Sheng Y, Bean HD, Mamajanova I, Hud NV, Leszczynski J (2009) A comprehensive investigation of the energetics of pyrimidine nucleoside formation in a model prebiotic reaction. J Am Chem Soc 131:16088–16095

Sheng J, Li L, Engelhart AE, Gan JH, Wang JW, Szostak JW (2014) Structural insights into the effects of 2′-5′ linkages on the RNA duplex. Proc Natl Acad Sci U S A 111:3050–3055

Shubin N (2008) Your inner fish: a journey into the 3.5-billion-year history of the human body. Pantheon, New York

Stoks PG, Schwartz AW (1979) Uracil in carbonaceous meteorites. Nature 282:709–710

Strecker A (1850) Ueber die künstliche Bildung der Milchsäure und einen neuen, dem Glycocoll homologen Körper. Justus Liebigs Ann Chem 75:27–45

Sun ZH, Chen DL, Lan T, McLaughlin LW (2002) Importance of minor groove functional groups for the stability of DNA duplexes. Biopolymers 65:211–217

Watson JD, Crick FHC (1953) Molecular structure of nucleic acids: a structure for deoxyribose nucleic acid. Nature 171:737–738

Weber AL (1992) Prebiotic sugar synthesis: hexose and hydroxy acid synthesis from glyceraldehyde catalyzed by iron(III) hydroxide oxide. J Mol Evol 35:1–6

Westheimer FH (1987) Why nature chose phosphates. Science 235:1173–1178

Wimberly BT, Brodersen DE, Clemons WM, Morgan-Warren RJ, Carter AP, Vonrhein C, Hartsch T, Ramakrishnan V (2000) Structure of the 30S ribosomal subunit. Nature 407:327–339

Woese C (1967) The genetic code. Harper & Row, New York, pp 179–195

Wulff G, Clarkson G (1994) On the synthesis of *C*-glycosyl compounds containing double-bonds without the use of protecting groups. Carbohydr Res 257:81–95

Yang Y-W, Zhang S, McCullum EO, Chaput JC (2007) Experimental evidence that GNA and TNA were not sequential polymers in the prebiotic evolution of RNA. J Mol Evol 65:289–295

Yu Y, Nakamura D, DeBoyace K, Neisius AW, McGown LB (2008) Tunable thermoassociation of binary guanosine gels. J Phys Chem B 112:1130–1134

Chapter 6
The Origin of the Ionized Linker: Geochemical Predestination for Phosphate?

Matthew A. Pasek

Abstract A major event in the origin of life on the earth must have been the formation of self-replicating polymers [e.g., Gilbert (Nature 319(6055):618, 1986)]. It is likely that any robust self-replicating polymer would have needed an ionized linker to slow hydrolysis and prevent diffusion. In modern life, the ionized linker is phosphate. In this chapter, I consider other alternatives to phosphate as linkers prior to the evolution of modern RNA/DNA. From a chemical and geological perspective phosphate is suggested to be the most likely molecule capable of performing the key activities of an ionized linker within a nucleic acid.

6.1 Polymers in the Origin of Life

A major event in the origin of life on the earth must have been the formation of self-replicating polymers (e.g., Gilbert 1986). A self-replicating polymer, if also bearing some catalytic ability, could have allowed selection from materials otherwise governed by organic chemistry and chemical evolution, leading to survival based on fitness and biological, Darwinian evolution. In modern life, the self-replicating polymers are a combination of DNA, RNA, and protein, which either store information as DNA, or provide the basis for chemical selection (protein) or act as the intermediary between the two (RNA) in a role that is likely ancient.

A major problem in our understanding of the origin of life is the formation of polymers such as these. Polymers form from repeating monomer units, such as proteins from amino acids, nucleic acids from nucleotides, and polysaccharides from sugars. These three polymers are the most ubiquitous across all modern forms of life.

The initial success of the Miller-Urey experiment (1959) provided a prebiotic synthesis of amino acids from high energy discharge in a reducing atmosphere, and

M. A. Pasek (✉)
School of Geosciences, University of South Florida, Tampa, FL, USA
e-mail: mpasek@usf.edu

© Springer International Publishing AG, part of Springer Nature 2018
C. Menor-Salván (ed.), *Prebiotic Chemistry and Chemical Evolution of Nucleic Acids*, Nucleic Acids and Molecular Biology 35,
https://doi.org/10.1007/978-3-319-93584-3_6

hence proteins were viewed as possibly being the earliest polymer present at the origin of life (Fox 1969). In part this perception—of the primacy of proteins—was a product of its time. Proteins and the structures of enzymes had been known for many years in biochemistry and were well-known to play key roles in life. In contrast, the Watson-Crick DNA model (1953), which showed how DNA could structurally hold genetic information, was contemporaneous with Miller's work, published only 20 days apart from the report of Miller's first experiment (1953).

In the 1980s it was shown that RNA could also serve as an enzyme (Cech and Bass 1986). This finding demonstrated that the key role of proteins—catalysis—could be accomplished by a material that looked almost identical to DNA. With this finding, in addition to an accumulation of evidence suggesting that RNA was a key part in numerous biochemical functions as coenzymes (White 1976), the RNA world gained traction as a leading hypothesis in the origin of life, carrying with it an idea that nucleic acids were the critical polymer for prebiotic chemistry.

Thus two types of polymers have competed for the role of original biopolymer: the nucleic acids and proteins. Modern life also includes polysaccharides; however, these are principally structural in function and have little catalytic or genetic potential; hence it is unlikely they were part of the prebiotic chemistry-biochemistry transition. Of proteins and nucleic acids, the formation of monomers of the proteins appears to be easier. The synthesis of amino acids requires reduced nitrogen compounds such as NH_4^+ and CN^-, reacting with an aldehyde via the Strecker synthesis to give an amino acid. All of these compounds are produced in CH_4/N_2 atmospheres with water when a high-energy driver—such as a spark discharge—is added to overcome kinetic barriers (Miller 1953).

In contrast, nucleotide formation occurs at the intersection of formose chemistry to make ribose and reduced nitrogen chemistry to make the nucleobases, and reports of their synthesis rely on compounds with varying carbon oxidation state (Powner et al. 2009). When formaldehyde and ammonium compounds are added together, amino acid synthesis is the major pathway, instead of nucleotide synthesis. In addition, these reactions must also occur in an environment where phosphorylation is also plausible, an environment typically viewed as being low in water activity.

The formation of polymers from amino acids or nucleotides also experiences numerous difficulties with side reactions. Synthesizing polymers of RNA and DNA in modern life is done by the reaction of a nucleotide triphosphate with the 3-′-hydroxyl group of the polymer, building nucleic acids. Before the enzymes that promote this reaction arose, other condensation routes were likely active, including cyclization of the phosphate, or attachment to other OH groups. Such routes may not always occur through 5′, 3′-linkages and can react by the 2′-linkage of ribose as well. The formation of polypeptides by amino acid condensation also struggles with the reactivity of amino acids to form cyclic dimers called diketopiperazines, demonstrating that problematic side reactions are not unique to nucleic acids.

Peptides do not perform replication at the sequence level, and what replication that does appear to occur with peptides is limited to changing other peptides (e.g., Prions, Prusiner 1991). Thus nucleic acids are likely the original polymer storing genetic information. However, due to the numerous difficulties in the abiotic

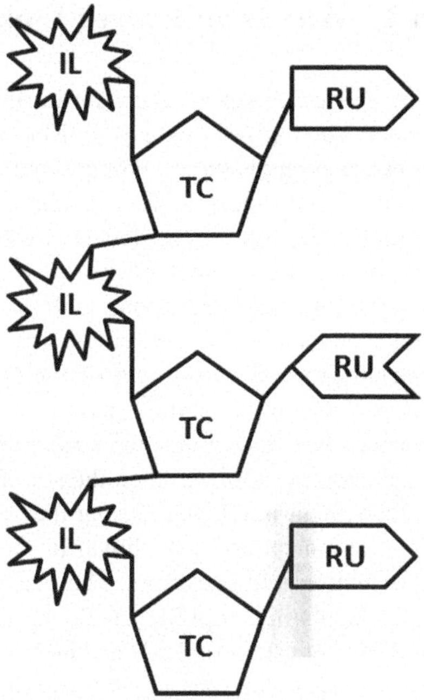

Fig. 6.1 Schematic representation of a nucleic acid. RU is recognition unit (e.g., nucleobase), TC is trifunctional connector (e.g., ribose), and IL is ionized linker (e.g., phosphate)

synthesis of RNA and DNA (e.g., Orgel 2004), it is plausible that an alternative to RNA and DNA—similar to these molecules structure and form—preceded them.

Hud et al. (2013) broke down genetic material into three major parts (Fig. 6.1). The "Recognition Units" (RU) are the nucleobases in RNA and DNA. These are the materials that carry genetic information via coding and can be stabilized by Watson-Crick base pairing. Alternatives to A, G, C, and U/T have been considered in the prebiotic literature for some time (Benner and Sismour 2005; Menor-Salván et al. 2009; Callahan et al. 2011; Cafferty et al. 2016a). Furthermore, biochemical nucleic acids are rather tolerant of alternative nucleobases, with variations common to RNA (e.g., pseudouridine, inosine, and methylguanosine), thus suggesting that other nucleobases could have been used prior to the locking in of codon translation.

The other two units of genetic material are the "Trifunctional Connector" (TC) and "Ionized Linker" (IL). Both of these form the backbone of nucleic acids and consist of ribose/deoxyribose and phosphate, respectively. The trifunctional connector serves to form a bridging polymer together with the ionized linker, and the trifunctional connector also binds the nucleobase through a glycosidic bond. In modern biology the trifunctional connector consists only of ribose and deoxyribose, with no other sugars filling this role.

The ionized linker is exclusively phosphate in modern biology, though some bacteria are known to have thiophosphate as a modified linker, formed after replication by unusual enzymes that modify the phosphate backbone to make thiophosphate (Wang et al. 2007). Such a modification is attributed as a defensive mechanism that protects DNA from nucleases (Chen et al. 2010) and remains the only known modification of the ionized linker in modern life that occurs naturally.

6.2 Why Is an Ionized Linker Necessary?

An ionized linker is likely an intrinsic part of nucleic acids that are dissolved in water. The ionized linker establishes the fundamental structure of a nucleic acid. The ionization repels the monomeric units within the backbone away from each other to promote base-pairing, and the charge prevents diffusion through lipid membranes, protects the nucleic acid from hydrolysis, and, at the monomer level, promotes dissolution of the nucleotides.

The ionized linker, along with the trifunctional connector, forms the backbone of nucleic acids. Since the ionized linker carries a charge, and since like charges repel, the backbone of two strands of nucleic acids also repel. When a strand of nucleic acid meets is complementary base-pair sequence, then hydrogen bonding from the nucleobases determines the structure of the nucleic acid, making helices. Were the backbone not ionized, then the structure of a nucleic acid would be driven by other noncovalent bonds, increasing the “messiness” of the nucleic acid.

The ionization of the linker has a further major benefit: it prevents hydrolysis of the nucleic acid backbone. The negative charge confers stability toward hydrolysis, specifically nucleophilic attack, an important feature of a nucleic acid that may have millions of ester bonds (Westheimer 1987). As long as the rate of hydrolysis of ester bonds is less than the number of bonds divided by the life span of an organism, the nucleic acid can be considered to be sufficiently stable as genetic material.

Another important property of having a charged nucleic acid is the fact that charge decreases the diffusivity of a nucleic acid across a membrane (Westheimer 1987). Since nucleic acids get their name from being found within a cell nucleus, the retention of genetic information is critical to biological evolution. Were genetic information capable of diffusing through a cell boundary, evolution would not have the ability to select upon unique sequences manifested in individuals. In aqueous solution, a charged polymer cannot diffuse through an organic membrane composed of lipids; hence, individual identity is retained along with the chance of inheritance.

At the monomer level, ionization promotes dissolution in water. For instance, the solubility at room temperature of adenosine is about 0.07 g/L in water (Merck 1996), whereas adenosine monophosphate is soluble to about 8 g/L (Wang et al. 2009). This is due to the polar character of water, and the enhanced solubility better allows the formation of nucleic acids in solution.

As highlighted by these characteristics, an ionized linker such as phosphate in nucleic acids may be presumed to be necessary if life develops in a polar solvent, such as water. By enhancing the stability of the nucleic acid, increasing the solubility of its monomeric units, stabilizing the structure of the nucleic acid, and preventing its diffusion out of a membrane, the ionized linker is a critical part of the structure of nucleic acids. Given the presumed ubiquity of water as a solvent throughout the universe, it is likely that ionization is universally an important facet of biological genetic material.

6.3 Phosphate as the Ionized Linker

Why then did life specifically chose phosphorus as phosphate as its ionized linker? Several works have addressed the reasons for phosphate, including Westheimer (1987), Kamerlin et al. (2013) and Wohlgemuth et al. (2017). The findings of these papers will be reviewed at the end of this chapter. Phosphate is capable of doing things that, at first glance, might be replaceable with other molecules, for instance arsenate.

Most alternatives to phosphates seek to replace phosphate with either something similar to phosphate or, if suitably chemically different, something much more abundant than phosphate. Why then is phosphate suspected as presenting difficulties to the prebiotic chemistry of nucleic acid synthesis? Why might phosphate be questioned as the original ionized linker?

The difficulties faced by phosphate are threefold: the first is that phosphate is rare. Compared to the other biogenic elements, phosphate is most rare of the elements used in nucleic acids with respect to cosmic abundance and ocean chemistry (Pasek and Lauretta 2005). The former is due to the nucleosynthetic processes that resulted in the elemental distribution of the solar system (and elsewhere). Phosphorus, being an odd atomic number element, is difficult to synthesize by helium atom fusion, which forms even atomic number elements preferentially. The second issue, of solubility, is due to the high stability of common phosphate minerals, such as the mineral apatite, $Ca_5(PO_4)_3(OH,F,Cl)$. These minerals are poorly soluble in water, and hence phosphate is buffered by their low dissolution. In the modern ocean, phosphate has concentrations between 10^{-8} and 10^{-6} molar. Such low concentrations are inhibitory for phosphorylation reactions (e.g., Pasek 2017).

Thirdly, phosphate is not reactive. The formation of both organic monoesters and diesters of phosphate requires the addition of energy and/or the loss of water, which can prevent the phosphorylation of organics to form the key constituents of nucleic acids. Multiple studies, including that of Burcar et al. (2016), have attempted to find solutions to this issue, and though significant progress has been made, especially in the spontaneous formation of monoesters, formation of diesters under plausible prebiotic conditions is still considered difficult. Given that orthophosphate diesters are the backbone of nucleic acids, this highlights the issues of phosphate as the original ionized linker.

The backbone of RNA and DNA consists of six repeating bonds: -O-P-O-C-C-C-. If a nucleic acid similar to RNA or DNA were present on the early earth, then it is plausible, if not likely, that the polymeric chain would have also consist of between five and seven repeating bonds, due to several molecular constraints (Cafferty et al. 2016b). In a search for an alternative to phosphate as an ionized linker, these repeating polymeric units should form the basis of any plausible polymer.

6.4 Alternatives to Phosphate as Ionized Linkers

There have been several proposed replacements of phosphate with other substances, in an attempt to circumvent (1) the low solubility of phosphates and (2) the poor reactivity of phosphates toward organics to make diester nucleic acids. Broadly, these replacements may be considered to fall into three main groupings: inorganic ion replacements, organic replacements, and phosphorus oxyacid replacements. The first two were examined in part by Westheimer (1987), and the last one is new to this work.

6.4.1 Inorganic Replacements

6.4.1.1 Nitrogen

Replacing the phosphate with other elements within its column can be viewed through the lens of science fiction, with silicon-based lifeforms being encounter by future space explorers, and is based in the argument of element periodicity: elements within the same column have similar chemical behaviors. Phosphorus has two contenders: immediately above it is nitrogen and immediately below is arsenic.

The exchange of nitrogen with phosphorus results in much more significant change than the exchange of phosphorus with arsenic, both in terms of biological functionality and in terms of potential toxicity (e.g., Hughes 2002). Nitrogen as nitrate is hard-pressed to replace phosphate. Nitrate does not form monoesters readily, and it does not form diesters while retaining a charge. This is due to nitrogen lacking accessible d orbitals. Nitrogen is unable to have more than four bonds under typical conditions, whereas phosphorus as a default has five bonds. In this respect, fundamental chemistry prevents nitrate from substituting for phosphate: it is unable to form diesters while retaining a charge.

An alternative route may be to consider nitrogen as an amine instead of nitrate. In this case, the ionized linker would be a diamine. Diamine compounds could feasibly form an ionized backbone at neutral pH, albeit with opposite charge from present day. Westheimer (1987) notes that positive charge may be detrimental as positive charges increase the rate of nucleophilic attack, and hence decrease the stability of the nucleic acid. Additionally, other backbone modifications would be necessary, as condensation with ammonium would result in a four bond -C-C-C-N- repeating unit, likely too short for a typical nucleic acid. The trifunctional connector would either have to connect differently or would have to be significantly modified to allow NH^+ to be an adequate ionized linker. An example of a successful nucleic acid based on nitrogen is peptide nucleic acids, which do away completely with both sugar and phosphate and have been considered as precursors to RNA and DNA (Nelson et al. 2000). Such nucleic acids are not charged and hence may not gain the many benefits of charge as outlined in the prior sections.

6.4.1.2 Arsenic

The possibility of replacing phosphate with arsenate in nucleic acids received its greatest support from Wolfe-Simon et al. (2011). This study purported to show the incorporation of arsenic into the DNA backbone of a microbe from Mono Lake, California, USA. Response to the paper was significant and rapid, mostly arguing that the findings were interpreted incorrectly (e.g., Benner 2011; Schoepp-Cothenet et al. 2011), and, eventually, most researchers concluded that the findings were incorrect (Erb et al. 2012; Reaves et al. 2012; Elias et al. 2012).

On paper, the idea seems reasonable enough. The pK_as of arsenic acid (H_3AsO_4) are very close to the pK_as of phosphoric acid (H_3PO_4), and both ions are similar in size. Arsenate does substitute for phosphate in some enzymes (Mukhopadhyay et al. 2002); such interchangeability is the root of some of its toxicity. If arsenate were to replace phosphate in the backbone of a nucleic acid, then it should be presumed that, absent other factors, the nucleic acid would be similar to RNA/DNA.

These other factors do indeed play a significant role, though. Arsenate esters (Fekry et al. 2011) are significantly less stable than phosphate esters (e.g., Williams and Wyman 2001). Fekry et al. (2011) estimate the hydrolytic half-life of DNA as about 30×10^6 years, compared to 0.06 seconds for the corresponding arsenate-replaced DNA. Arsenic oxyacids also readily exchange oxygen with water, especially in contrast to phosphorus oxyacids (Fekry et al. 2011 and references therein), suggesting low stability of arsenate esters.

In addition to the low kinetic stability of arsenate esters, arsenic is also redox sensitive over the range of terrestrial redox-pH conditions that are commonly found on the surface of the earth (see Sect. 6.5). Arsenate reduces to arsenite, which has significantly different properties compared to arsenate.

6.4.1.3 Borate

Borate as an ionized linker could be justified due to its strong propensity to react with sugars. This propensity has been used in prebiotic syntheses of the current trifunctional connector, ribose (Ricardo et al. 2004; Benner et al. 2012). In these reactions, borate spontaneously links to the 2′ and 3′ hydroxyl groups on ribose, forming a cyclic borate ester. Borates are also known to spontaneously link across natural carbohydrates, forming borate ester polysaccharides (e.g., O'Neill et al. 1996). Borate also acts as a phosphate mimic in some enzymes, replacing phosphate in dihydroxyacetone-phosphate aldolases (Sugiyama et al. 2006).

Boronic acid nucleotides have been used to replace nucleotides in nucleic acid with some success (Martin et al. 2011, 2013). The synthesis of these molecules, however, invokes chemistry unlikely to be present on the early earth, due to the use of solvents other than water and boron reagents other than borate (e.g., diisopinocampheylborane; see Martin et al. 2009).

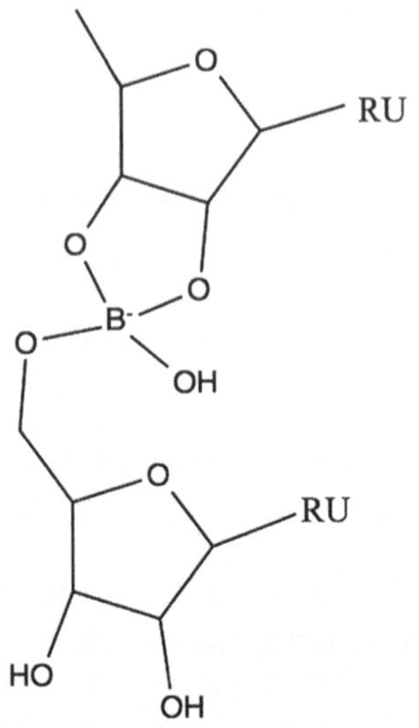

Fig. 6.2 Borate-based nucleic acid, formed of cross-linked ribose with borate. RU is the recognition unit

Borate spontaneously forms cyclic esters more easily than phosphate and other ions due to its small size. However, forming nucleic acids from ribose and borate can be envisioned (Fig. 6.2), though the conformation of such a nucleic acid may be rather strained and has a longer backbone than current nucleic acids. Additionally, given the extent of boron research in prebiotic chemistry, the fact that such structures have never been reported indicates these polymers do not form spontaneously.

6.4.1.4 Aluminate

Beneath boron on the periodic table is aluminum, with aluminate, $Al(OH)_4^-$, being a natural analog of borate $H_2BO_3^-$, might be expected to behave similarly to boron, and might overcome some difficulties with the low abundance of boron in crustal rocks. However, aluminate does not appear to form esters to any extent, likely due to their low stability toward hydrolysis.

6.4.1.5 Silicon

In contrast to aluminum, silicon as silicic acid (H_4SiO_4) readily forms esters, many of which form staples of organic chemistry labs (e.g., tetramethylsilane). Silicon esters of organics are described as quick to hydrolyze (Westheimer 1987) and do not appear to form ionized linkages at neutral pH. The first deprotonation of H_4SiO_4 occurs under strongly alkaline conditions; hence, silicic acid is an ionized linker only at high pH.

6.4.1.6 Sulfur

The sulfate ion, in contrast to arsenate and vanadate, is stable over most of the conditions that may have been present on the developing earth. Additionally, sulfur is the only element that is known to substitute for phosphorus in phospholipids, which are the material that comprise cell membranes. Sulfolipids bear a C-SO_3^- moiety and are found principally in phosphorus-limited ecosystems (Van Mooy et al. 2006). Additionally, the reactions associated with sulfur redox in biological systems (from sulfate to sulfide) proceed through linking a sulfate molecule to ATP, indicating a relationship between the building blocks of nucleic acids and sulfate. However, the ATP is acting in this chemistry in its metabolic role, as opposed to its nucleic acid building block role. Phosphatase enzymes are also known to hydrolyze organosulfates (and vice versa with sulfatases), hinting that there is some biochemical similarity between the two molecules (Pabis et al. 2016).

However, sulfate is unlikely to act as an ionized linker as the diester of sulfate bears no charge. Furthermore, biologic organosulfur compounds are typically thiols, as opposed to oxyanions, and hence there's little role for sulfur in nucleic acids as an ionized linker.

6.4.1.7 Vanadate

Oft-forgotten but of significant interest as a phosphate replacement is the ion vanadate (VO_4^{3-}). Vanadium is known to replace phosphate in biochemical reactions (Lopez et al. 1976; Bornscheuer and Kazlauskas 2004), as the vanadate ion behaves similarly to phosphate and has similar pK_as and ionic sizes, akin to arsenate. The esters of vanadate appear to be generally stable over a short term and occur in equilibrium with alcohols (Tracey and Gresser 1988; Tracey et al. 1988).

Potential problems with vanadate are twofold and include the redox instability of vanadate over a large redox region and the low abundance of vanadium. In this respect, vanadate is akin to arsenate, suffering from many of the same problems with redox, abundance, and possibly long-term ester instability.

6.4.2 *Organic Replacements*

6.4.2.1 Citrate

Westheimer (1987) first suggested citrate as an organic replacement for phosphate, though it was dismissed in that work, as the ionization K_As were too close to have a major positive effect on preventing hydrolysis. Furthermore, the large size of the citrate molecule minimizes the benefits of having a negative charge, as the negative charge is further away from the linker and doesn't help prevent hydrolysis as well.

6.4.2.2 Glyoxylate

Bean et al. (2006) explored the possibility of replacing phosphate with glyoxylate ($O_2C\text{-}CH(OH)_2^-$). This study found that glyoxylate spontaneously links nucleosides, while retaining the negative charge from the $-CO_2^-$ group. The proximity of the $-CO_2^-$ group to the linked center also overcomes the issue of distance between ester bonds and ionized charged that citrate face and means the $-CO_2^-$ group helps prevent hydrolysis, in contrast to citrate where the ionized group is too far from the linked chain.

The spontaneous formation of nucleosides linked by glyoxylate occurs because the acetal bond formation is exothermic. The reaction occurs most easily when the activity of water is low, and products include some dinucleotides, and appears to require metal cations such as Mg^{2+}. Not all nucleosides reacted with glyoxylate; this was attributed to differences in nucleoside solubility.

Unlike most of the other ionized linkers discussed, glyoxylate is prochiral. Prochiral linkers result in a new chiral center on reaction, and this effect was addressed by Bean et al. (2006), who suggested that, as long as the chirality was random, there would be little net effect on nucleic acid structure.

A prebiotic route to glyoxylate was identified by Mohammed et al. (2017), consisting of a transamination of glycine and formaldehyde at 50–70 °C, and this route addresses the question of whether glyoxylate was present on the early earth. Some amount of glyoxylate would presumably have been present, provided there were both glycine and formaldehyde, which are presumed to be relatively abundant prebiotic feedstock molecules for the early earth.

6.4.3 Phosphorus Oxyanion Replacements

Phosphate, while amenable to being the modern ionized linker, may not have been the only phosphorus oxyanion present on the early earth. Other phosphorus oxyanions may have also been present (Pasek and Lauretta 2005), including hypophosphite ($H_2PO_2^-$), phosphite (HPO_3^{2-}), pyrophosphate ($P_2O_7^{4-}$), and hypophosphate ($P_2O_6^{4-}$). Substitution of these ions for phosphate may have overcome some of the difficulties typically associated with prebiotic phosphorylation, including the low solubility and reactivity of phosphate.

6.4.3.1 Thiophosphate

The only known actual substitute for phosphate in DNA is thiophosphate (SPO_3^{3-}). Thiophosphate is formed by enzymes that act specifically on the DNA backbone of some organisms, selectively removing one oxygen atom and replacing it with a sulfur atom (Wang et al. 2007). DNA appears to suffer no significant conformational changes with this replacement, which appears to be a defensive mechanism against environmental oxidants such as H_2O_2, preventing DNA break damage (Xie et al. 2012).

Thiophosphate, as such, should be a relatively successful replacement for phosphate as an ionized linker. The issue with thiophosphate is its abundance. There are no known thiophosphate minerals, and outside of some biochemical sources, thiophosphate is not generated naturally by any known natural reactions.

6.4.3.2 Phosphite

Gulick (1955) first proposed that alternatives to phosphate—phosphite and hypophosphite—could have provided reactive P for the first biological organisms. Building on this hypothesis, De Graaf and Schwartz (2005) demonstrated the synthesis of nucleoside phosphonates (bearing a HPO_3^{2-} ion instead of a HPO_4^{2-} ion). However, Peyser and Ferris (2001) had earlier demonstrated that nucleic acids formed of phosphorous acid (H_3PO_3) were susceptible to rapid hydrolysis and breakdown. Additionally, once formed the nucleic acids are not charged as phosphite lacks a third hydroxide group compared to phosphate.

Phosphite has some clear benefits over phosphate as it is more soluble and it is more reactive than phosphate. The demonstration that phosphite is a natural product of the corrosion of meteoritic mineral schreibersite (Pasek et al. 2007) and can be found in natural samples (Pasek and Block 2009; Pasek et al. 2013) indicates that the ion is geochemically available. If phosphite nucleotides could be formed and then subsequently quickly oxidized as suggested by De Graaf and Schwartz (2005), then phosphite nucleosides would work well as a phosphite replacement.

6.4.3.3 Phosphonates

In contrast to the phosphites, which utilize inorganic phosphite as the phosphorus oxyanion, an alternate route to synthesizing nucleic acids is to incorporate C-P bonds into the ring structure of molecules of interest. Two such possibilities are discussed in the literature. De Graaf et al. (1998) investigate sugar phosphonates formed by the reaction of phosphonoacetaldehyde with formaldehyde via the formose reaction. The products of this reaction are six-membered rings with two phosphite groups. A nucleic acid formed from this material was envisioned by De Graaf et al. (1998) but necessarily has diphosphonate linkages (Fig. 6.3). Such a system ties the ionized linker and trifunctional connector together. Diphosphonate linkages likely are as stable to hydrolysis as pyrophosphate, which has hydrolytic life spans of 1000 years or so (Pasek et al. 2008).

Another route to forming nucleic acids using phosphonates comes from Bryant et al. (2010). By reacting hypophosphite ($H_2PO_2^-$) with pyruvate via a phospho-aldol addition, a lactone phosphinate is produced that incorporates phosphorus into the ring structure. Although it is unlikely that such a structure could form a nucleic

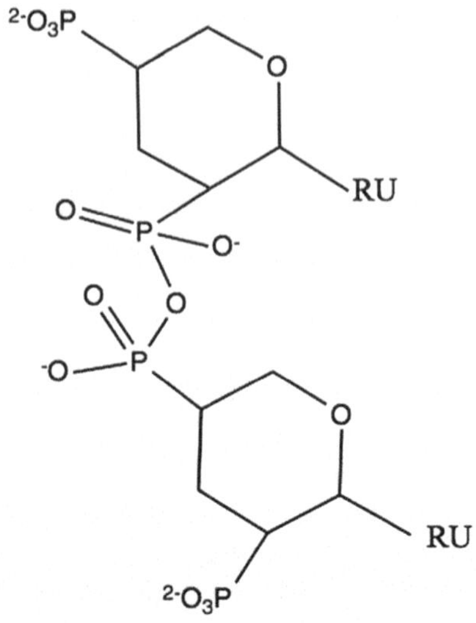

Fig. 6.3 De Graaf et al. (1998) proposed phosphonate ribose-ring structure

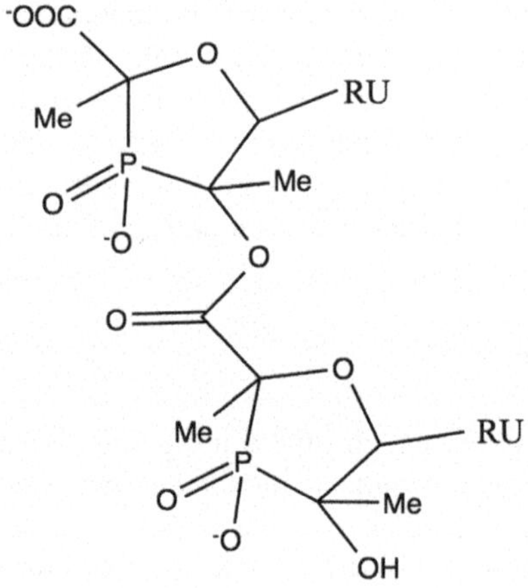

Fig. 6.4 A lactone-phosphinate compound was reported in Bryant et al. (2010), from reaction of hypophosphite with pyruvate. If the lactone is reduced at the carbonyl, and the reduction provides as a site to attach a recognition unit, then the nucleic acid below could be envisioned. As in the De Graaf et al. (1998) phosphonate bond, such a structure merges the ionized linker with the trifunctional connector

acid, as linking such a structure to a nucleobase would not proceed through an obvious route, if the carbonyl formed by this reaction is reduced, it may provide a good site for a glycosidic bond (Fig. 6.4). Although the rate of hydrolysis of esters is faster than that of phosphoesters, the proximity of the ionized P to the ester may help stabilize this structure. However, as of yet, this structure has not been synthesized and serves mostly to provide another example of a potential phosphonate chemistry that could combine the ionized linker with the trifunctional connector, akin to the De Graaf et al. (1998) model.

In both cases, these phosphonates are reaction products of inorganic reduced phosphorus compounds (phosphite and hypophosphite) with organic reagents. A high abundance on the early earth of these compounds is not guaranteed and would be contingent on concentrating the reagents in a suitable environment. Esters formed from these compounds could be linked by a P-O-P bond, which can be remarkably stable.

6.4.3.4 Hydroxymethylphosphonate

A specific phosphonate, hydroxymethylphosphonate or HMP, can be envisioned as an ionized linker that expands the length of the linker. Such a system could work well if the trifunctional connector were smaller, for instance, a replacement of ribose with threose as the backbone sugar. HMP also overcomes some of the issues with phosphites in that the linker remains ionized when forming nucleic acids, and it does not have the problems associated with the phosphonates as the linker is not part of the trifunctional connector. However, the linkage through the hydroxymethyl group is not an ester and hence may not be as stable as the phosphoester bond.

HMP is formed relatively readily by reaction of reduced P compounds, such as phosphite, with formaldehyde (Pasek et al. 2007). It is one of the simplest organophosphonates to form from inorganic phosphorus compounds and may have been present in high concentrations in some systems. The stability of these compounds is not well known, though esters of alpha-hydroxyphosphonic acids tend to be pretty stable (Wuggenig and Hammerschmidt 1998).

6.4.3.5 Hypophosphate

The hypophosphate ion ($H_2P_2O_6^{2-}$) is an unusual ion formed by reaction of metal phosphides with water and is typically the third of fourth most common ion in these experimental solutions (Pasek et al. 2007). It has a P-P bond and is one of the few phosphorus compounds with this linkage that occurs (presumably) in nature on the surface of meteorites. Its presence is indicative of phosphite radicals (PO_3^{2-}) as reactive intermediates (Pasek et al. 2015). Hypophosphate is stable in solution and in solid form for periods lasting years or more.

Much akin to HMP reactions, hypophosphate could form a linkage between a smaller trifunctional connector such as threose (Fig. 6.5). The backbone of such a nucleic acid would be quite different compared to RNA/DNA and would likely be

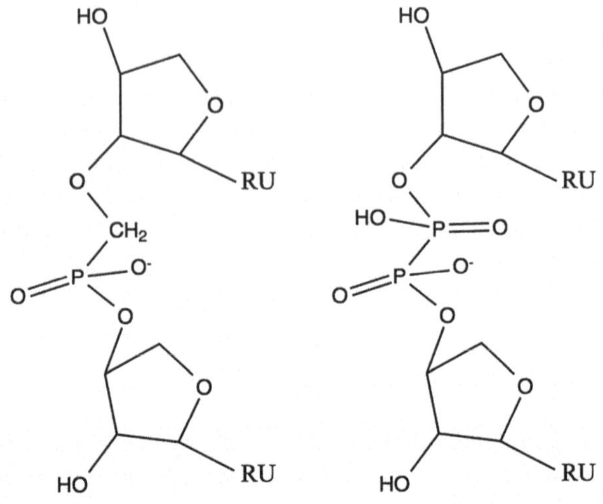

Fig. 6.5 Potential alternative nucleic acids employing (**a**) hydroxymethylphosphonate and (**b**) hypophosphate ionized linkers. The trifunctional connector may change to threose to have the same bond number per repeating unit as RNA and DNA: six in RNA-CCCOPO vs (**a**) CCOCPO or (**b**) CCOPPO. RU is the recognition unit

even more negatively charged, though the final pK_a of hypophosphate is rather large (12–14).

The stability of hypophosphate esters is unknown, though in this case it is not a question of their low stability but likely of the relatively low research that has been done on this ion. Hypophosphate does not form esters easily, in contrast to phosphite, and hence it is possible that the esters of this compound may not be stable. In biochemical studies, hypophosphate can interfere significantly with phosphate chemistry due to the difficulty of breaking the P-P bond (Pawlowska et al. 2016).

6.5 Why Then Phosphate?

Phosphate may be the preferred ionized linker for several key reasons. Phosphate is more abundant than many of these alternatives. Hence, if a replacement is possible, it may not be likely if the replacement is rare. Furthermore, several of these potential replacements are not stable over the varied terrestrial conditions potentially present on the early earth. They may become oxidized or reduced depending on the environment. Additionally, empirical evidence suggests that replacing the phosphate with different materials just doesn't work so well—the esters may be or have been shown to be quite unstable. Finally, some of the ions are prochiral: they induce the nucleic acid to have another chiral center, which adds to the molecular complexity of their synthesis. Whether or not this is an issue is unclear for the development of life on the earth.

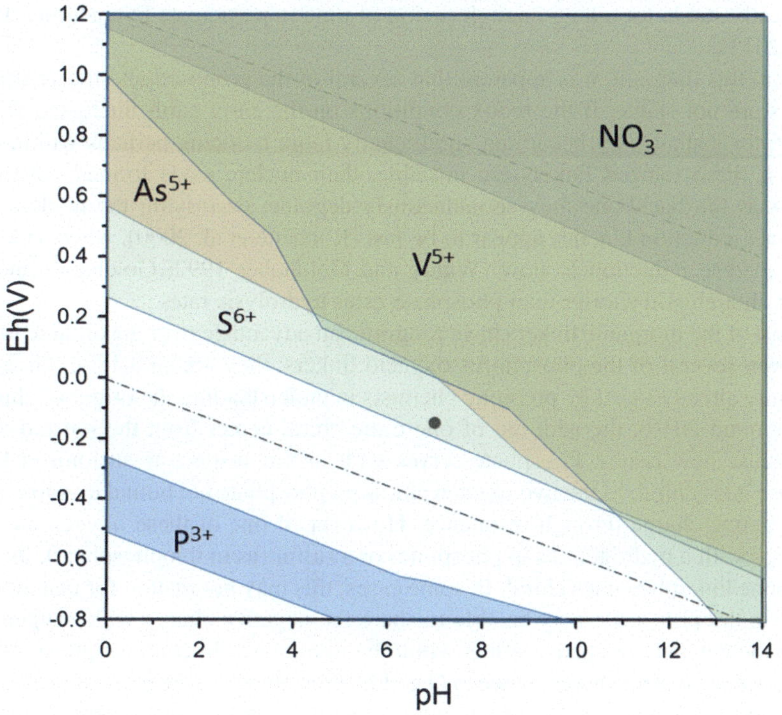

Fig. 6.6 Redox and pH conditions at 298 K necessary for the stability of some of the proposed ionized linkers. The black dot represents the most oxidizing conditions for ammonium at pH 7. For sulfate (S^{6+}), arsenate (As^{5+}), vanadate (V^{5+}), and nitrate, the stability zone technically extends to the most oxidizing conditions as well; hence, these are stable in air and overlap. The two dashed lines represent the typical oxidizing conditions of surface water (top, in contact with O_2 in air) and the presumed reducing conditions where water breaks down to H_2. Stability fields come from Takeno (2005)

The stability of phosphate over a large redox (Eh) and pH range is a key feature of why phosphate may have been employed as a nucleic acid ionized linker. The stability fields for nitrate, vanadate, arsenate, phosphite, and sulfate are shown in Fig. 6.6. Borate and silicic acid are stable over this entire range, with the exception of the possibility of forming BH_4^- under very low redox conditions (unlikely on the early earth). Included on this diagram is a likely boundary point for prebiotic chemistry—the point where ammonium is dominant (vs. N_2) at pH 7. Given the role of reduced nitrogen species in modern life, it is likely that NH_4^+ was an important species that should have been stable on the surface of the earth. It is apparent that nitrate is too oxidizing to likely have been present on the early earth and that arsenate and vanadate may be problematic as they also require more oxidizing conditions relative to the reducing conditions necessary for NH_4^+. Phosphite requires extremely reducing conditions, but, if formed, appears to be kinetically stable for a long enough period of time to participate in reactions (Pasek et al. 2013).

From this diagram, it is apparent that several of the proposed alternative ionized linkers are not stable, if the redox conditions on the early earth fluctuated significantly, for instance, by becoming significantly more reducing perhaps within early cells. If these ionized linkers are unstable, then nucleic acids formed with these linkers as the backbone may spontaneously degrade, destroying the nucleic acid. Arsenate reduction kinetics appear to be fast (Rochette et al. 2000), whereas sulfate and vanadate reduction is slow (Wanty and Goldhaber 1992; Goldhaber and Orr 1995), though still shorter than phosphate ester hydrolysis rates.

Most of the inorganic linkers have a significant advantage over the organic linkers and even several of the phosphorus oxyacid linkers: they are all achiral. Given the difficulty already faced by prebiotic chemists in understanding the origin of chirality (Blackmond 2010), the addition of one more chiral center from the ionized linker may cause new issues. Phosphate serves well in this respect as an ionized linker because it is achiral, as the two oxygen atoms on phosphate not bound to ribose share the negative charge through resonance. However, if one of these oxygen atoms is replaced with a hydrogen (as in phosphite) or a sulfur (as in thiophosphate), then the backbone linker becomes chiral. In some cases, this may not matter, for instance, the sulfur in thiophosphate may be able to share the negative charge with oxygen, and hence the net electronegative effect would be minimized. In contrast, phosphite has no chance of sharing charge between O and H; thus, there will be electronic effects to replacing an O with an H, which would result in structural changes. Structural changes may not be important for single strands of nucleic acid, but if duplex were to form, they might cause significant issues.

Chirality may be especially problematic for organic replacements as, like phosphite, they are not capable of resonance stabilization of charge between the two unlinked units. As such, these linkers may induce folding of the nucleic acid, especially if they all assume the same chirality, perhaps if formed enzymatically. Bean et al. (2006) argued that, in the absence of a chiral selecting mechanism, random assembly will not significantly affect duplex formation and neither would alternating assembly (S, R, S, R, etc.).

A summary of the characteristics of various ionized linkers is provided in Table 6.1. The rates of hydrolysis of esters of these ionized linkers demonstrate that phosphate is unique in its ability to form stable diesters. Most other esters of ions are unstable, with diesters or polymers (when data is available) lasting only for seconds (arsenic) up to years. These fast hydrolysis rates are due to large differences in the ionized characteristics of various esters, and those without charge (e.g., boric and silicic esters) do not bear charge at neutral pH. The ionization helps prevent hydrolytic attack of the ester.

The relative abundances of the proposed replacement ions are shown in Fig. 6.7. These calculations omit the organic replacements, whose abundance would depend on prebiotic environment, and omit phosphorus oxyanion alternatives, which should be a subset of the total P abundance. It is clear from these graphs that phosphorus is much more abundant than some of the linkers, principally arsenate, vanadate, and borate, but is less abundant than Al and Si, the major rock-forming elements on the

Table 6.1 pK_as, bond lengths (Å) ester half-lives, and prochirality of the proposed alternative linkers (and phosphate)

Linker	Prochiral	Atom-O bond length	pK_a 1	pK_a 2	pK_a 3	Ester half-lives	Reference
H_3AsO_4	N	1.7	2.26	6.76	11.29	Seconds	Fekry et al. (2011)
H_3VO_4	N	1.7	3.8	8.3	13.1	Seconds	Borden et al. (2006)
H_3BO_3	N	1.4	9.24	>14		Minutes	Steinberg and Hunter (1957) (triesters)
$H_4Al(OH)_4$	N	1.7	4.8	5.1	6	Not measured	
H_3PO_4	N	1.52	2.15	7.2	12.35	Millennia	Westheimer (1987)
H_4SiO_4	N	1.6	9.9	11.8	12	Hours	Guthrie (1978), Ossenkamp et al. (2001)
H_2SO_4	N	1.49	<0	2.15		Hours to days	Guthrie (1978)
HNO_3	N	1.25	<0			Years	Guthrie (1978)
Phosphite	Y	1.5	1.3	6.7		Hours to weeks	Peyser and Ferris (2001), Mitchell et al. (1998)
Phosphonate	Y	1.5	2	7		Hours to years	Niemi et al. (1999)
HMP	N	1.5	0	8		Unknown, likely years	
$H_4P_2O_6$	N	1.5	2.1	6.8	9.5	Unknown, likely years	
Glyoxylate	Y	1.4	3.3			Likely years	Westheimer (1987)
Citrate	Y	1.4	2.92	4.28	5.21	Years	Westheimer (1987)
H_3PSO_3	Y	1.5	1.2	5.6	11.5	Millennia	

earth's crust. Although such calculations could be repeated for ocean abundances, the abundance of phosphate in the modern ocean is strongly affected by biological processes, and the composition of the ocean prior to the oxygenation of the atmosphere is unknown.

Given that neither aluminate nor silicate is effective at forming esters due to low stability, this leads us to a conclusion about phosphate: that it is the most abundant ionized linker capable of doing its job. Alternatives to phosphate are either too rare, add too much complexity, or are too reactive to fill the job as well as phosphate can. That is not to say that some of original linkers may have been playing a role in early nucleic acid synthesis, but once nucleic acids started to become selected for stability and ease of formation (including linker rarity and lack of chirality), then phosphate is likely to have taken over. Whether this happened before the onset of Darwinian evolution—and hence the origin of life—is unknown. If it happened during the chemical selection/evolution stage, then phosphate may have been the first ionized linker in life.

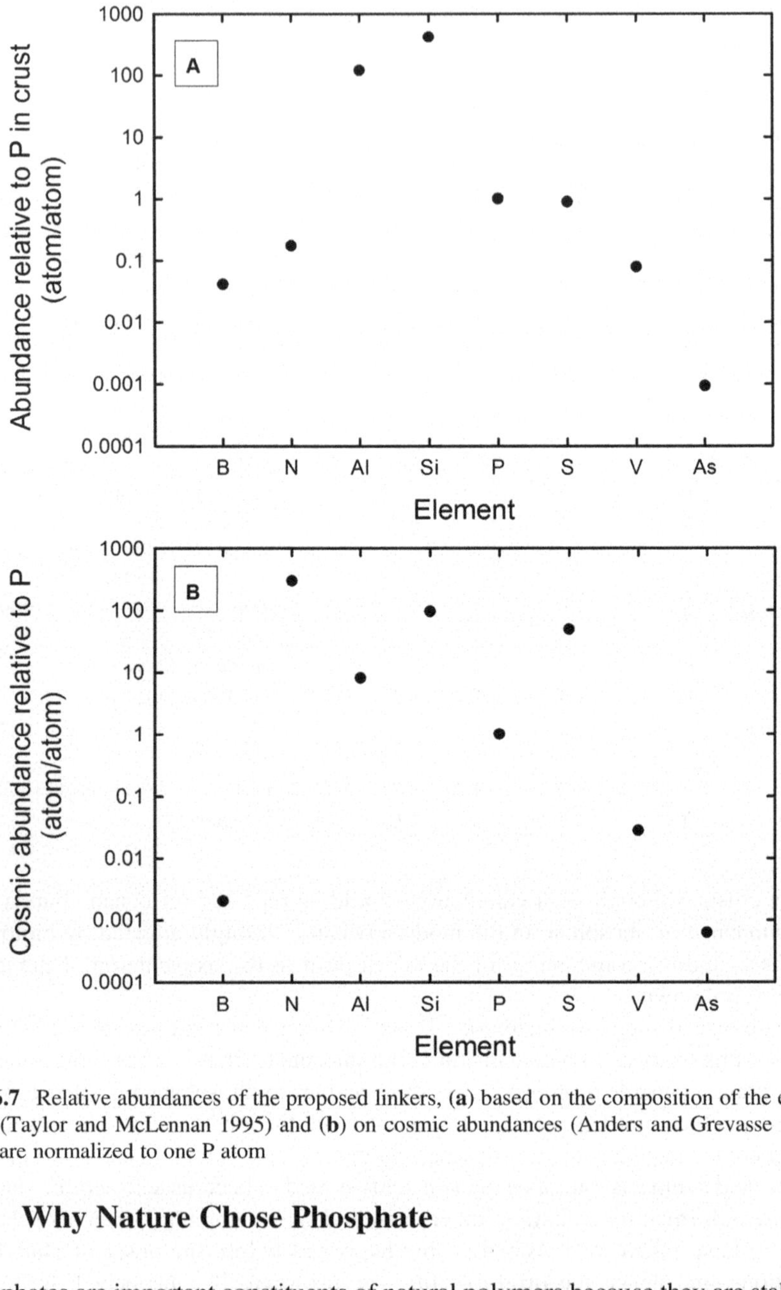

Fig. 6.7 Relative abundances of the proposed linkers, (**a**) based on the composition of the earth's crust (Taylor and McLennan 1995) and (**b**) on cosmic abundances (Anders and Grevasse 1989). Both are normalized to one P atom

6.6 Why Nature Chose Phosphate

Phosphates are important constituents of natural polymers because they are stable to hydrolysis, they are achiral, and they are ionized. Phosphate is the most common potential ionized linker that might occur in water; hence, its choice as the ionized linker appears to have been geochemically predestined.

This is not to say there couldn't be alternatives in early polymers. Glyoxylate, borate, and possibly other linkers have several significant advantages over phosphate in that they spontaneously link to sugars and may form dimers with some ease. In contrast, phosphate does not form polymers spontaneously. It is possible that one of these linkers preceded phosphate, and then natural or chemical selection pushed polymers linked by ionized species such as phosphate to dominate.

The principal difficulty of phosphate as an ionized linker is dimer formation. One of the successes of prebiotic chemistry of the past 15 years has been the identification of a number of routes to forming organophosphates monomers from simpler starting reagents. Although nucleic acids have yet to be synthesized with any of these routes, nucleotides have been synthesized both through use of energetic phosphorus-bearing minerals (Gull et al. 2015) and by dissolution in low water activity solvents (Burcar et al. 2016). These results indicate that formation of phosphate monoesters is not problematic. Formation of the diesters remains a challenge and, if such a reaction was prebiotic, likely proceeded through some sort of activation of the monoester, for instance, by linking a second or third phosphate to the molecule, such as by nucleotide triphosphates.

Indeed, the prevalence of phosphate in metabolic molecules belies a potential reason for its dominance: it is part of the main energy storage molecule in life, ATP. The triphosphate linkage in ATP carries chemical energy for metabolic reactions (as do other nucleotide triphosphates, such as GTP, though with much less frequency). The triphosphate group of ATP and other nucleotide triphosphates is the business end of the molecule. Energy is stored within the phosphoanhydride bond. This has led some to propose that nucleotide triphosphates, were they prebiotic, would have produced nucleic acids by assembling these monomers (e.g., Yamagata 1999). The difficulty with this assumption is that nucleotide triphosphates have highly complex organic molecules attached to the relatively simple triphosphate. These include molecules with specific stereochemistry (ribose) and specificity of the nucleobases. In contrast, simplifying the triphosphate to an energetic polyphosphate is consistent with modern metabolic energy storage molecules in some primitive organisms (Achbergerová and Nahálka 2011). The storage of energy in polyphosphates again is something that appears to be unique to phosphate, as polyarsenates and polyvanadates are not energy-storing (Klemperer et al. 1992).

For these reasons it is apparent that phosphate is the best at what it does: making stable nucleic acids. If early in life's history instability could have been useful (as per Bean et al. 2006), then phosphate may not have been part of the first nucleic acids. However, since there is such interplay between metabolic reactions involving polyphosphate and the polyphosphate-based building blocks of nucleic acids, phosphate was likely incorporated early in the history of life. Routes to forming phosphorylated nucleic acids should still be a goal of prebiotic chemistry research, as there appears to be an adequate ability to make phosphate monoesters. Ideally, a prebiotic route that could demonstrate nucleic acid formation from energetic phosphorus molecules (e.g., triphosphate) could provide strong evidence for forming the original nucleic acids.

References

Achbergerová L, Nahálka J (2011) Polyphosphate-an ancient energy source and active metabolic regulator. Microb Cell Factories 10(1):63

Anders E, Grevesse N (1989) Abundances of the elements: meteoritic and solar. Geochim Cosmochim Acta 53(1):197–214

Bean HD, Anet FA, Gould IR, Hud NV (2006) Glyoxylate as a backbone linkage for a prebiotic ancestor of RNA. Orig Life Evol Biosph 36(1):39–63

Benner SA (2011) Comment on "A bacterium that can grow by using arsenic instead of phosphorus". Science 332:1149

Benner SA, Sismour AM (2005) Synthetic biology. Nat Rev Genet 6(7):533–543

Benner SA, Kim HJ, Carrigan MA (2012) Asphalt, water, and the prebiotic synthesis of ribose, ribonucleosides, and RNA. Acc Chem Res 45(12):2025–2034

Blackmond DG (2010) The origin of biological homochirality. Cold Spring Harb Perspect Biol 2 (5):a002147

Borden J, Crans DC, Florián J (2006) Transition state analogues for nucleotidyl transfer reactions: structure and stability of pentavalent vanadate and phosphate ester dianions. J Phys Chem B 110 (30):14988–14999

Bornscheuer UT, Kazlauskas RJ (2004) Catalytic promiscuity in biocatalysis: using old enzymes to form new bonds and follow new pathways. Angew Chem Int Ed 43(45):6032–6040

Bryant DE, Marriott KE, Macgregor SA, Kilner C, Pasek MA, Kee TP (2010) On the prebiotic potential of reduced oxidation state phosphorus: the H-phosphinate–pyruvate system. Chem Commun 46(21):3726–3728

Burcar B, Pasek M, Gull M, Cafferty BJ, Velasco F, Hud NV, Menor-Salván C (2016) Darwin's warm little pond: a one-pot reaction for prebiotic phosphorylation and the mobilization of phosphate from minerals in a urea-based solvent. Angew Chem Int Ed 55(42):13249–13253

Cafferty BJ, Fialho DM, Khanam J, Krishnamurthy R, Hud NV (2016a) Spontaneous formation and base pairing of plausible prebiotic nucleotides in water. Nat Commun 7

Cafferty BJ, Musetti C, Kim K, Horowitz ED, Krishnamurthy R, Hud NV (2016b) Small molecule-mediated duplex formation of nucleic acids with 'incompatible'backbones. Chem Commun 52 (31):5436–5439

Callahan MP, Smith KE, Cleaves HJ, Ruzicka J, Stern JC, Glavin DP et al (2011) Carbonaceous meteorites contain a wide range of extraterrestrial nucleobases. Proc Natl Acad Sci 108 (34):13995–13998

Cech TR, Bass BL (1986) Biological catalysis by RNA. Annu Rev Biochem 55(1):599–629

Chen S, Wang L, Deng Z (2010) Twenty years hunting for sulfur in DNA. Protein Cell 1(1):14–21

De Graaf RM, Schwartz AW (2005) Thermal synthesis of nucleoside H-phosphonates under mild conditions. Orig Life Evol Biosph 35(1):1–10

De Graaf RM, Visscher J, Schwartz AW (1998) Prebiotic chemistry of phosphonic acids: products derived from phosphonoacetaldehyde in the presence of formaldehyde. Orig Life Evol Biosph 28(3):271–282

Elias M, Wellner A, Goldin-Azulay K, Chabriere E, Vorholt JA, Erb TJ, Tawfik DS (2012) The molecular basis of phosphate discrimination in arsenate-rich environments. Nature 491 (7422):134–137

Erb TJ, Kiefer P, Hattendorf B, Günther D, Vorholt JA (2012) GFAJ-1 is an arsenate-resistant, phosphate-dependent organism. Science 337(6093):467–470

Fekry MI, Tipton PA, Gates KS (2011) Kinetic consequences of replacing the internucleotide phosphorus atoms in DNA with arsenic. ACS Chem Biol 6(2):127–130

Fox SW (1969) Self-ordered polymers and propagative cell-like systems. Naturwissenschaften 56 (1):1–9

Gilbert W (1986) Origin of life: The RNA world. Nature 319(6055)

Goldhaber MB, Orr WL (1995) Kinetic controls on thermochemical sulfate reduction as a source of sedimentary H_2S. ACS Symp Ser 612:412–425

Gulick A (1955) Phosphorus as a factor in the origin of life. Am Sci 43(3):479–489

Gull M, Mojica MA, Fernández FM, Gaul DA, Orlando TM, Liotta CL, Pasek MA (2015) Nucleoside phosphorylation by the mineral schreibersite. Sci Rep 5:17198–17198. https://doi.org/10.1038/srep17198

Guthrie JP (1978) Hydrolysis of esters of oxy acids: pKa values for strong acids; Brønsted relationship for attack of water at methyl; free energies of hydrolysis of esters of oxy acids; and a linear relationship between free energy of hydrolysis and pKa holding over a range of 20 pK units. Can J Chem 56(17):2342–2354

Hud NV, Cafferty BJ, Krishnamurthy R, Williams LD (2013) The origin of RNA and "my grandfather's axe". Chem Biol 20(4):466–474

Hughes MF (2002) Arsenic toxicity and potential mechanisms of action. Toxicol Lett 133(1):1–16

Kamerlin SC, Sharma PK, Prasad RB, Warshel A (2013) Why nature really chose phosphate. Q Rev Biophys 46(1):1

Klemperer WG, Marquart TA, Yaghi OM (1992) New directions in polyvanadate chemistry: from cages and clusters to baskets, belts, bowls, and barrels. Angew Chem Int Ed Engl 31(1):49–51

Lopez V, Stevens T, Lindquist RN (1976) Vanadium ion inhibition of alkaline phosphatase-catalyzed phosphate ester hydrolysis. Arch Biochem Biophys 175(1):31–38

Martin AR, Barvik I, Luvino D, Smietana M, Vasseur JJ (2011) Dynamic and programmable DNA-templated boronic ester formation. Angew Chem Int Ed 50(18):4193–4196

Martin AR, Mohanan K, Luvino D, Floquet N, Baraguey C, Smietana M, Vasseur JJ (2009) Expanding the borononucleotide family: synthesis of borono-analogues of dCMP, dGMP and dAMP. Org Biomol Chem 7(21):4369–4377

Martin AR, Vasseur JJ, Smietana M (2013) Boron and nucleic acid chemistries: merging the best of both worlds. Chem Soc Rev 42(13):5684–5713

Menor-Salván C, Ruiz-Bermejo D, Guzmán MI, Osuna-Esteban S, Veintemillas-Verdaguer S (2009) Synthesis of pyrimidines and triazines in ice: implications for the prebiotic chemistry of nucleobases. Chem Eur J 15(17):4411–4418

Merck Index (1996) 12th ed. Merck, Whitehouse Station, NJ, p 152

Miller SL (1953) A production of amino acids under possible primitive earth conditions. Science 117(3046):528–529

Miller SL, Urey HC (1959) Organic compound synthesis on the primitive earth. Science 130 (3370):245–251

Mitchell MC, Taylor RJ, Kee TP (1998) On the hydrolysis of dimethyl-H-phosphonate. An 18O-labelling and 31P-NMR study. Polyhedron 17(4):433–442

Mohammed FS, Chen K, Mojica M, Conley M, Napoline JW, Butch C, Pollet P, Krishnamurthy R, Liotta CL (2017) A plausible prebiotic origin of glyoxylate: nonenzymatic transamination reactions of glycine with formaldehyde. Synlett 28(01):93–97

Mukhopadhyay R, Rosen BP, Phung LT, Silver S (2002) Microbial arsenic: from geocycles to genes and enzymes. FEMS Microbiol Rev 26(3):311–325

Nelson KE, Levy M, Miller SL (2000) Peptide nucleic acids rather than RNA may have been the first genetic molecule. Proc Natl Acad Sci 97(8):3868–3871

Niemi R, Vepsäläinen J, Taipale H, Järvinen T (1999) Bisphosphonate prodrugs: synthesis and in vitro evaluation of novel acyloxyalkyl esters of clodronic acid. J Med Chem 42 (24):5053–5058

O'Neill MA, Warrenfeltz D, Kates K, Pellerin P, Doco T, Darvill AG, Albersheim P (1996) Rhamnogalacturonan-II, a pectic polysaccharide in the walls of growing plant cell, forms a dimer that is covalently cross-linked by a borate ester in vitro conditions for the formation and hydrolysis of the dimer. J Biol Chem 271(37):22923–22930

Orgel LE (2004) Prebiotic chemistry and the origin of the RNA world. Crit Rev Biochem Mol Biol 39(2):99–123

Ossenkamp GC, Kemmitt T, Johnston JH (2001) New approaches to surface-alkoxylated silica with increased hydrolytic stability. Chem Mater 13(11):3975–3980

Pabis A, Duarte F, Kamerlin SC (2016) Promiscuity in the enzymatic catalysis of phosphate and sulfate transfer. Biochemistry 55(22):3061–3081

Pasek MA (2017) Schreibersite on the early earth: scenarios for prebiotic phosphorylation. Geosci Front 8:329–335

Pasek M, Block K (2009) Lightning-induced reduction of phosphorus oxidation state. Nat Geosci 2 (8):553–556

Pasek MA, Dworkin JP, Lauretta DS (2007) A radical pathway for organic phosphorylation during schreibersite corrosion with implications for the origin of life. Geochim Cosmochim Acta 71 (7):1721–1736

Pasek MA, Harnmeijer JP, Buick R, Gull M, Atlas Z (2013) Evidence for reactive reduced phosphorus species in the early Archean ocean. Proc Natl Acad Sci 110(25):10089–10094

Pasek M, Herschy B, Kee TP (2015) Phosphorus: a case for mineral-organic reactions in prebiotic chemistry. Orig Life Evol Biosph 45(1–2):207–218

Pasek MA, Kee TP, Bryant DE, Pavlov AA, Lunine JI (2008) Production of potentially prebiotic condensed phosphates by phosphorus redox chemistry. Angew Chem Int Ed 47(41):7918–7920

Pasek MA, Lauretta DS (2005) Aqueous corrosion of phosphide minerals from iron meteorites: a highly reactive source of prebiotic phosphorus on the surface of the early Earth. Astrobiology 5 (4):515–535

Pawlowska R, Korczynski D, Nawrot B, Stec WJ, Chworos A (2016) The α-thio and/or β-γ-hypophosphate analogs of ATP as cofactors of T4 DNA ligase. Bioorg Chem 67:110–115

Peyser JR, Ferris JP (2001) The rates of hydrolysis of thymidyl-3′, 5′-thymidine-H-phosphonate: the possible role of nucleic acids linked by diesters of phosphorous acid in the origins of life. Orig Life Evol Biosph 31(4):363–380

Powner MW, Gerland B, Sutherland JD (2009) Synthesis of activated pyrimidine ribonucleotides in prebiotically plausible conditions. Nature 459(7244):239–242

Prusiner SB (1991) Molecular biology of prion diseases. Science 252(5012):1515–1523

Reaves ML, Sinha S, Rabinowitz JD, Kruglyak L, Redfield RJ (2012) Absence of detectable arsenate in DNA from arsenate-grown GFAJ-1 cells. Science 337(6093):470–473

Ricardo A, Carrigan MA, Olcott AN, Benner SA (2004) Borate minerals stabilize ribose. Science 303(5655):196–196

Rochette EA, Bostick BC, Li G, Fendorf S (2000) Kinetics of arsenate reduction by dissolved sulfide. Environ Sci Technol 34(22):4714–4720

Schoepp-Cothenet B, Nitschke W, Barge LM, Ponce A, Russell MJ, Tsapin AI (2011) Comment on “A bacterium that can grow by using arsenic instead of phosphorus”. Science 332:1149

Steinberg H, Hunter DL (1957) Preparation and rate of hydrolysis of boric acid esters. Ind Eng Chem 49(2):174–181

Sugiyama M, Hong Z, Whalen LJ, Greenberg WA, Wong CH (2006) Borate as a phosphate ester mimic in aldolase-catalyzed reactions: practical synthesis of L-fructose and L-Iminocyclitols. Adv Synth Catal 348(18):2555–2559

Takeno, N. (2005). Atlas of Eh-pH diagrams. Geological survey of Japan open file report, 419, p 102

Tracey AS, Gresser MJ (1988) The characterization of primary, secondary, and tertiary vanadate alkyl esters by 51V nuclear magnetic resonance spectroscopy. Can J Chem 66(10):2570–2574

Tracey AS, Galeffi B, Mahjour S (1988) Vanadium (V) oxyanions. The dependence of vanadate alkyl ester formation on the pKa of the parent alcohols. Can J Chem 66(9):2294–2298

Taylor SR, McLennan SM (1995) The geochemical evolution of the continental crust. Rev Geophys 33(2):241–265

Van Mooy BA, Rocap G, Fredricks HF, Evans CT, Devol AH (2006) Sulfolipids dramatically decrease phosphorus demand by picocyanobacteria in oligotrophic marine environments. Proc Natl Acad Sci 103(23):8607–8612

Wang D, Li ZJ, Ying HJ (2009) Solubility of adenosine 5′-monophosphate in different solvents from (288.15 to 330.15) K. J Chem Eng Data 55(2):992–993

Wang L, Chen S, Xu T, Taghizadeh K, Wishnok JS, Zhou X et al (2007) Phosphorothioation of DNA in bacteria by *dnd* genes. Nat Chem Biol 3(11):709–710

Wanty RB, Goldhaber MB (1992) Thermodynamics and kinetics of reactions involving vanadium in natural systems: accumulation of vanadium in sedimentary rocks. Geochim Cosmochim Acta 56(4):1471–1483

Watson JD, Crick FHC (1953) Molecular structure of nucleic acids: a structure for deoxyribose nucleic acid. Nature 171:737–738

Westheimer FH (1987) Why nature chose phosphates. Science 235(4793):1173–1178

White HB (1976) Coenzymes as fossils of an earlier metabolic state. J Mol Evol 7(2):101–104

Williams NH, Wyman P (2001) Base catalysed phosphate diester hydrolysis. Chem Commun (14):1268–1269

Wolfe-Simon F, Blum JS, Kulp TR, Gordon GW, Hoeft SE, Pett-Ridge J et al (2011) A bacterium that can grow by using arsenic instead of phosphorus. Science 332(6034):1163–1166

Wohlgemuth R, Liese A, Streit W (2017) Biocatalytic phosphorylations of metabolites: past, present, and future. Trends Biotechnol 35(5):452–465

Wuggenig F, Hammerschmidt F (1998) Enzymes in organic chemistry VI [1]. Enantioselective hydrolysis of 1-chloroacetoxycycloalkylmethylphosphonates with lipase AP 6 from Aspergillus niger and chemoenzymatic synthesis of chiral, nonracemic 1-aminocyclohexylmethylphosphonic acids. Monatsh Chem Chem Mon 129(4):423–436

Xie X, Liang J, Pu T, Xu F, Yao F, Yang Y et al (2012) Phosphorothioate DNA as an antioxidant in bacteria. Nucleic Acids Res 40(18):9115–9124

Yamagata Y (1999) Prebiotic formation of ADP and ATP from AMP, calcium phosphates and cyanate in aqueous solution. Orig Life Evol Biosph 29(5):511–520

Chapter 7
Template-Directed Replication of Nucleic Acids Mediated by Viscous Environments

Isaac Gállego and Christine He

Abstract Many hypotheses concerning the nature of early life assume that genetic information was once transferred through the template-directed synthesis of RNA, prior to the evolution of genetically encoded protein synthesis. However, attempts at identifying the earliest mechanism for the protein-free, template-directed replication of nucleic acids remain an elusive goal. A fundamental biophysical problem known as strand inhibition limits copying of a nucleic acid duplex: transferring information from a template sequence in the presence of its complementary strand is inhibited by the stability of the template duplex. This chapter discusses state-of-the-art strategies and a novel method which uses viscous solvents to overcome strand inhibition during template copying, one of the most challenging problems in polymer self-replication.

7.1 Introduction

All living organisms have the ability to replicate their own genomes, and transfer genetic information to their offspring. Hence, any plausible prebiotic scenario must address how a chemical system capable of self-replication and transfer of information evolved.

Nucleic acids store and transfer genetic information in all living organisms (Alberts et al. 2008). Many models of early evolution assume that the first informational polymer was also a nucleic acid, based on the ability of nucleic acids to robustly transfer sequence information through Watson-Crick base pairing (Watson and Crick 1953). In the 1980s it was discovered that RNA, in addition to its role as an informational polymer, possesses catalytic properties and carries out a range of key biochemical reactions (Kruger et al. 1982). Later on, it was discovered that the linking of amino acids within the ribosome—arguably the most ancient

I. Gállego (✉)
MRC-Laboratory of Molecular Biology, Cambridge, UK

C. He
University of California, Berkeley, CA, USA

© Springer International Publishing AG, part of Springer Nature 2018

C. Menor-Salván (ed.), *Prebiotic Chemistry and Chemical Evolution of Nucleic Acids*, Nucleic Acids and Molecular Biology 35,
https://doi.org/10.1007/978-3-319-93584-3_7

macromolecular structure still present in life today (Woese 2001)—is carried out entirely by RNA (Cech 2000). These discoveries, along with ongoing research which indicates that RNA's cellular role is more expansive than previously thought (Storz 2002; Ghildiyal and Zamore 2009; Breaker 2011; Breaker and Joyce 2014), suggest that the template-directed synthesis of RNA was a critical process in the transition from prebiotic chemistry to life. This evidence motivated the development of an RNA World hypothesis, which proposes that RNA was the key biopolymer in early evolution, serving as both a genetic information carrier and catalyst prior to the adoption of these roles by DNA and proteins, respectively (Gilbert 1986).

However, despite more than half a century of research into the chemical origins of nucleic acids, a robust pathway from simple chemical precursors to self-replicating nucleic acid polymers has not been identified (Joyce 2002; Cafferty and Hud 2014; Szostak 2012). Open questions remain about every stage in the abiotic formation of nucleic acids, including synthesis of mononucleotides (Hud et al. 2013), non-templated polymerization of mononucleotides into oligonucleotides (Szostak 2012), and information transfer from the resulting nucleic acid templates (Bell and Dutta 2002) (see Fig. 7.1). Discussion of mono- and oligonucleotide formation is extensively reviewed in previous chapters within this book.

In this chapter, we focus explicitly on the problem of replication (Fig. 7.1 step c): how was sequence information transferred from a nucleic acid sequence to a complementary strand? Identifying a robust, prebiotically plausible mechanism for template-directed nucleic acid synthesis remains an elusive goal. In contemporary biology, DNA replication is a complex process, orchestrated by specialized, highly evolved enzymes working in a concerted, stepwise fashion (Kunkel and Bebenek 2000; Bell and Dutta 2002). In contrast, replication of an early informational polymer must have occurred by a simpler, protein-free mechanism that may have borne little resemblance to the process of DNA replication in life today.

In this chapter, we summarize the strategies that have been proposed for driving replication of nucleic acids in prebiotic conditions and the limitations of these models. Additionally, we introduce a proposed model for information transfer that is based on employing alternative environments to circumvent one of the main bottlenecks in polymer replication, the strand inhibition problem.

7.2 Primer-Dependent Information Transfer

Much of the research focused on identifying a prebiotic route for nucleic acid replication has employed a mechanism that resembles extant polymerase enzymes, where a short oligonucleotide sequence—a primer—is annealed to a template strand and then extended at its 3′ end by sequential addition of mononucleotides (Fig. 7.2a). However, different laboratories have pursued different strategies for template-directed information transfer. These strategies can be divided into those that employ and do not employ a catalyst to promote primer extension.

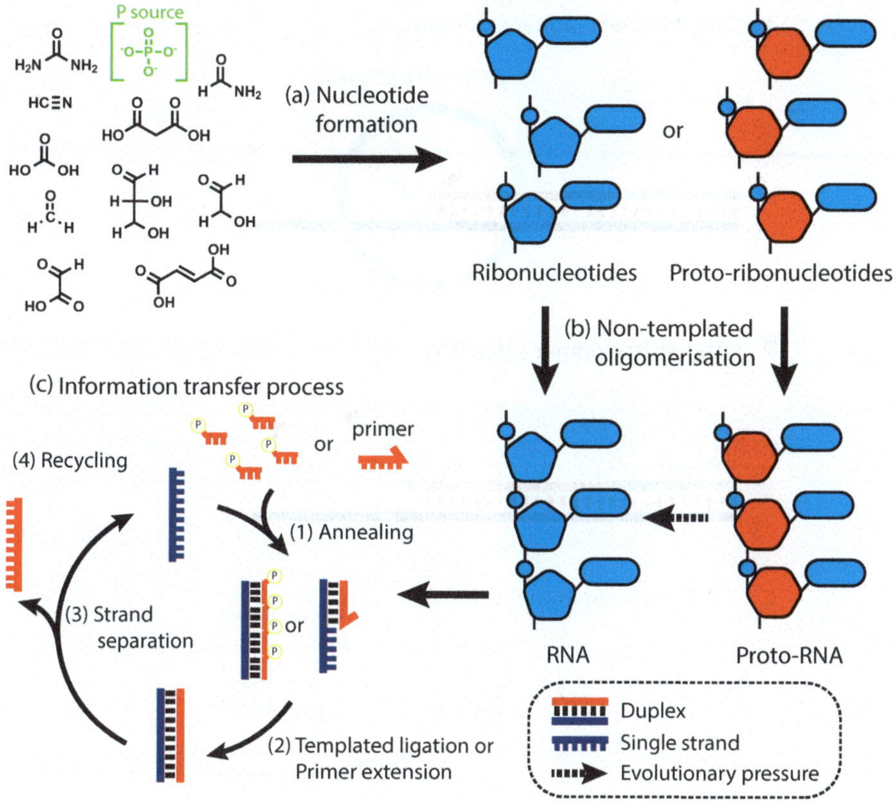

Fig. 7.1 Step-by-step synthesis of (proto-)nucleotides (step a), non-templated oligomerization of proto-nculeotides (step b), and template-based transfer of information from oligonucleotides under prebiotic conditions (step c). The scheme shows two possible pathways to the formation of RNA, where extant ribonucleotides could either be synthesized directly (step a, left side) or preceded by proto-ribonucleotides that oligomerize into proto-RNA (step a, right side). Subsequently, evolutionary pressure (chemically or biologically driven) would convert proto-RNA into extant RNA

7.2.1 *Catalyst-Free Primer Extension*

Without catalysts to drive phosphodiester bond formation, primer extension requires chemical modification of the mononucleotide building blocks. In a method pioneered by the Orgel laboratory (Inoue and Orgel 1981) and then thoroughly explored by the Szostak laboratory (Mansy et al. 2008; Mansy and Szostak 2009), the 5′ phosphate of the mononucleotides is chemically activated with nitrogenous heterocycles (Kervio et al. 2016)—such as 2-methylimidazolide or oxyazabenzotriazole—that serve as reactive leaving groups during phosphodiester bond formation (Fig. 7.2b). Recently, this primer extension method has been used to synthesize functional RNA enzymes from non-functional shorter fragments lacking specific regions (Adamala et al. 2015). The Szostak laboratory has integrated the

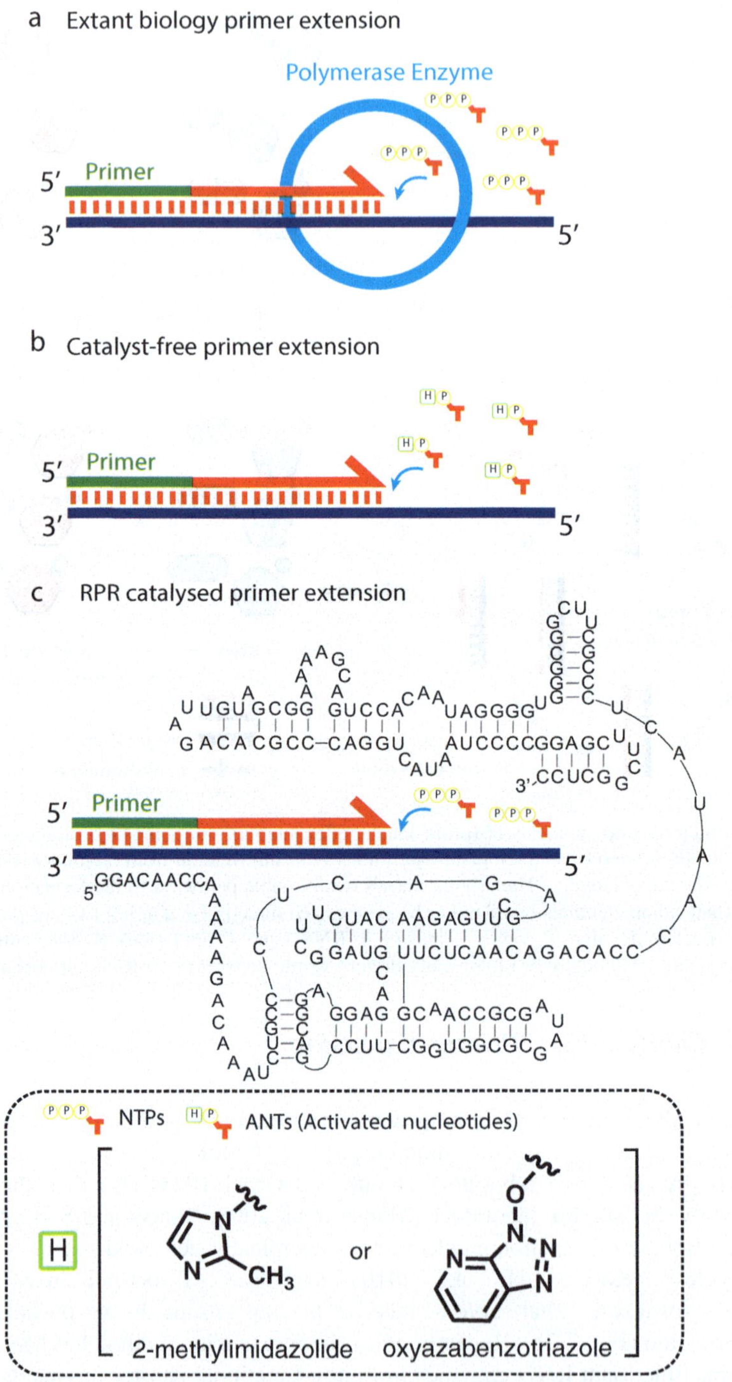

Fig. 7.2 Methods of information transfer by primer extension on a template

primer enzyme-free extension method within fatty acid-based compartmentalization systems (Mansy et al. 2008), with the aim of emulating a protocell that is capable of homeostasis (Engelhart et al. 2016).

However, several problems limit this model of template-directed copying, including weak association of mononucleotides with the primer-template complex and competition for template binding by inactivated mononucleotides (resulting from hydrolysis of the activated phosphate group) (Kervio et al. 2016). In addition, the primer extension reaction itself proceeds at slow rates, in low yields, and in a strongly sequence-dependent manner. In order to extend a primer with all four nucleobases, downstream "helper" oligonucleotides are utilized to provide an additional stacking surface that stabilizes the association of the incoming mononucleotide with the primer-template complex (Deck et al. 2011), and can also catalyse upstream bond formation (Prywes et al. 2016).

7.2.2 Primer Extension Mediated by Ribozymes

Given the difficulties associated with catalyst-free extension of RNA primers, other groups have pursued approaches which employ an RNA catalyst (ribozyme). According to the RNA World hypothesis, RNA once served as both genetic information carrier and primary catalyst for key reactions—including, potentially, its own replication.

To address the question of whether RNA might have once catalysed its own replication, researchers have used in vitro evolution to generate ribozymes capable of primer extension. This strategy initially led to the development of a ribozyme that served as an RNA ligase (Ekland and Bartel 1995; Shechner et al. 2009). Further in vitro evolution generated an RNA-templated polymerase ribozyme (RPR) capable of extending an RNA primer that is annealed to short templates (Johnston et al. 2001). In this case, the RPR extends the 3′ extreme of the primer oligonucleotide (Fig. 7.2c), and the 5′ extreme of the free mononucleotides is activated by a triphosphate—similar to the mechanism used by protein polymerases. More recently, the Holliger laboratory evolved RPRs with higher fidelity and processivity (Wochner et al. 2011; Attwater et al. 2013a) that can synthesize complementary strands from longer (~200 nt) templates. Generally, primer extension is more difficult on templates with internal structure (hairpins); recently an RPR was developed by Horning and Joyce with some structure tolerance (Horning and Joyce 2016). Another interesting approach has been the use of small ribozyme ligases to synthesize a functional RPR by stitching together fragments of the RPR from a pool of sequences (Mutschler et al. 2015). More recently, the Holliger laboratory has also developed a triplet-based ribozyme that is capable of going through templates with much more complex internal structure, and that has a fidelity-recognition mechanism similar to that of the small ribosomal subunit (Attwater et al. 2018).

7.3 The Strand Inhibition Problem

In the previous section, we discussed different strategies that have been used to drive information transfer from an RNA template strand. However, the aforementioned systems face a problem after the synthesis of the complementary strand has proceeded. The result of synthesizing a complementary strand from a template strand is a duplex that must be separated into single strands before another round of template-directed synthesis can occur. In the absence of enzymes, elevated temperatures can drive the separation of a duplex into single strands. However, for copying to proceed, the temperature must then be lowered to the point at which mono- or oligonucleotide substrates can stably bind to their complementary sites on the template strands. Reformation of the template duplex is kinetically and thermodynamically favoured over template-substrate binding, so that once the temperature drops, the template duplex reforms and substrate binding is inhibited (Fig. 7.3). This long-standing problem in enzyme-free, template-directed nucleic acid synthesis is known as strand inhibition (Grossmann et al. 2008; Fernando et al. 2007).

Few strategies have been developed to explicitly address the strand inhibition problem, and existing approaches are far from robust or self-sustaining, requiring manual washing steps to separate copy strands from template strands (Deck et al. 2011; Luther et al. 1998) (Fig. 7.4a). Other approaches employ non-prebiotic chemical modifications to specific sites on the template or substrates that, after ligation, produce an unstable duplex and facilitate strand separation (Dose et al. 2006; Zhan and Lynn 1997; Kausar et al. 2011; Abe and Kool 2004) (Fig. 7.4b). These limitations motivate the continued search for more general, non-enzymatic processes that could have facilitated copying from long template strands of arbitrary sequence before the emergence of polymerase enzymes (Adamala et al. 2015; Mutschler et al. 2015; Walker et al. 2012; Kreysing et al. 2015).

7.4 Alternative Environments and Their Potential Role in the Origins of Nucleic Acid Replication

In identifying a mechanism for enzyme-free nucleic acid replication, it is important to consider the physical environment and solvent conditions in which such a process might have transpired. The physico-chemical properties of the environment dictate

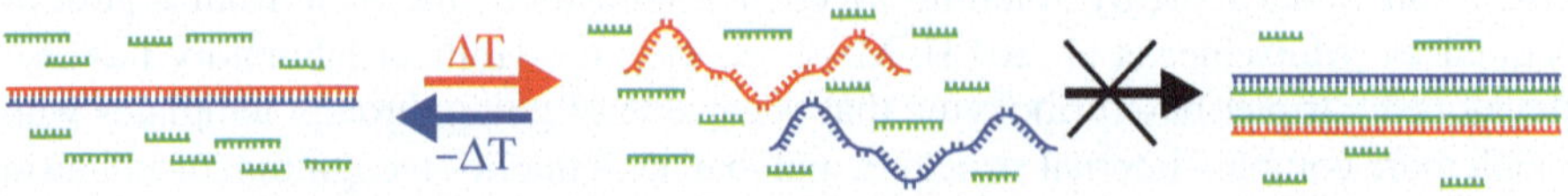

Fig. 7.3 Illustration of the strand inhibition problem for long complementary template strands in a low viscosity solvent such as water

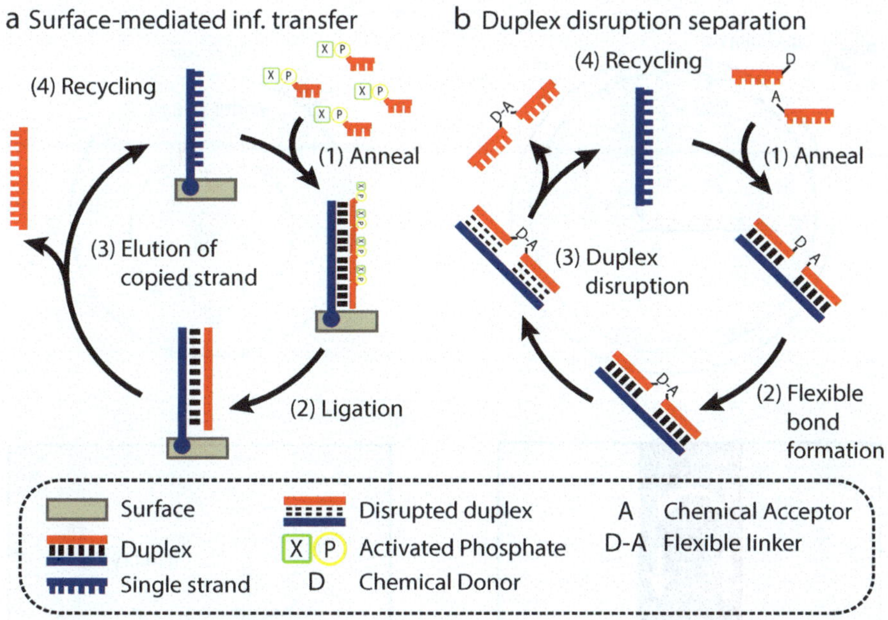

Fig. 7.4 Proposed strategies to solve the strand inhibition problem. (**a–b**) Panels showing stepwise processes by which the strand inhibition problem can be overcome during templated copying. (**a**) Method that exploits the anchorage of the template strand to a surface. (**b**) Strategy in which a flexible linker/duplex disruptor is introduced within the newly synthesized complementary strand, promoting duplex separation. A similar cycle can be applied where the template contains the flexible linker to overcome strand inhibition. These strategies were developed earlier than the method we propose in this chapter (see Sect. 7.5.1)

the chemical stability and conformation of molecules—critical factors in any nucleic acid self-assembly process.

Historically, most work in the field of prebiotic chemistry has been performed in environments mimicking the aqueous or highly hydrated milieu utilized by current biology. However, important advances have been made by using the environment as a means to control and drive prebiotic processes, mainly by compartmentalizing or confining molecules in specific microenvironments. Compartmentalization—be it physical or temporal—was a key evolutionary development during the origins of life and integral to the transition from prebiotic chemistry to cellular life. Compartmentalization offers many advantages to a prebiotic system including limited diffusion, concentration of scarce molecules and resources, and protection from chemical agents or external fluctuations that may affect replication or catalytic processes (Stoeger et al. 2016). A clear example of compartmentalization was demonstrated by the protocell work of Szostak and co-workers (Fig. 7.5a), where compartmentalization by fatty acid vesicles provides specific environments in which reactants and products can be concentrated to facilitate information transfer from RNA templates (Mansy et al. 2008; Mansy and Szostak 2009), as well as drive basic cellular-like

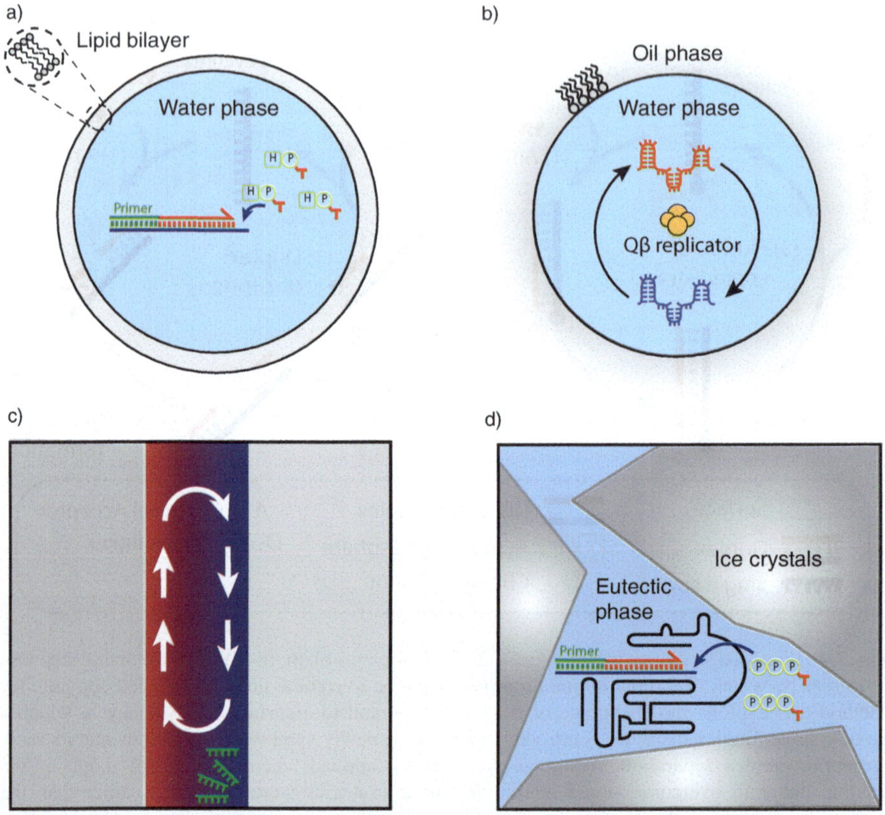

Fig. 7.5 Methods for compartmentalization of nucleic acid molecules and information transfer processes. (**a**) Protocell-like compartmentalization. (**b**) Compartmentalization within a water-in-oil emulsion system. (**c**) Thermophoresis method for local concentration of nucleic acids. (**d**) In-ice compartmentalization through eutectic ice phase formation. Figure adapted from Wachowius et al. (2017)

processes (Engelhart et al. 2016). Here the semipermeable lipid bilayer acts as a physical barrier, creating isolated, thermally stable microenvironments where template-directed synthesis occurs (Mansy and Szostak 2008). However, at the same time, this barrier hinders the transport of most reactants within the compartments, hindering protometabolism processes.

Although not as widely considered in the origins of life community, compartmentalization using emulsions has been utilized to study the linkage between genotype and phenotype. An example of this phenomenon is water-in-oil (W/O) emulsions, where an inert oil phase contains a suspension of aqueous cell-like droplets that are transiently stable (Fig. 7.5b). W/O emulsions have been used to explore the impact of compartmentalization in the Qβ replication system (Ichihashi et al. 2013; Matsumura et al. 2016). These studies find that the replicase phenotype can only outlast quickly replicating, shorter sequences (i.e. parasites) when the process is, at least transiently, compartmentalized.

Alternatively, temperature has also been used as a driving force for confinement, concentration, and size-dependent selection of molecules. Braun and co-workers have demonstrated that temperature gradients within a specific capillary-shaped space promote thermophoresis (Fig. 7.5c), locally concentrating molecules within certain regions of the capillary (Mast et al. 2013). This process can also be used to physically separate nucleic acids in a length-dependent fashion (Kreysing et al. 2015). This group has proposed that this system mimics the rock pores in hydrothermal vents and could have been a driving force for nucleic acid replication on the prebiotic Earth. On the other hand, Holliger and co-workers have proposed a scenario in which lower temperatures (−7 °C) would promote the formation of a eutectic phase within ice (Fig. 7.5d). This phase is liquid, highly rich in ions, and locally concentrates molecular species (including nucleic acids) while preserving the chemical stability of RNA due to low temperature. Holliger and co-workers have demonstrated that the catalytic activity (Attwater et al. 2010) and development of RPRs with enhanced activity (Attwater et al. 2013b) are promoted in a eutectic ice system. They have also demonstrated that freeze-thaw cycles promote the formation of active RPRs stitched together from constitutive RNA fragments (Mutschler et al. 2015).

In addition to driving compartmentalization of molecules, temperature fluctuations can also promote enhanced reactivity of chemical species. The driving force for such temperature changes could have been periodic cycles of hot/cold, dry/wet conditions, driven by day/night or seasonal transitions on the early Earth. During periods of high temperature, water would have evaporated from pools on dry land. Rehydration—through water uptake from a humid environment, rain, or tides—would bring the pool back to a hydrated state. Hud and co-workers have utilized alternating hot/cold, dry/wet cycles to promote condensation of sugars with analogue bases (Chen et al. 2014), condensation reactions to form polyesters (Mamajanov et al. 2014), and small peptide synthesis through ester intermediates (Forsythe et al. 2015).

An important result of a hydration/dehydration cycle is the concentration of small organic molecules as water evaporates, resulting in transient formation of nonaqueous solvents. Such alternative solvents may possess physico-chemical properties that could promote nucleic acid replication and overcome limiting steps present in aqueous buffer. Herein, we present a model system focused on viscosity as a mechanism for facilitating the template-directed synthesis of nucleic acids (Fig. 7.6). Viscous environments could have been generated on the early Earth during periodic cycles of hot/cold, dry/wet conditions, the same type of process that enables dehydration-based chemistry (Mamajanov et al. 2014; Chen et al. 2014; Forsythe et al. 2015). In this case, periods of high temperature would evaporate water from pools, concentrating organic compounds or salts into a viscous solution. Rehydration would bring the pool back to a hydrated, low-viscosity state. We propose that similar fluctuations of water activity, temperature, and viscosity could have facilitated the copying of a nucleic acid duplex in prebiotic conditions.

7.4.1 *Viscous Solvents as a Means to Solve the Strand Inhibition Problem*

In a viscous environment, diffusion of nucleic acid polymers is slowed in a size-dependent manner. A highly viscous solvent can significantly retard the movement of long template strands, slowing the association of a long template strand with its complementary strand to form a template duplex. Additionally, intramolecular secondary structure (e.g. hairpins) on the template strands forms quickly, serving to further kinetically trap the template as single strands. Meanwhile, mononucleotides and short oligonucleotides remain relatively mobile. The difference in mobility between template strands and oligonucleotide substrates in a viscous environment creates a time window for assembly of oligonucleotide substrates on their complementary template strands, overcoming the problem of strand inhibition and enabling template-directed nucleic acid synthesis. Figure 7.6 illustrates the different steps of this replication cycle.

In addition to providing a mechanism to circumvent the strand inhibition problem, viscosity provides a potential mechanism for the selection and amplification of catalytically active RNA sequences (ribozymes) from a prebiotic pool of diverse sequences—a key process in the emergence of a putative RNA World. In contrast to

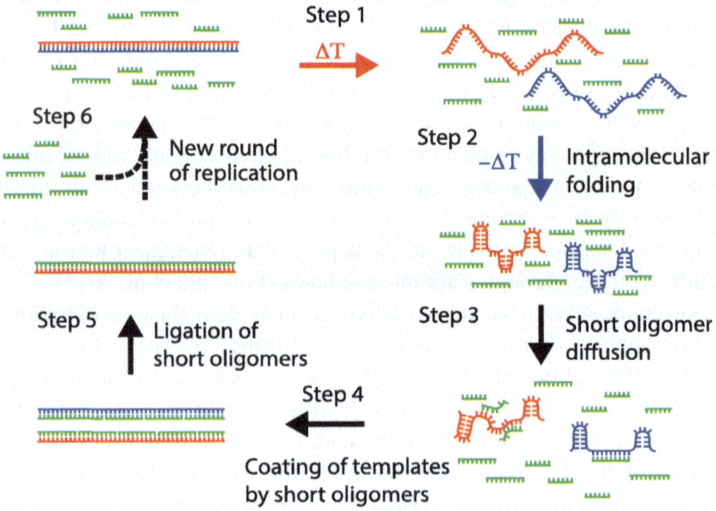

Fig. 7.6 Proposed process for duplex replication in a viscous solvent. Step (1): Heating causes thermal denaturation of the template duplex and reduced viscosity of the solvent, allowing the template single strands to diffuse apart. Step (2): Upon cooling, solvent viscosity increases, kinetically trapping internal hairpins on the template single strands. Step (3): Oligonucleotide substrates, which are more mobile than the much longer templates, diffuse and bind to complementary sites on the template strands. Step (4): Eventually, the oligonucleotides completely coat and unfold the trapped template strands. Step (5): Ligation (chemical or enzymatic) of the bound substrate oligonucleotides completes the process of duplex replication. Step (6): Another round of viscosity-mediated replication can begin

aqueous environments, viscous environments may promote the replication of longer, highly structured nucleic acid sequences—which are less mobile and more readily kinetically trapped—over shorter, unstructured sequences. Concerns over how copying of short template sequences can "outcompete" long sequences have existed since the 1960s, when Spiegelman demonstrated that short sequences are enzymatically replicated more quickly than longer sequences in an extracellular replication system, resulting in shortening of template sequences and loss of genetic information over time (Mills et al. 1967). This trend can be reversed in certain continuous flow capillary systems, where longer template sequences are kinetically trapped, allowing sequence copying by an protein polymerase (in a PCR-like fashion), while shorter strands are washed out of the capillary (Mast et al. 2013). Viscosity, too, has the potential to promote the replication of longer template sequences over shorter ones, as long as template strands are more easily kinetically trapped (contain more internal structure and are less mobile than shorter templates). Viscous environments also represent a potential solution to the "replicator-catalyst" paradox of the RNA World, which arises from the fact that sequences with well-folded intramolecular structures—a requirement for catalytic activity—are less accessible as templates for replication (Ivica et al. 2013; Attwater et al. 2013b). In a viscous environment, the formation of a stable intramolecular structure on a template sequence might be highly beneficial for replication by preventing the reannealing of complementary template strands into a duplex. Therefore, viscous environments have the potential to provide a novel route to enzyme-free nucleic acid replication that may explain the emergence of desirable template features that are not favoured in aqueous environments.

7.4.2 *Eutectic Solvents as the "Thickener" in the Prebiotic Soup*

Testing the proposed self-replication process required a viscous solvent in which nucleic acid duplexes are stable at room temperature but denature at low enough temperatures that the nucleic acids are not irreversibly altered (e.g. depurination) at denaturing temperatures. We chose to focus on eutectic solvents, a class of solvents closely related to ionic liquids (Abbott et al. 2003, 2011; Smith et al. 2014). These solvents are mixtures of two or more components whose melting point is significantly lower than that of either individual component. Figure 7.7 shows some examples of small molecule, hydrogen bond donors that form eutectic solvents in combination with a halide salt such as choline chloride.

Previous studies performed by the Hud laboratory have shown that DNA and RNA form stable secondary structures—duplex, triplex, and quadruplex—in anhydrous and hydrated eutectic solvents, though the thermodynamic stability and exact secondary structure of nucleic acid sequences may be altered compared to aqueous buffer (Mamajanov et al. 2010; Lannan et al. 2012). Later studies demonstrated that

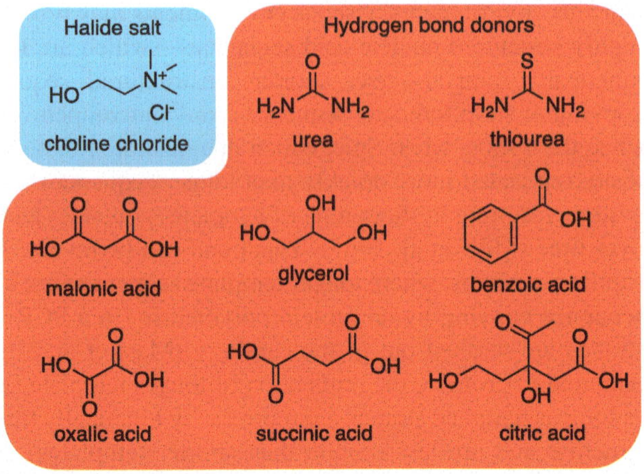

Fig. 7.7 Structure of choline chloride and some hydrogen bond donors used in the formation of eutectic solvents. The hydrogen bond donors shown here are considered to be prebiotically plausible

highly hydrated ionic liquids are also a suitable milieu to maintain secondary structure of nucleic acids (Tateishi-Karimata and Sugimoto 2012) and that some of these solvents may alter the effect of GC content on the thermal stability of the double helix (Tateishi-Karimata and Sugimoto 2012; Portella et al. 2014). Eutectic solvents are miscible with water, hygroscopic, stable upon heating, and often composed of small molecules, making them attractive as potential solvents for day/night, hydration/evaporation cycles on the prebiotic Earth. Simulated day/night cycles in eutectic solvents have been shown to promote the formation of organophosphates (Gull et al. 2014; Burcar et al. 2016), overcoming the prebiotic "phosphorylation problem" present in aqueous conditions.

Based on its ability to promote folding of DNA origami (Gállego et al. 2015), we chose glycholine as a eutectic solvent, composed of a 4:1 molar ratio of glycerol and choline chloride that has a room temperature viscosity of 437 cP. The melting temperatures (T_m's) of three distinct DNA structures (a 17 nt hairpin, 32 bp duplex, and a 3 kb long duplex) were determined to be significantly lower and closer to each other in anhydrous glycholine than those measured in aqueous buffer (Table 7.1).

7.5 The Longer the Better: Breaking a Long-Standing Paradigm in Nucleic Acid Replication

Replication and transfer of genetic information is a defining feature of extant biology. At some point on the early Earth, a process capable of copying nucleic acid polymers long enough to carry functional information must have evolved. However, for many years, origins of life researchers have faced a bottleneck in which very long polymers inhibit self-copying, limiting the synthesis of polymers long enough to carry complex information and serve as a templates for the next rounds of copying at the same time (Ivica et al. 2013). Based on these assumptions,

Table 7.1 Melting temperatures of DNA duplex species in glycholine and in aqueous buffer (20 mM Tris pH 7.5, 0.1 M NaCl)

		T_m (°C)	
Species	Base pairs	Aqueous buffer	Glycholine
17 nt hairpin	7 bp stem	79.5	44.7
32 bp duplex	32	72.2	49.0
3 kb duplex	2957	88.0	50.7

and using length-dependent trapping of viscous solvents such as glycholine, we propose a model system process for viscosity-enabled information transfer from nucleic acids (He et al. 2017). This strategy breaks a long-standing paradigm in the origins of life field and provides a scenario by which longer and more complex polymers are preferentially replicated over shorter, simpler sequences. In the following section, we implement a model system to transfer information from a gene-length (>300 nt) RNA duplex and discuss the implications of our approach for origins of life.

7.5.1 A Model System to Enable Information Transfer From Gene-Length Polymers

The process we propose is divided into Steps 1–5 of a replication cycle, with Step 6 representing the return of the system to its original state (Fig. 7.6). We apply our solvent-mediated replication process to generate a complementary sequence from a 352 nucleotide (nt) region of a 3 kilobase (kb) pair DNA template by utilizing eleven 32 nt oligomers that are complementary to specific, flanking regions within the template (see Fig. 7.9a). Because our aim was to develop a general approach to replicate arbitrary sequences of gene-length size, a linearized 3 kb bacterial plasmid was chosen as a model for a mixed, heterogeneous sequence, i.e. a sequence that has not been specifically designed for the purpose of overcoming strand inhibition.

Steps 1 and 2. Viscosity Enables Kinetic Trapping of Single-Stranded Templates

After heating, the long templates separate, but when cooling, our proposed process promotes the intramolecular folding of the single-stranded templates (Fig. 7.6, Step 2). This process should be increasingly favoured over duplex reformation as solvent viscosity increases due to progressive cooling of the solvent. The formation of internal structure (such as hairpin-like structures) is kinetically favoured, since it is a unimolecular process, as compared to a slower, duplex formation which is a bimolecular process involving the diffusion of single strands through the solvent (Hagen 2010). The kinetics of secondary structure reformation was studied after heat cycling (thermal denaturation and cooling to 20 °C) for a 17 nt hairpin, 32 bp duplex, and 3 kb duplex DNA in aqueous buffer and in glycholine (Fig. 7.8). This analysis demonstrated that duplex formation kinetics of a short duplex is about 15 times slower in viscous solvents. In contrast, intramolecular hairpin formation was not

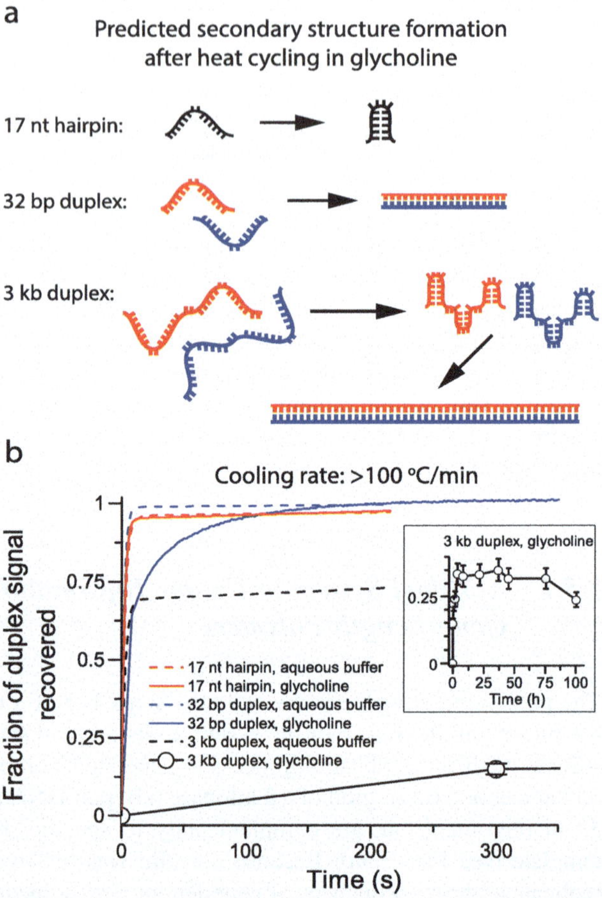

Fig. 7.8 (**a**) Predicted secondary structures of the 17 nt hairpin, 32 bp duplex, and 3 kb duplex. (**b**) Kinetics of duplex formation after thermal denaturation and cooling, as measured by UV-Vis absorbance and agarose gel electrophoresis (3 kb duplex)

affected by viscous environments as compared with low viscosity in a water-based solvent.

The effect of glycholine on the reformation rate of the 3 kb duplex was striking, with only ~25% of the full length duplex reformed after 100 h, compared to ~70% duplex reformation after 20 s in aqueous buffer (Fig. 7.8b). It is well known that genome-length, single-stranded DNA (ssDNA) generated via rapid cooling (e.g. >48 °C/min) from a thermally denatured state contains substantial intramolecular structure (Doty et al. 1960). Thus, reformation of the 3 kb DNA duplex in glycholine is likely delayed both by slowed diffusion of the ssDNA through the viscous solvent and by the persistence of intramolecular structures, the latter presenting an additional kinetic barrier to duplex formation (Viasnoff et al. 2006; Zhang et al. 2014). Figure 7.8a shows a schematic of predicted secondary structure formation after heat cycling for the 17 nt hairpin, 32 nt duplex, and 3 kb duplex DNA in glycholine.

A more detailed experiment demonstrated a direct correlation between the amount of kinetically trapped, single-stranded 3 kb duplex and the cooling rate, mostly due to internal structure formation. This experiment found that a cooling rate of 4 °C/min maximized the trapping differences between glycholine as compared with aqueous milieu, and could be easily achieved in a prebiotic scenario (He et al. 2017).

Steps 3 and 4. Oligonucleotides Bind and Cooperatively Open Up a Gene-Length Portion of the Kinetically Trapped Templates

During cooling in glycholine, short oligonucleotides diffuse faster than the trapped templates (Step 3), enabling full coating of a continuous section of a single-stranded template (Fig. 7.6, Step 4). To test this process, we designed a system in which a 32 nt oligonucleotide (F0), which binds to a site on the antisense template strand, is flanked by ten 32 nt oligonucleotides. Five of the ten flanking sequences bind directly upstream of F0 on the template strand (named L1 through L5), and five bind directly downstream (named R1 through R5). When assembled on the 3 kb antisense strand and ligated together, this series of 11 oligonucleotides—with F0 in the central position—forms a continuous 352 nt strand (Fig. 7.9). F0 was labeled with fluorescein (FAM) at its 5′ extreme to monitor the binding process of F0 to the template and the relative mobility of the template that contained the bound oligonucleotides.

In this experiment, glycholine samples containing the 3 kb DNA duplex and F0 only or the entire set of eleven 32-mer oligonucleotides (L5 through R5) were thermally cycled (denatured at 95 °C and cooled to 20 °C), and the olignuclueotide binding kinetics at 20 °C were then monitored. Reformation of the 3 kb duplex over time was followed by the recovery in the intensity of the less mobile duplex band [Fig. 7.9b; also see supporting information in He et al. (2017)]. After thermal cycling we observed that F0 binds to its ssDNA target and remained bound for over 10 days (Fig. 7.9c). In the EtBr image, two bands were observed with similar electrophoretic mobilities indicative of 2961 nt ssDNA. The FAM channel image revealed that F0 is bound to the more mobile of these two bands (Fig. 7.9b, asterisk), indicating that these upper and lower bands correspond, respectively, to 2961 nt ssDNA with and without the bound oligonucleotides.

When all eleven 32-mer oligonucleotides were present, the extent of F0 binding increased relative to when only F0 is present in the system (Fig. 7.9c). The increase in F0 binding with the binding of additional oligonucleotides suggests that L5 through R5 cooperatively unfold intramolecular structures along the single-stranded template via toehold-mediated strand displacement (Panyutin and Hsieh 1994; Green and Tibbetts 1981; Radding et al. 1977), increasing the accessibility of target sites in the same region—including the F0 binding site. This result illustrates how the binding of oligonucleotides to a double-stranded template can be greatly enhanced by thermal cycling in a viscous solvent, effectively overcoming strand inhibition.

Step 5: Viscosity-Enabled Information Transfer from a 3 kb DNA Duplex

As the final step in demonstrating information transfer from a duplex template (Fig. 7.6, Step 5), the eleven 32-mers bound to the 2961 nt ssDNA template were

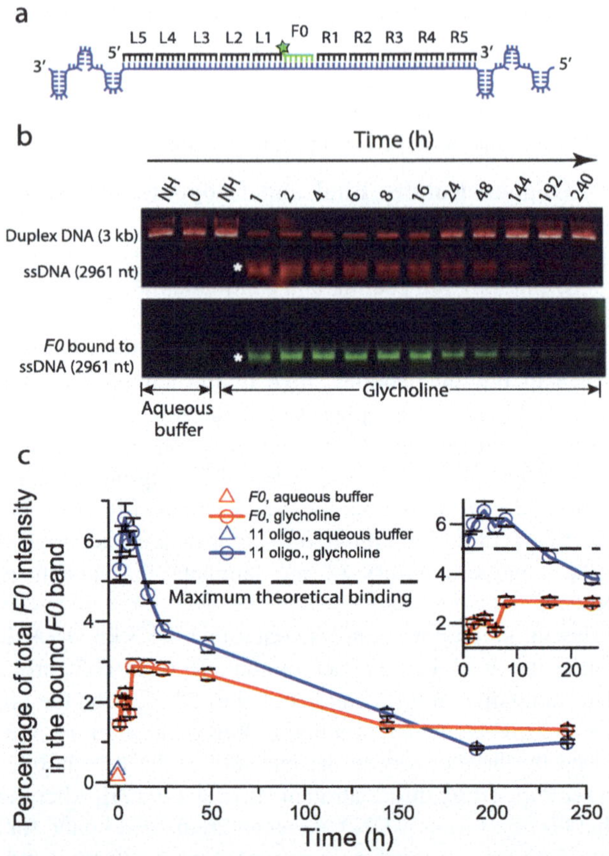

Fig. 7.9 Kinetics of oligonucleotide binding to a denatured 3 kb DNA duplex. (**a**) A schematic illustrating the binding of eleven oligonucleotides (32 nt each) to the 2961 nt template strand. F0, the central oligonucleotide, is labeled with fluorescein (FAM), represented by a green star. (**b**) Agarose gel (2%) showing the results of thermal cycling L5 through R5 with 3 kb duplex DNA in glycholine. The gel was stained with ethidium bromide (EtBr) and then imaged through EtBr-specific (red) and FAM-specific (green) channels. EtBr image shows the kinetics of 3 kb DNA annealing, while FAM image shows that F0 is binding to the ssDNA. (**c**) The percentage of F0 bound to the 3 kb DNA over time. Error bars account for all known sources of error. The maximum theoretical limit of 5% bound F0 is based on a 20:1 molar ratio of F0 to 3 kb template

linked using T4 DNA ligase, a non-prebiotic but robust ligation method. Importantly, we note that ligase was added after samples were thermally cycled and maintained at 20 °C for 4 h. Therefore, the assembly of oligonucleotides on its kinetically trapped template strand was enabled by the viscous solvent, without the aid of a protein enzyme (He et al. 2017).

After ligation, when all eleven 32-mer substrates were present with 3 kb DNA, 11 product bands were observed (Fig. 7.10b), which correspond to all potential polymers produced by ligation of the 32-mer substrates (Fig. 7.10a). The maximum length product possible (352 nt), formed by ligation of all eleven 32-mers, is

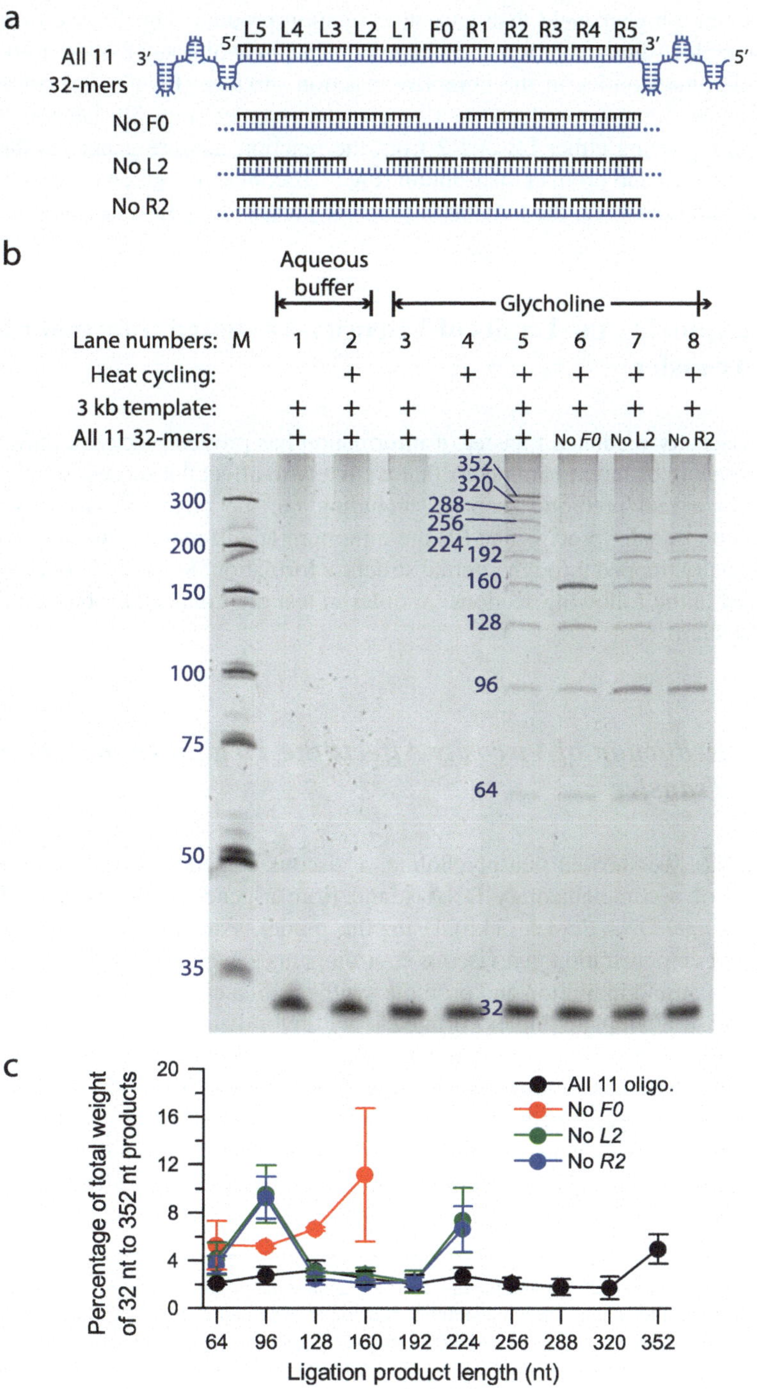

Fig. 7.10 Viscosity-enabled gene-length information transfer from a 3 kb duplex. (**a**) Schematic of the predicted products of systems composed of the 3 kb duplex and different combinations of oligonucleotide substrates that bind to an internal 352 nt region. (**b**) Denaturing polyacrylamide gel

produced in a higher yield than any other ligation product (Fig. 7.10c). Consistent with a template-directed process, the removal of a single oligonucleotide from the set of 11 oligonucleotides in the complete reaction mixture (Fig. 7.10a) limited the length of the ligation products in a predictable manner (Fig. 3b, lanes 6–8). For example, removing either L2 or R2 from the reaction mixture generates the same banding pattern and product distribution (Fig. 7.10c, blue and green curves) because both situations truncate the maximum possible product length in the same way.

7.6 Exploring the Limits of Viscosity-Mediated Information Transfer

While viscosity-mediated transfer of information has proven successful, it is necessary to explore different parameters that are likely to affect the success of this model system in a real prebiotic scenario including (1) the effect of viscosity in the information transfer process and (2) the minimum length required for a template to be kinetically trapped through internal structure formation. Some of these aspects are discussed in the following sections, in order to test the limits and robustness of our model system.

7.6.1 Reduction of Viscosity Affects the Yield of Information Transfer

In Sect. 7.6, we showed that glycholine, a viscous eutectic solvent, facilitates the synthesis of a complementary DNA strand from a gene-length region (>350 nt) within a longer DNA duplex (3 kb). Using this model system, we have made a proof-of-concept demonstration that viscous environments can be utilized to overcome the problem of strand inhibition and promote synthesis of a complementary strand from one strand of a template duplex. If nucleic acid hybridization is indeed a diffusion-limited process in viscous environments, then increased viscosity should promote more efficient information transfer from template strands. More viscous conditions would lead to longer kinetic trapping of intramolecular structures on the template strands, as well as a greater mobility difference between long template strands and short oligonucleotides. However, in principle, a solvent of sufficiently high viscosity

Fig. 7.10 (continued) (10%) stained with SYBR Gold showing products of thermal cycling the 3 kb duplex with complementary oligonucleotides. Reactions run in lanes 6, 7, and 8 were lacking F0, L2, and R2, respectively. When the 3 kb DNA and the 32-mers are heat cycled in aqueous buffer, no ligation products are observed (lane 2). (**c**) Densitometry analysis showing the percentage of the total mass in each ligation product band. Data is averaged over three experimental repeats, with error bars indicating standard deviations

will hinder the diffusion of the short oligonucleotides to the point where association with the template strands is hindered. Therefore, it is logical to expect that the yield of information transfer process increases with increasing viscosity up to a certain point and then decreases as the diffusion of all nucleic acid polymers is severely hindered.

To investigate the viscosity dependence of the method, experiments of thermal cycling of the 3 kb DNA duplex and oligonucleotides L5 through R5 were carried out in different aqueous mixtures of glycholine (Fig. 7.11a) which span a large viscosity range from 34 cP (80% glycholine by weight) to 437 cP (100% glycholine). After ligation by T4 DNA ligase, the resulting products yields of the viscosity-mediated information transfer process were analysed (Fig. 7.11b). The results indicated that, within aqueous mixtures of glycholine, the yield is strongly correlated with the viscosity of the solvent and decreases as the viscosity drops.

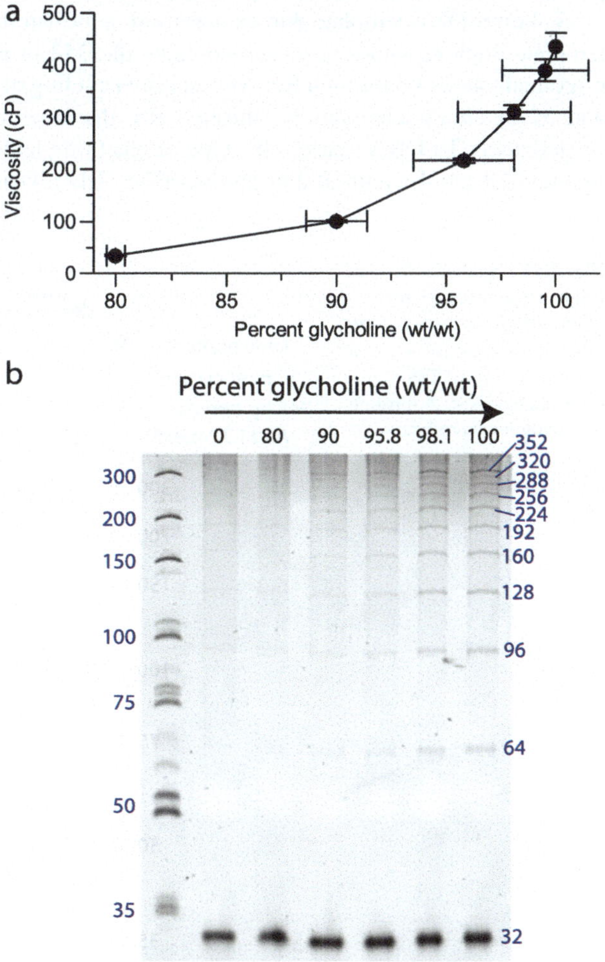

Fig. 7.11 Yield of viscosity-mediated information transfer as a function of viscosity in aqueous mixtures of glycholine. (**a**) Viscosity measurements of aqueous mixtures of glycholine. Data points shown are averaged from five measurements. (**b**) Denaturing polyacrylamide gel (10%, 8 M urea) showing the results of heat cycling and ligation of a mixture of the 3 kb DNA template and eleven 32-mer DNA oligonucleotides (L5–R5) in different aqueous mixtures of glycholine

7.6.2 *Shorter Templates Can Serve as Support for Information Transfer*

While the template utilized in our system provided a model of a mixed, arbitrary sequence of gene length, a length of 3 kb is arguably beyond the limits of prebiotic feasibility for a template. Additionally, the viscosity-mediated information transfer process should favour the replication of long templates over short templates, since long template strands are likely to form more intramolecular duplex structure—enhancing their kinetic trapping—and are less mobile (i.e. will reanneal with the complementary template strand more slowly) in a viscous solvent than short sequences. In theory, templates that are too short (i.e. lacking in intramolecular structure and/or too mobile) will not be efficiently copied in a viscous environment. To prove this theory, the model information transfer was tested using a shorter template whose size is on the same order of a minimal RNA-directed RNA polymerase ribozyme (~200 nt minimum length) (Wochner et al. 2011).

A shorter DNA template duplex was used: a 545 bp fragment of the original 3 kb template duplex, whose sequence contains the 352 nt region for binding of DNA oligonucleotides L5 through R5. After thermal cycling of the 545 bp template duplex and 32 nt oligonucleotides L5 through R5, the assembled oligonucleotides were ligated using T4 DNA ligase. This type of reaction yielded all 11 possible products, including the full-length 352 nt product (Fig. 7.12). Densitometry analysis showed

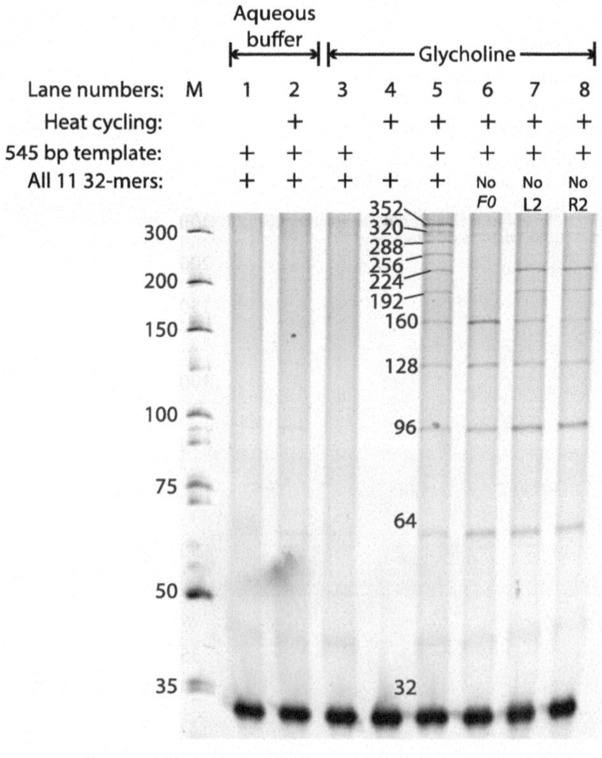

Fig. 7.12 Analysis of information transfer in glycholine utilizing a 545 bp DNA template duplex and a 4:1 molar ratio of DNA oligonucleotides L5 through R5. Samples were heat cycled and incubated with T4 DNA ligase

that the similar yields were obtained from both the 545 bp and 3 kb templates (He et al. 2017). These results indicated that a template whose length is on the same order as a ribozyme can become efficiently kinetically trapped in glycholine, enabling template-directed information transfer.

7.7 Viscosity-Mediated Information Transfer Is Possible Using RNA as a Genetic Polymer

Many researchers believe that, prior to the evolution of DNA and coded proteins, RNA played a more central role in both information storage and catalysis. However, DNA and RNA—besides the use of thymidine by the former, and uridine by the later—differ in that RNA contains an additional hydroxyl moiety in the 2′ position of the ribose ring, as compared to DNA. This hydroxyl group has a profound effect on the biophysical and chemical properties of RNA when compared to DNA—it changes the thermal stability and rigidity of the duplex (Roberts and Crothers 1992; Wang and Kool 1995) and makes single-stranded RNA much more chemically unstable due to base-catalysed hydrolysis (Li and Breaker 1999). Hence, changing the viscosity and chemical composition of the milieu could have a different outcome in viscosity-mediated information transfer based on RNA as genetic polymer.

Therefore, the approach was also tested by using a system composed of a RNA template duplex with the same sequence as the 545 bp DNA template duplex described in our model system (r545 bp duplex) and eleven 32 nt RNA oligonucleotides (rL5 through rR5) that have the same sequences as the eleven 32 nt DNA oligonucleotides described earlier.

As seen with the DNA system, thermal cycling of the r545 bp template duplex and oligonucleotides rL5 through rR5 in glycholine, and subsequent ligation with T4 RNA ligase 2, resulted in formation of all 11 possible products (Fig. 7.13, lane 5). As in the DNA system, the removal of a single RNA oligonucleotide from the reaction mixture limited the length of the ligation products in a predictable manner (Fig. 7.13, lanes 6–8). These results indicated that viscosity promoted information transfer from both DNA and RNA polymers, suggesting that our approach is a robust and potentially general strategy for overcoming strand inhibition for a range of informational polymers.

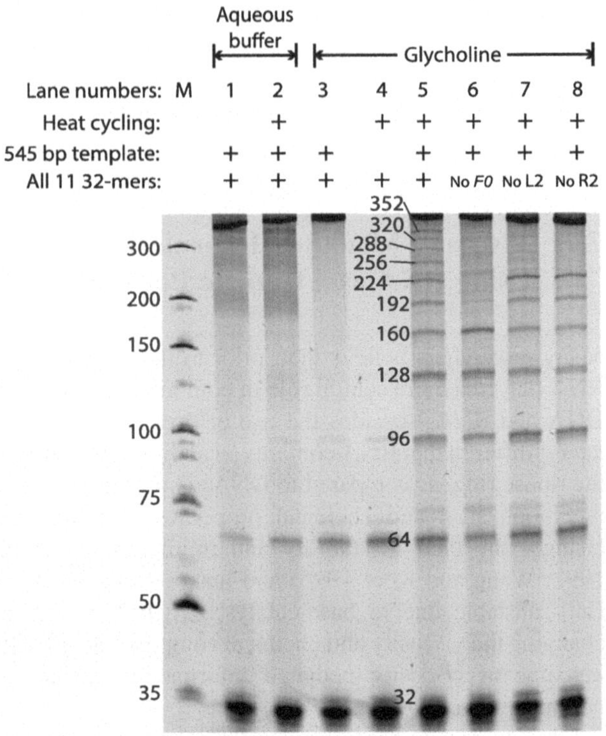

Fig. 7.13 Analysis of information transfer in glycholine utilizing a 545 bp RNA template duplex and a 4:1 molar ratio of RNA oligonucleotides rL5 through rR5. Samples were heat cycled and incubated with T4 RNA ligase 2

7.8 Conclusion and Perspectives of the Use of Thick, Viscous Environments on the Origin of Life Scenario

While significant advances have been made in recent years towards elucidating a prebiotic route to mononucleotides and nucleic acid polymers (Powner et al. 2009; Patel et al. 2015; Cafferty et al. 2013, 2016; Chen et al. 2014; Burcar et al. 2016), demonstrating enzyme-free replication of nucleic acid polymers remains an unsolved challenge (Szostak 2012). Our results have demonstrated that hot/cool cycles in viscous environments promote copying from gene-length, mixed sequence nucleic acid duplexes. We have shown that biophysical challenges associated with copying naked nucleic acids in the absence of proteins—such as the strand inhibition problem (He et al. 2017)—may be overcome through the use of viscous solvents, which can dramatically alter the thermodynamics and kinetics of nucleic acid duplex formation compared to aqueous buffer (Mamajanov et al. 2010; Lannan et al. 2012). These results also highlight the important role that geochemical factors and the physical environment likely played in the emergence and evolution of nucleic acids.

Further exploration of the prebiotic role of viscous environments may centre on their ability to promote nucleic acid replication in a length- and structure-dependent manner, providing a pressure for the selection of certain sequence features over multiple rounds of replication. It is currently unclear how a catalytically active RNA sequence would have been selected from a complex pool of sequences and persisted in a prebiotic environment. However, a recent study utilizing short RNA randomers suggests that active ribozymes could have emerged from such pools (Mutschler el al. 2018). It has been known since the 1960s that long genomic sequences are outcompeted by the faster replication of shorter parasitic sequences, resulting in shortening of the average sequence length over multiple rounds of replication (Mills et al. 1967). Studies have demonstrated that the loss of genetic information during replication in prebiotic conditions can be avoided in specific environments that promote selective accumulation of long sequences, such as a flow system that simulates a hydrothermal vent pore (Kreysing et al. 2015) or repeated cycling between nucleic acid compartmentalization in W/O droplets and mixing in bulk solution (Matsumura et al. 2016).

Viscous environments have a similar effect as compartmentalization, but rather than physically separating sequences by length or replication rate, long sequences are more readily kinetically trapped in the single-stranded template state. Therefore, viscosity should provide a means of enhancing replication of long, gene-length sequences over shorter sequences. Viscous environments could have been generated on the early Earth by evaporation of water from terrestrial pools and provide a more general solution than previous approaches (Kreysing et al. 2015) to maintain sequences separated long enough to constitute a genetically viable system.

While our results indicate that our information transfer process is indeed viscosity-mediated in glycholine (and its aqueous mixtures), an important question is whether other viscous environments can also promote information transfer. Glycerol is arguably a molecule which would have been available on the early Earth (Kaiser et al. 2014). However, the presence of choline on the prebiotic Earth is questionable. As such, glycholine is a model viscous environment, and a natural next question is whether this information transfer process is dictated by viscosity or by other features of the solvent that may alter the kinetics of nucleic acid hybridization.

It is important to note that glycholine is a small molecule solvent, whose viscosity arises from the hydrogen bonding network between the individual components that comprise the eutectic phase (Smith et al. 2014). The effects of large, macromolecular viscogens such as PEG are distinct from that of small molecule solvents, since high viscosity is also accompanied by molecular crowding effects, which increase the rate of reactions such as template duplex reformation from single strands (Zhou et al. 2008), and increased thermal stability (Mutshcler et al. 2015), accentuating the problem of strand inhibition.

Therefore, future work should explore the potential for viscosity-mediated information transfer in other viscous organic solvents. Examining information transfer across a range of solvents will not only allow us to determine whether viscosity is a general physical property—regardless of the specific solvent identity—that could have promoted replication but will also allow deeper exploration of the relationship between viscosity and efficiency of the information transfer process.

References

Abbott AP, Capper G, Davies DL, Rasheed RK, Tambyrajah V (2003) Novel solvent properties of choline chloride/urea mixtures. Chem Commun (Camb) 1:70–71. https://doi.org/10.1039/b210714g

Abbott AP, Harris RC, Ryder KS, D'Agostino C, Gladden LF, Mantle MD (2011) Glycerol eutectics as sustainable solvent systems. Green Chem 13(1):82–90. https://doi.org/10.1039/C0GC00395F

Abe H, Kool ET (2004) Destabilizing universal linkers for signal amplification in self-ligating probes for RNA. J Am Chem Soc 126(43):13980–13986. https://doi.org/10.1021/ja046791c

Adamala K, Engelhart AE, Szostak JW (2015) Generation of functional RNAs from inactive oligonucleotide complexes by non-enzymatic primer extension. J Am Chem Soc 137(1):483–489. https://doi.org/10.1021/ja511564d

Alberts B, Johnson A, Lewis J, Raff M, Roberts K, Walter P (2008) Molecular biology of the cell, 5th edn. Graland Science, New York, NY

Attwater J, Wochner A, Pinheiro VB, Coulson A, Holliger P (2010) Ice as a protocellular medium for RNA replication. Nat Commun 1:76. https://doi.org/10.1038/ncomms1076

Attwater J, Tagami S, Kimoto M, Butler K, Kool ET, Wengel J, Herdewijn P, Hirao I, Holliger P (2013a) Chemical fidelity of an RNA polymerase ribozyme. Chem Sci 4(7):2804. https://doi.org/10.1039/c3sc50574j

Attwater J, Wochner A, Holliger P (2013b) In-ice evolution of RNA polymerase ribozyme activity. Nat Chem 5(12):1011–1018. https://doi.org/10.1038/nchem.1781

Attwater J, Raguram A, Morgunov AS, Gianni E, Holliger P (2018) Ribozyme-catalysed RNA synthesis using triplet building blocks. eLife 7:e35255

Bell SP, Dutta A (2002) DNA replication in eukaryotic cells. Annu Rev Biochem 71:333–374. https://doi.org/10.1146/annurev.biochem.71.110601.135425

Breaker RR (2011) Prospects for riboswitch discovery and analysis. Mol Cell 43(6):867–879. https://doi.org/10.1016/j.molcel.2011.08.024

Breaker RR, Joyce GF (2014) The expanding view of RNA and DNA function. Chem Biol 21(9):1059–1065. https://doi.org/10.1016/j.chembiol.2014.07.008

Burcar B, Pasek M, Gull M, Cafferty BJ, Velasco F, Hud NV, Menor-Salván C (2016) Darwin's warm little pond: a one-pot reaction for prebiotic phosphorylation and the mobilization of phosphate from minerals in a urea-based solvent. Angew Chem Int Ed 55(42):13249–13253. https://doi.org/10.1002/anie.201607773

Cafferty BJ, Hud NV (2014) Abiotic synthesis of RNA in water: a common goal of prebiotic chemistry and bottom-up synthetic biology. Curr Opin Chem Biol 22:146–157. https://doi.org/10.1016/j.cbpa.2014.09.015

Cafferty BJ, Gállego I, Chen MC, Farley KI, Eritja R, Hud NV (2013) Efficient self-assembly in water of long noncovalent polymers by nucleobase analogues. J Am Chem Soc 135(7):2447–2450. https://doi.org/10.1021/ja312155v

Cafferty BJ, Fialho DM, Khanam J, Krishnamurthy R, Hud NV (2016) Spontaneous formation and base pairing of plausible prebiotic nucleotides in water. Nat Commun 7:11328

Cech TR (2000) The ribosome is a ribozyme. Science 289(5481):878–879. https://doi.org/10.1126/science.289.5481.878

Chen MC, Cafferty BJ, Mamajanov I, Gállego I, Khanam J, Krishnamurthy R, Hud NV (2014) Spontaneous prebiotic formation of a beta-ribofuranoside that self-assembles with a complementary heterocycle. J Am Chem Soc 136(15):5640–5646. https://doi.org/10.1021/ja410124v

Deck C, Jauker M, Richert C (2011) Efficient enzyme-free copying of all four nucleobases templated by immobilized RNA. Nat Chem 3(8):603–608

Dose C, Ficht S, Seitz O (2006) Reducing product inhibition in DNA-template-controlled ligation reactions. Angew Chem Int Ed 45:5369–5373

Doty P, Marmur J, Eigner J, Schildkraut C (1960) Strand separation and specific recombination in deoxyribonucleic acids: physical chemical studies. Proc Natl Acad Sci USA 46(4):461–476

Ekland EH, Bartel DP (1995) The secondary structure and sequence optimization of an RNA ligase ribozyme. Nucleic Acids Res 23(16):3231–3238. https://doi.org/10.1093/nar/23.16.3231

Engelhart AE, Adamala KP, Szostak JW (2016) A simple physical mechanism enables homeostasis in primitive cells. Nat Chem 8(5):448–453. https://doi.org/10.1038/nchem.2475

Fernando C, Gv K, Szathmary E (2007) A stochastic model of nonenzymatic nucleic acid replication: "elongators" sequester replicators. J Mol Evol 64:572–585

Forsythe JG, Yu S-S, Mamajanov I, Grover MA, Krishnamurthy R, Fernández FM, Hud NV (2015) Ester-mediated amide bond formation driven by wet–dry cycles: a possible path to polypeptides on the prebiotic earth. Angew Chem Int Ed 54(34):9871–9875. https://doi.org/10.1002/anie.201503792

Gállego I, Grover MA, Hud NV (2015) Folding and imaging of DNA nanostructures in anhydrous and hydrated deep-eutectic solvents. Angew Chem Int Ed 54(23):6765–6769. https://doi.org/10.1002/anie.201412354

Ghildiyal M, Zamore PD (2009) Small silencing RNAs: an expanding universe. Nat Rev Genet 10 (2):94–108

Gilbert W (1986) Origin of life – the RNA world. Nature 319(6055):618–618. https://doi.org/10.1038/319618a0

Green C, Tibbetts C (1981) Reassociation rate limited displacement of DNA strands by branch migration. Nucleic Acids Res 9(8):1905–1918. https://doi.org/10.1093/nar/9.8.1905

Grossmann TN, Strohbach A, Seitz O (2008) Achieving turnover in DNA-templated reactions. ChemBioChem 9(14):2185–2192. https://doi.org/10.1002/cbic.200800290

Gull M, Zhou MS, Fernandez FM, Pasek MA (2014) Prebiotic phosphate ester syntheses in a deep eutectic solvent. J Mol Evol 78(2):109–117. https://doi.org/10.1007/s00239-013-9605-9

Hagen SJ (2010) Solvent viscosity and friction in protein folding dynamics. Curr Protein Pept Sci 11(5):385–395

He C, Gállego I, Laughlin B, Grover MA, Hud NV (2017) A viscous solvent enables information transfer from gene-length nucleic acids in a model prebiotic replication cycle. Nat Chem 9 (4):318–324. https://doi.org/10.1038/nchem.2628

Horning DP, Joyce GF (2016) Amplification of RNA by an RNA polymerase ribozyme. Proc Natl Acad Sci USA 113(35):9786–9791. https://doi.org/10.1073/pnas.1610103113

Hud Nicholas V, Cafferty Brian J, Krishnamurthy R, Williams Loren D (2013) The origin of RNA and "my grandfather's axe". Chem Biol 20(4):466–474. https://doi.org/10.1016/j.chembiol.2013.03.012

Ichihashi N, Usui K, Kazuta Y, Sunami T, Matsuura T, Yomo T (2013) Darwinian evolution in a translation-coupled RNA replication system within a cell-like compartment. Nat Commun 4:2494. https://doi.org/10.1038/ncomms3494

Inoue T, Orgel LE (1981) Substituent control of the poly(C)-directed oligomerization of guanosine 5'-phosphoroimidazolide. J Am Chem Soc 103(25):7666–7667. https://doi.org/10.1021/ja00415a051

Ivica NA, Obermayer B, Campbell GW, Rajamani S, Gerland U, Chen IA (2013) The paradox of dual roles in the RNA world: resolving the conflict between stable folding and templating ability. J Mol Evol 77(3):55–63. https://doi.org/10.1007/s00239-013-9584-x

Johnston WK, Unrau PJ, Lawrence MS, Glasner ME, Bartel DP (2001) RNA-catalyzed RNA polymerization: accurate and general RNA-templated primer extension. Science 292 (5520):1319–1325. https://doi.org/10.1126/science.1060786

Joyce GF (2002) The antiquity of RNA-based evolution. Nature 418(6894):214–221

Kaiser RI, Maity S, Jones BM (2014) Synthesis of prebiotic glycerol in interstellar ices. Angew Chem 127(1):197–202. https://doi.org/10.1002/ange.201408729

Kausar A, McKay RD, Lam J, Bhogal RS, Tang AY, Gibbs-Davis JM (2011) Tuning DNA stability to achieve turnover in template for an enzymatic ligation reaction. Angew Chem Int Ed 50:8922–8926

Kervio E, Sosson M, Richert C (2016) The effect of leaving groups on binding and reactivity in enzyme-free copying of DNA and RNA. Nucleic Acids Res. https://doi.org/10.1093/nar/gkw476

Kreysing M, Keil L, Lanzmich S, Braun D (2015) Heat flux across an open pore enables the continuous replication and selection of oligonucleotides towards increasing length. Nat Chem 7 (3):203–208. https://doi.org/10.1038/nchem.2155

Kruger K, Grabowski PJ, Zaug AJ, Sands J, Gottschling DE, Cech TR (1982) Self-splicing RNA: autoexcision and autocyclization of the ribosomal RNA intervening sequence of tetrahymena. Cell 31(1):147–157. https://doi.org/10.1016/0092-8674(82)90414-7

Kunkel TA, Bebenek R (2000) DNA replication fidelity. Annu Rev Biochem 69:497–529. https://doi.org/10.1146/annurev.biochem.69.1.497

Lannan FM, Mamajanov I, Hud NV (2012) Human telomere sequence DNA in water-free and high-viscosity solvents: G-quadruplex folding governed by Kramers rate theory. J Am Chem Soc 134 (37):15324–15330. https://doi.org/10.1021/ja303499m

Li Y, Breaker RR (1999) Kinetics of RNA degradation by specific base catalysis of transesterification involving the 2'-hydroxyl group. J Am Chem Soc 121(23):5364–5372. https://doi.org/10.1021/ja990592p

Luther A, Brandsch R, Gv K (1998) Surface-promoted replication and exponential amplification of DNA analogues. Nature 396:245–248

Mamajanov I, Engelhart AE, Bean HD, Hud NV (2010) DNA and RNA in anhydrous media: duplex, triplex, and g-quadruplex secondary structures in a deep eutectic solvent. Angew Chem Int Ed 49(36):6310–6314. https://doi.org/10.1002/anie.201001561

Mamajanov I, MacDonald PJ, Ying JY, Duncanson DM, Dowdy GR, Walker CA, Engelhart AE, Fernandez FM, Grover MA, Hud NV, Schork FJ (2014) Ester formation and hydrolysis during wet-dry cycles: generation of far-from-equilibrium polymers in a model prebiotic reaction. Macromolecules 47(4):1334–1343. https://doi.org/10.1021/ma402256d

Mansy SS, Szostak JW (2008) Thermostability of model protocell membranes. Proc Natl Acad Sci USA 105(36):13351–13355. https://doi.org/10.1073/pnas.0805086105

Mansy SS, Szostak JW (2009) Reconstructing the emergence of cellular life through the synthesis of model protocells. Cold Spring Harb Symp Quant Biol 74:47–54. https://doi.org/10.1101/sqb.2009.74.014

Mansy SS, Schrum JP, Krishnamurthy M, Tobe S, Treco DA, Szostak JW (2008) Template-directed synthesis of a genetic polymer in a model protocell. Nature 454(7200):122–125. https://doi.org/10.1038/nature07018

Mast CB, Schink S, Gerland U, Braun D (2013) Escalation of polymerization in a thermal gradient. Proc Natl Acad Sci USA 110(20):8030–8035. https://doi.org/10.1073/pnas.1303222110

Matsumura S, Kun A, Ryckelynck M, Coldren F, Szilagyi A, Jossinet F, Rick C, Nghe P, Szathmary E, Griffiths AD (2016) Transient compartmentalization of RNA replicators prevents extinction due to parasites. Science 354(6317):1293–1296. https://doi.org/10.1126/science.aag1582

Mills DR, Peterson RL, Spiegelman S (1967) An extracellular darwinian experiment with a self-duplicating nucleic acid molecule. Proc Natl Acad Sci 58(1):217–224. https://doi.org/10.1073/pnas.58.1.217

Mutschler H, Wochner A, Holliger P (2015) Freeze–thaw cycles as drivers of complex ribozyme assembly. Nat Chem 7(6):502–508. https://doi.org/10.1038/nchem.2251

Mutschler H, Taylor AI, Lightowlers A, Houlihan G, Abramov M, Herdewijn P, Holliger P (2018) Innate potential of random genetic oligomer pools for recombination. bioRxiv 320499. https://doi.org/10.1101/320499

Panyutin IG, Hsieh P (1994) The kinetics of spontaneous DNA branch migration. Proc Natl Acad Sci USA 91(6):2021–2025

Patel BH, Percivalle C, Ritson DJ, DuffyColm D, Sutherland JD (2015) Common origins of RNA, protein and lipid precursors in a cyanosulfidic protometabolism. Nat Chem 7(4):301–307. https://doi.org/10.1038/nchem.2202

Portella G, Germann MW, Hud NV, Orozco M (2014) MD and NMR analyses of choline and TMA binding to duplex DNA: on the origins of aberrant sequence-dependent stability by alkyl cations in aqueous and water-free solvents. J Am Chem Soc 136(8):3075–3086. https://doi.org/10.1021/ja410698u

Powner MW, Gerland B, Sutherland JD (2009) Synthesis of activated pyrimidine ribonucleotides in prebiotically plausible conditions. Nature 459(7244):239–242

Prywes N, Blain JC, Del Frate F, Szostak JW (2016) Nonenzymatic copying of RNA templates containing all four letters is catalyzed by activated oligonucleotides. eLife 5:e17756. https://doi.org/10.7554/eLife.17756

Radding CM, Beattie KL, Holloman WK, Wiegand RC (1977) Uptake of homologous single-stranded fragments by superhelical DNA: IV. Branch migration. J Mol Biol 116(4):825–839. https://doi.org/10.1016/0022-2836(77)90273-X

Roberts R, Crothers D (1992) Stability and properties of double and triple helices: dramatic effects of RNA or DNA backbone composition. Science 258(5087):1463–1466. https://doi.org/10.1126/science.1279808

Shechner DM, Grant RA, Bagby SC, Koldobskaya Y, Piccirilli JA, Bartel DP (2009) Crystal structure of the catalytic core of an RNA-polymerase ribozyme. Science 326(5957):1271–1275. https://doi.org/10.1126/science.1174676

Smith EL, Abbott AP, Ryder KS (2014) Deep eutectic solvents (DESs) and their applications. Chem Rev 114(21):11060–11082. https://doi.org/10.1021/cr300162p

Stoeger T, Battich N, Pelkmans L (2016) Passive noise filtering by cellular compartmentalization. Cell 164(6):1151–1161. https://doi.org/10.1016/j.cell.2016.02.005

Storz G (2002) An expanding universe of noncoding RNAs. Science 296(5571):1260–1263. https://doi.org/10.1126/science.1072249

Szostak JW (2012) The eightfold path to non-enzymatic RNA replication. J Syst Chem 3:2. https://doi.org/10.1186/1759-2208-3-2

Tateishi-Karimata H, Sugimoto N (2012) A-T base pairs are more stable than G-C base pairs in a hydrated ionic liquid. Angew Chem Int Ed Eng 51(6):1416–1419. https://doi.org/10.1002/anie.201106423

Viasnoff V, Meller A, Isambert H (2006) DNA nanomechanical switches under folding kinetics control. Nano Lett 6(1):101–104. https://doi.org/10.1021/nl052161c

Wachowius F, Attwater J, Holliger P (2017) Nucleic acids: function and potential for abiogenesis. Q Rev Biophys 50. https://doi.org/10.1017/s0033583517000038

Walker SI, Grover MA, Hud NV (2012) Universal sequence replication, reversible polymerization and early functional biopolymers: a model for the initiation of prebiotic sequence evolution. PLoS One 7(4):e34166. https://doi.org/10.1371/journal.pone.0034166

Wang S, Kool ET (1995) Origins of the large differences in stability of DNA and RNA helixes: C-5 methyl and 2'-hydroxyl effects. Biochemistry 34(12):4125–4132. https://doi.org/10.1021/bi00012a031

Watson JD, Crick FC (1953) Molecular structure of nucleic acids: a structure for deoxyribose nucleic acids. Nature 171:737–738

Wochner A, Attwater J, Coulson A, Holliger P (2011) Ribozyme-catalyzed transcription of an active ribozyme. Science 332(6026):209–229. https://doi.org/10.1126/science.1200752

Woese CR (2001) Translation: in retrospect and prospect. RNA 7:1055–1067. https://doi.org/10.1017/S1355838201010615

Zhan Z-YJ, Lynn DG (1997) Chemical amplification through template-directed synthesis. J Am Chem Soc 119:12420–12421

Zhang X, Qu Y, Chen H, Rouzina I, Zhang S, Doyle PS, Yan J (2014) Interconversion between three overstretched DNA structures. J Am Chem Soc 136(45):16073–16080. https://doi.org/10.1021/ja5090805

Zhou H-X, Rivas G, Minton AP (2008) Macromolecular crowding and confinement: biochemical, biophysical, and potential physiological consequences. Annu Rev Biophys 37:375–397. https://doi.org/10.1146/annurev.biophys.37.032807.125817

Chapter 8
Folding and Catalysis Near Life's Origin: Support for Fe^{2+} as a Dominant Divalent Cation

C. Denise Okafor, Jessica C. Bowman, Nicholas V. Hud, Jennifer B. Glass, and Loren Dean Williams

Abstract There is broad consensus that during and immediately following the origin of life, RNA was the single biopolymer or was among a small group of cooperating biopolymers. During the origin of life, the Archean Earth was anoxic; Fe^{2+} was abundant and relatively benign. We hypothesize that RNA used Fe^{2+} as a cofactor instead of, or along with, Mg^{2+} during the inception and early phases of biology, until the Great Oxidation Event (GOE). In this model, RNA participated in a metal substitution during the GOE, whereby Mg^{2+} replaced Fe^{2+} as the dominant RNA cofactor. A GOE-induced Fe^{2+} to Mg^{2+} substitution predicts that under 'early Earth' (anoxic) conditions, Fe^{2+} can participate in a variety of functions, including mediation of RNA folding and catalysis by ribozymes and proteins. Understanding the influence of Fe^{2+} on nucleic acid structure and function could provide an important link between the geological record and the ancestral biological world. This review focuses on experimental work investigating the interactions and functions of RNA and nucleic acid processing proteins with Fe^{2+} under anoxic, early Earth conditions.

8.1 Introduction

Cations play complex and essential roles in the folding of RNA into compact native states, in which negatively charged phosphate groups of the backbone are forced into close proximity (Brion and Westhof 1997; Hsiao et al. 2008). In association with folded RNAs, cations occupy a continuum of states distinguished by extent of direct

---

C. D. Okafor · J. C. Bowman · N. V. Hud · L. D. Williams (✉)
School of Chemistry and Biochemistry, NSF-NASA Center for Chemical Evolution, Georgia Institute of Technology, Atlanta, GA, USA
e-mail: loren.williams@chemistry.gatech.edu

J. B. Glass
School of Earth and Atmospheric Sciences, Georgia Institute of Technology, Atlanta, GA, USA

© Springer International Publishing AG, part of Springer Nature 2018
C. Menor-Salván (ed.), *Prebiotic Chemistry and Chemical Evolution of Nucleic Acids*, Nucleic Acids and Molecular Biology 35,
https://doi.org/10.1007/978-3-319-93584-3_8

coordination by RNA, thermodynamic contributions to folding, rates and dimensionalities of diffusion, and influence on specific structural states and on populations (Bowman et al. 2012). We partition this continuum of states into four classes: free, condensed, glassy, and chelated cations.

Condensed divalent cations are electrostatically linked to RNA and are found within envelopes that extend well beyond the van der Waals surface of the collapsed RNA. Condensed cations are fully hydrated, with near bulk-like diffusion properties. Glassy cations, with mobility restricted to one or two dimensions, are closely associated with RNA and are often partially dehydrated. Chelated cations, with two to four first-shell RNA ligands, can stabilize specific conformational states of the RNA. Chelated cations are less frequent than glassy ions, which are less frequent than condensed ions.

Cations assist with RNA catalysis by stabilizing transition states (Butcher 2011; Johnson-Buck et al. 2011) and/or by participating in catalysis (Hanna and Doudna 2000). Mg^{2+} is considered the dominant cofactor in ribozyme catalysis and in nucleic acid processing enzymes, and is one of the most abundant divalent cations in vivo (Draper 2004).

We have developed a model in which Fe^{2+}, in association with RNA and with nucleic acid processing enzymes, was a dominant divalent metal ion in life on the early Earth. We further postulate that Fe^{2+} complexes have been retained in certain extant anaerobes. Our recent results support a model in which iron, possibly along with magnesium, was a critical cofactor for nucleic acids during the Archean Eon. The anoxic and iron-rich (in combination, "ferruginous") conditions of the Archean would have inhibited destructive iron-mediated processes such as Fenton chemistry (Prousek 2007; Kozlowski et al. 2014), allowing Fe^{2+} to become deeply embedded in biological chemistry (Theil and Goss 2009).

Iron was abundant, soluble, and benign when life originated and first proliferated (Anbar 2008; Hazen and Ferry 2010; Holland 2006; Johnson et al. 2008). For around two billion years, the anoxic conditions of the ancient Earth sustained soluble Fe^{2+} rather than insoluble Fe^{3+} that is present in seawater today at picomolar concentrations. The abundance of iron in Archean seawater is evident from the extraordinary banded iron formations that span two billion years of the geologic rock record (Reinhard et al. 2017). It was originally assumed that in Archean seawater Fe^{2+} would be buffered at ~0.1 mM by equilibrium with ferrous carbonate (siderite, $FeCO_3$) (Holland 1973, 1984; Drever 1974). However, subsequent work has shown that siderite could have been strongly supersaturated in ancient oceans due to the slow kinetics of siderite precipitation. Instead, Derry (2015) suggests that ancient Fe^{2+} concentrations would have been controlled by the solubility of ferrous phosphate (vivianite), $3Fe^{2+} + 2HPO_4^{2-} + 8H_2O = Fe_3(PO_4)_2 \cdot 8H_2O + 2H^+$. The most recent estimates of Archean HPO_4^{2-} is 0.04–0.13 μM (Jones et al. 2015), which would yield $Fe^{2+} > 1$ mM (Derry 2015). Much less is known about seawater Mg^{2+} in the Archean, except that concentrations would have been lower than in modern oceans (~10 mM vs 52 mM Mg^{2+} in modern oceans) due to enhanced hydrothermal activity, which strips Mg^{2+} from seawater during hot-water-rock interactions in the ocean crust (Izawa et al. 2010). It is important to note that there

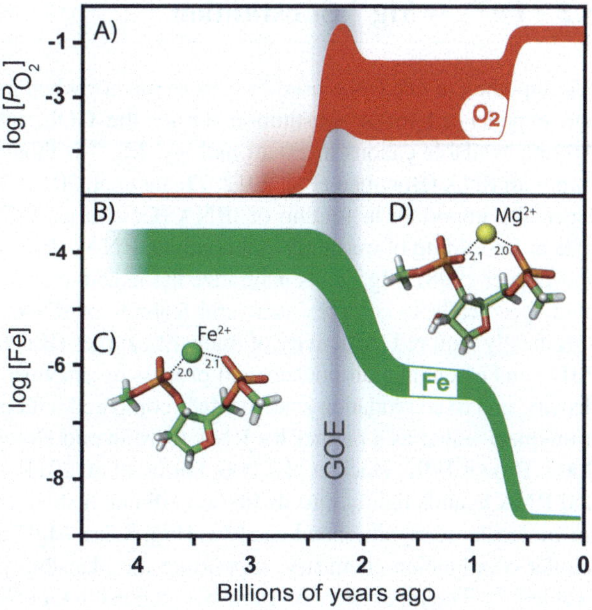

Fig. 8.1 Fe^{2+} was abundant in the biosphere before the Great Oxidation Event (GOE), when O_2 began to accumulate in the atmosphere. Approximate amount of (**a**) atmospheric O_2 and (**b**) dissolved iron in Earth's oceans over time. RNA coordinating (**c**) Fe^{2+} or (**d**) Mg^{2+}. Adapted from Lyons et al. (2014) and Anbar (2008). Vertical thickness of lines indicating O_2 and Fe^{2+} concentrations represent uncertainty in historical levels

are order-of-magnitude uncertainties for the ion content of Archean seawater (Holland et al. 2003), but the latest estimates suggest that Fe^{2+} and Mg^{2+} could have approached equimolar concentrations in the low mM in Archean seawater, while in modern oxic waters, Mg^{2+} is ~10 orders of magnitude more abundant than Fe^{2+}.

The Great Oxidation Event (GOE) forced the modern condition of iron scarcity and iron-mediated oxidative damage (Aguirre and Culotta 2012; Ushizaka et al. 2011; Martin and Imlay 2011; Cotruvo and Stubbe 2011; Wolfe-Simon et al. 2006; Anjem et al. 2009; Harel et al. 2014; Torrents et al. 2002). With the rise in oxygen, much of the Fe^{2+} of the early Earth was oxidized and geologically sequestered (Fig. 8.1). However, toxicity and vanishingly low concentrations of Fe^{2+} on the surface of the modern oxic Earth have not reversed the effects of iron's extensive evolutionary history and great utility in catalysis. Iron is the most abundant transition metal by far in human cells (Iyengar and Woittiez 1988).

The idea that Fe^{2+} played a crucial role in association with RNA on the early Earth is motivated by the theories that life may have originated (1) with RNA-based genetic and metabolic systems, i.e., the RNA world (Atkins et al. 2011), or (2) in a system of RNA-protein mutualism (Lanier et al. 2017). In both of these scenarios, and in essentially all models of the origin of life, RNA was an important component of early life in an anoxic, iron-rich environment. Therefore, understanding the influence of Fe^{2+} on nucleic acid structure and function could provide important links between the geological record and the ancestral biological world. This review focuses on work that has investigated the interactions of RNA, Fe^{2+}, and proteins under anoxic, early Earth conditions.

8.2 $Fe^{2+} \rightarrow Mg^{2+}$ Substitution

We hypothesize that RNA used Fe^{2+} as a cofactor when iron was benign and abundant, and experienced metal substitution during the GOE, whereby Mg^{2+} replaced Fe^{2+}. Among available cations in extant biology, Mg^{2+} is thought to have a special relationship with RNA (Bowman et al. 2012; Zheng et al. 2015). Mg^{2+} was seen early on to be especially important in folding of tRNA (Cole et al. 1972) and is now thought to be critical for folding of essentially all compact RNAs (Pyle 1993; Cate et al. 1996; Misra and Draper 1998). Mg^{2+} ions neutralize the negative charge of the RNA backbone and bind specifically to complex structural features of RNA (Petrov et al. 2012). Mg^{2+} is specifically required for activity of many ribozymes (Butcher 2011; Johnson-Buck et al. 2011) and essentially all nucleic acid processing enzymes. The small size, high charge density, and fixed oxidation state of Mg^{2+} compared with other biological cations makes it uniquely suited as a partner for RNA (Rashin and Honig 1985; Maguire and Cowan 2002; Brown 1992; Bock et al. 2006; Petrov et al. 2011). Mg^{2+} packs water molecules and RNA ligands tightly into its first coordination shell, orienting and polarizing water for molecular recognition and catalysis (Fig. 8.2a). Mg^{2+} and Fe^{2+} are characterized by similar coordination chemistry, supporting the plausibility of the $Fe^{2+} \rightarrow Mg^{2+}$ model (Table 8.1). The parameters of Table 8.1 suggest that Fe^{2+} and Mg^{2+} are similar in each of the classes of RNA association: free, condensed, glassy, and chelated. Our substitution hypothesis is based in part on analogy with metal substitution in some metalloproteins. The GOE resulted in decreases in Fe^{2+} concentrations and subsequent replacement in metabolic processes by other metals such as Cu^{2+}, Zn^{2+}, and Mn^{2+} (Aguirre and Culotta 2012; Ushizaka et al. 2011; Martin and Imlay 2011; Cotruvo and Stubbe 2011; Wolfe-Simon et al. 2006; Anjem et al. 2009; Harel et al. 2014; Torrents et al. 2002; Dupont et al. 2006, 2010).

Mg^{2+} coordinates RNA phosphate oxyanions with octahedral coordination geometry, bringing the first-shell ligands within ~2.1 angstroms of the metal ion. Quantum mechanics calculations show that the octahedral geometry of RNA-metal complexes and RNA conformation are conserved for Mg^{2+} or Fe^{2+} (Fig. 8.2) (Athavale et al. 2012).

A GOE-induced Fe^{2+} to Mg^{2+} substitution implies that under "early Earth" (anoxic) conditions, Fe^{2+} would have been able to perform a variety of functions, mediating RNA folding and acting as a cofactor for ribozyme and nucleic acid processing enzymes. In addition, because it is not limited to a single oxidation state, iron has the potential to confer redox functionality on RNA. In anoxia, Fe^{2+} would be unable to cleave RNA via the Fenton reaction (Prousek 2007; Kozlowski et al. 2014). For RNAs that are dependent on Mg^{2+} for folding or catalytic activity, structure and function should be conserved upon substitution of Mg^{2+} for Fe^{2+}. Likewise, nucleic acid processing proteins dependent on Mg^{2+} as a cofactor should be able to catalyze reactions with Fe^{2+}.

The $Fe^{2+} \rightarrow Mg^{2+}$ substitution hypothesis has been tested in vitro by reversing the putative substitution in anoxic conditions (Athavale et al. 2012; Okafor et al. 2017; Hsiao et al. 2013; Popovic et al. 2015). Mg^{2+} was removed and Fe^{2+} added to RNA

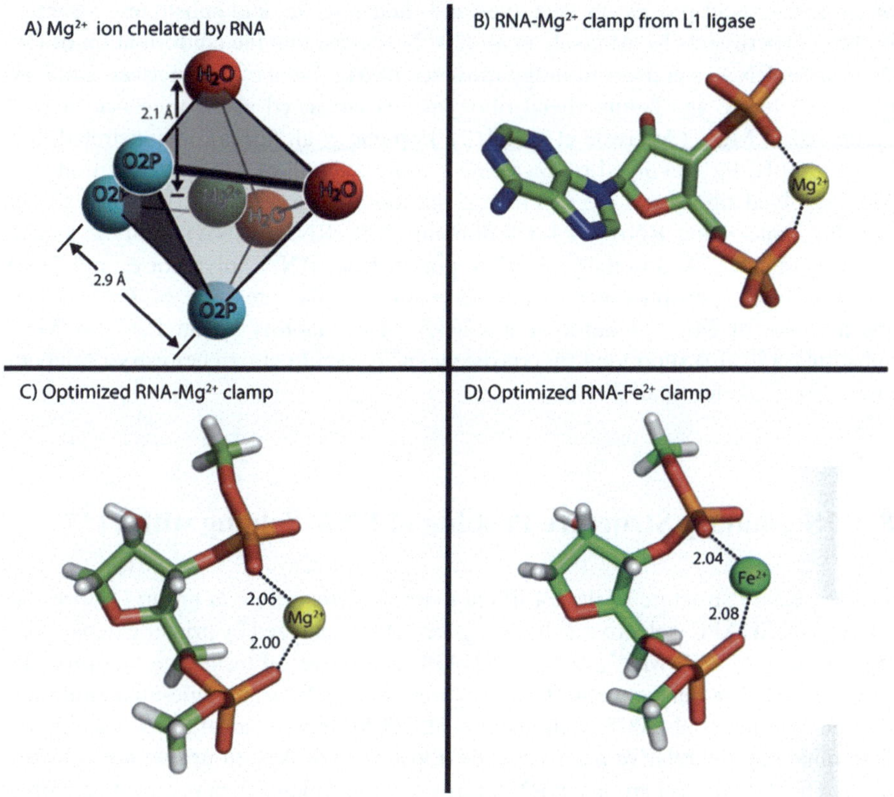

Fig. 8.2 (**a**) Mg^{2+} ion chelated by RNA. This Mg^{2+} ion (green sphere) is octahedral, with three first-shell phosphate oxygens of the rRNA (cyan) and three first-shell water oxygens (red). Mg^{2+}-oxygen distances are around 2.1 Å. (**b**) A first-shell RNA-Mg^{2+} complex: an RNA-Mg^{2+} clamp from the L1 ribozyme ligase (PDB 2OIU). (**c**) An RNA-Mg^{2+} clamp optimized by high-level QM calculations. (**d**) An optimized RNA-Fe^{2+} clamp. Each cation (Mg^{2+} or Fe^{2+}) is hexacoordinate. Mg^{2+} is shown as a yellow sphere, and Fe^{2+} is shown as a green sphere. Water molecules are omitted. Panel **a** adapted from Bowman et al. (2012). Panels **b**, **c**, and **d** from Athavale et al. (2012)

Table 8.1 Characteristics of Mg^{2+} and Fe^{2+}

	r (Å)[a]	AOCN[b]	$-\Delta H_{hyd}$[c]	pK_a[d]	ΔH[e]
Mg^{2+}	0.65	6	458[f]	11.4	
Fe^{2+}[g]	0.74	6	464[h]	9.5	−1.3

[a]Ionic radius (Brown 1988)
[b]Average observed coordination number (Brown 1988)
[c]Hydration enthalpy (kcal mol) (Brion and Westhof 1997)
[d]pK_a of $M^{2+}(H_2O)_6$ where $M^{2+} = Fe^{2+}$ or Mg^{2+} (Wulfsberg 1991)
[e]Interaction enthalpy relative to Mg^{2+} (kcal mol^{-1}) for RNA clamp formation (Dupont et al. 2010; Athavale et al. 2012). From Okafor et al. (2017)
[f]From Rashin and Honig (1985)
[g]High spin
[h]From Uudsemaa and Tamm (2004)

or to nucleic acid processing enzymes and their nucleic acid substrates. Structure probing experiments by Athavale et al. (2012) suggest that the conformation of P4–P6 domain RNA is conserved in the presence of Mg^{2+} or Fe^{2+}. Moreover, catalysis by the L1 ligase and hammerhead ribozymes is enhanced in the presence of Fe^{2+} compared to Mg^{2+} (Athavale et al. 2012). Popovic et al. (2015) demonstrated that, at neutral pH, Fe^{2+}-evolved ribozymes are more tolerant of Mg^{2+} substitution than Mg^{2+}-evolved ribozymes are of Fe^{2+} substitution. Hsiao et al. (2013) have shown that Fe^{2+} and some RNAs (P4–P6 domain, 23S rRNA, tRNA) catalyze single-electron transfer. Additionally, a DNA polymerase, RNA polymerase, and DNA ligase catalyze phosphodiester bond formation in the presence of Fe^{2+} and in the absence of Mg^{2+} (Okafor et al. 2017). The viability of an $Fe^{2+} \rightarrow Mg^{2+}$ substitution is also supported by consistencies in coordination chemistry (Athavale et al. 2012; Okafor et al. 2017).

8.3 Secondary Structure Probing of RNA Folding with Fe^{2+}

The secondary structure of the P4–P6 domain of *T. thermophila* group I intron was assayed by SHAPE (selective 2′-hydroxyl acylation analyzed by primer extension) in the presence of Na^+, Mg^{2+}, or Fe^{2+}. SHAPE is a powerful technique that provides secondary and tertiary structural information about RNAs at single-nucleotide resolution (Merino et al. 2005; Wilkinson et al. 2005). Briefly, local RNA flexibility is determined by the relative reactivities of ribose 2′-hydroxyl groups with an electrophile. The 2′-hydroxyl groups in RNA form adducts which, when reverse transcribed using fluorescently labeled primers, give truncated products. Resolution and visualization of fluorescent cDNA fragments using capillary electrophoresis permit the determination of RNA secondary structure based on the local flexibility of each nucleotide (Wilkinson et al. 2005, 2008).

In the presence of Na^+, the P4–P6 domain gave a SHAPE fingerprint consistent with the known secondary structure (Athavale et al. 2012; Cate et al. 1997). Upon addition of Mg^{2+}, SHAPE indicates that the structure of P4–P6 collapses to form long-range, tertiary interactions (Fig. 8.3). The structure obtained with Mg^{2+} is similar to previously reported SHAPE profiles for the same RNA in the presence of Mg^{2+} (Vicens et al. 2007).

The fingerprint of SHAPE reactivity for the P4–P6 domain RNA is conserved when Mg^{2+} is replaced by Fe^{2+} under anoxic conditions. This result suggests that tertiary interactions as well as specific and non-specific Mg^{2+}-RNA interactions are recapitulated by Fe^{2+} in the absence of oxygen. Fe^{2+} appears to be a true mimic for Mg^{2+} during RNA folding.

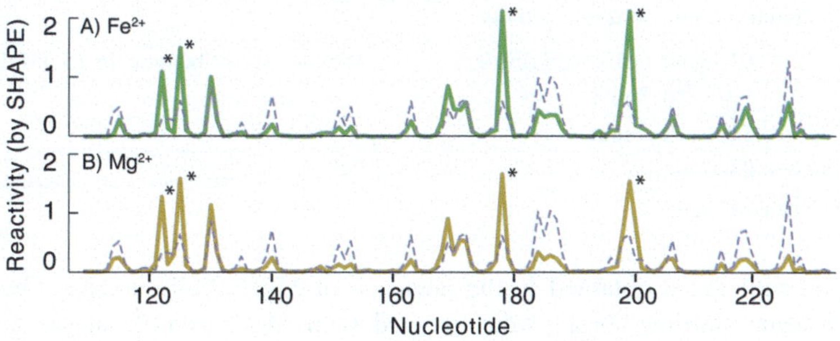

Fig. 8.3 Similar changes are observed in SHAPE reactivity with addition of Mg^{2+} or Fe^{2+}. (**a**) SHAPE profile of P4–P6 domain RNA in the presence of 250 mM NaCl and 2.5 mM Fe^{2+} (green) or 250 mM NaCl alone (purple, dashed). (**b**) SHAPE profile in the presence of 250 mM NaCl and 2.5 mM Mg^{2+} (yellow) or 250 mM NaCl alone (purple, dashed). Asterisks indicate sites with increased reactivity upon addition of divalent metal. Adapted from Athavale et al. (2012)

8.4 Ribozyme Activity with Fe^{2+}

We have also shown that Fe^{2+} is a good replacement for Mg^{2+} in mediating ribozyme catalysis (Athavale et al. 2012). The activity of two ribozymes was assayed in the presence of Mg^{2+} or Fe^{2+}. Both ribozymes displayed higher activity with Fe^{2+} than with Mg^{2+} (Table 8.2).

The L1 ribozyme ligase is an in vitro-selected, Mg^{2+}-dependent ribozyme that catalyzes formation of a phosphodiester linkage (Robertson and Scott 2007). Mg^{2+} coordinates non-bridging phosphate oxygens in the catalytic pocket, possibly stabilizing the negative charge developing in the transition state during catalysis.

The hammerhead ribozyme, which is widely distributed in the tree of life, cleaves the RNA backbone in a site-specific, reversible reaction, via nucleophilic attack by a 2′-hydroxyl group on the 3′-phosphorous atom (Scott 2007; Penedo et al. 2004). A cyclic 2′3′-phosphate is formed followed by departure of a phosphate oxygen as a 5′-hydroxyl. Divalent metal ions stabilize tertiary interactions and transition states in the ribozyme. Catalytic activity is observed in the folded ribozyme at low mM Mg^{2+} concentrations. While there is no evidence to show that divalent metals play a chemical role in catalysis, reduced activity is observed in the presence of monovalent ions only (Leclerc 2010). Divalent metal ions interacting with the hammerhead ribozyme likely remain hydrated.

The initial rate of ligation in L1 ligase was 25-fold higher in the presence of Fe^{2+} than Mg^{2+}, while the initial rate of hammerhead cleavage was 3.5-fold higher in the presence of Fe^{2+} than Mg^{2+} (Athavale et al. 2012). The observed enhancement in activities of these ribozymes indicates that Fe^{2+} substitutes for Mg^{2+} and facilitates catalysis not only in a case where the divalent metal directly participates in catalysis, as in the case of L1 ligase, but also in a case where Mg^{2+} may play non-specific supportive roles.

Popovic et al. (2015) used in vitro evolution to explore RNA function under early Earth conditions and to compare ribozyme populations that emerge under iron-rich

Table 8.2 Initial rates of ribozyme activity

	L1 ligase in 100 μM divalent	Hammerhead ribozyme in 25 μM divalent
Mg^{2+}	1.4×10^{-6} min^{-1}	1.1×10^{-2} min^{-1}
Fe^{2+}	3.5×10^{-5} min^{-1}	3.5×10^{-2} min^{-1}
k_{Fe}/k_{Mg}	25	3.5

From Athavale et al. (2012)

conditions with those obtained in the presence of Mg^{2+}. Self-cleaving ribozymes from the same starting library were evolved with Mg^{2+} or Fe^{2+} at pH 5 or pH 7 through four to seven selection steps and then evolved once more with the same or opposite ion. Evolved populations were compared, and the impact of metal ion identity and pH on ribozyme sequence examined. Both ion identity and pH were shown to favor evolution of specific ribozyme motifs. At pH 7, the dominant products of the selection were the same for both ions. However, the prevalence of hammerhead family motifs was greater in Fe^{2+} compared to Mg^{2+}. This result is consistent with prior work by Athavale et al. (2012) (Table 8.2) demonstrating enhanced hammerhead ribozyme activity in the presence of Fe^{2+} at pH 7.5. Ribozyme populations evolved in Fe^{2+} at pH 7 showed relatively little sensitivity to the substitution of Fe^{2+} for Mg^{2+}, but dominant motif families evolved in Mg^{2+} were perturbed by the substitution of Fe^{2+}. This observation appears to be consistent with the substitution of Fe^{2+} with Mg^{2+} over time in biopolymers, as the availability of Fe^{2+} declined in concert with the rise in atmospheric O_2. Populations evolved in Mg^{2+} or Fe^{2+} at pH 5 diverged from each other and those evolved at pH 7, suggesting that pH could have been an important factor in the ability of Mg^{2+} to replace Fe^{2+} in RNA function (Popovic et al. 2015).

8.5 Fe^{2+}-Mediated RNA Oxidoreduction

Fe^{2+} is able to confer oxidoreductase ability to RNA (Hsiao et al. 2013). In the absence of Mg^{2+} and under early Earth conditions, certain RNAs are observed to gain the ability to catalyze electron transfer in the presence of Fe^{2+}. We used a peroxidase assay, wherein H_2O_2 oxidizes a reducing agent, tetramethylbenzidine (TMB), to form a radical cation $TMB^{\bullet+}$ that absorbs at 652 nm (Josephy et al. 1982), to show that RNA-Fe^{2+} is a functional analog of horseradish peroxidase in catalyzing oxidoreduction. Oxidoreductase activity was observed in certain RNAs like P4–P6, 23S rRNA, the a-rRNA model of the ancestral ribosome, yeast $tRNA^{Phe}$, and Domain III of the 23S rRNA (Fig. 8.4). The activity was not observed in other RNAs such as a short duplex, a large unstructured RNA (satellite tobacco mosaic virus RNA), and a small single-stranded RNA. Activity was not observed with DNA or single nucleotides (ATP). Activity was also not observed in the absence of RNA or with any metal other than Fe^{2+}. In fact, the addition of other metals to the assay (e.g., Mg^{2+}, Na^{+}) was shown to attenuate oxidoreduction (Hsiao et al. 2013).

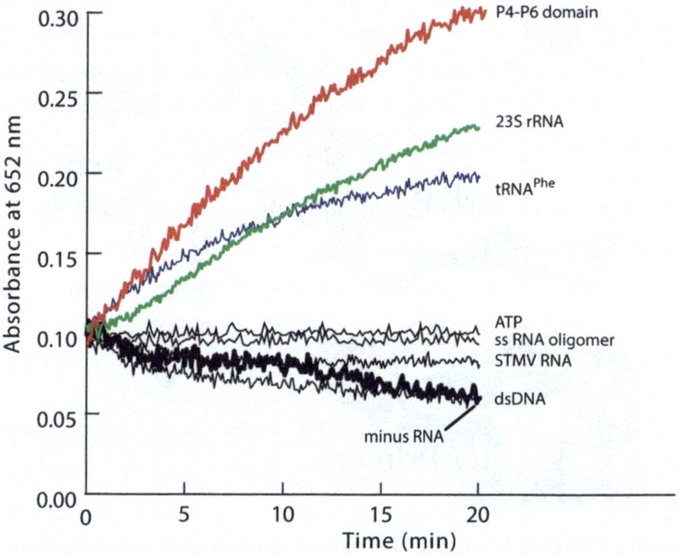

Fig. 8.4 Some RNAs (23S rRNA from *T. thermophilus*, P4–P6 domain RNA, yeast $tRNA^{Phe}$) in combination with Fe^{2+} catalyze single-electron transfer. Other nucleic acids (ATP, a short RNA oligomer, double-stranded DNA (dsDNA), and the RNA genome of STMV) are inefficient catalysts. All reactions were performed in the absence of O_2 and Mg^{2+} and in the presence of Fe^{2+}, tetramethylbenzidine, and H_2O_2. Adapted from Hsiao et al. (2013)

A Michaelis-Menten kinetic analysis of the oxidoreduction activity in RNA-Fe^{2+} complexes shows behavior consistent with true catalysis (Hsiao et al. 2013). Rate saturation was observed with increasing substrate concentration, and kinetic parameters were extracted from the data (Hsiao et al. 2013). While the RNA requirements for catalytic activity are not completely understood, the ability to coordinate Mg^{2+} ions is seen in all RNAs exhibiting catalysis, hinting at the importance of RNA coordination for Fe^{2+}-induced oxidoreductase activity.

It seems likely that chemical transformations such as oxidoreduction would have been required in primitive biological systems (Deamer and Weber 2010). RNA in combination with Fe^{2+} under pre-GOE conditions is indeed shown to catalyze oxidoreductase reactions, suggesting that the catalytic repertoire of RNA may have been much greater on an early Earth and was attenuated with the rise of O_2.

8.6 Fe^{2+} as a Mg^{2+} Substitute in Nucleic Acid Processing Enzymes

It is generally accepted that, in extant biology, nucleic acid processing enzymes such as ligases, polymerases, and kinases use a dual Mg^{2+} ion mechanism for catalysis (Fig. 8.5) (Rittié and Perbal 2008; Yang et al. 2006; Steitz 1999; Doherty and

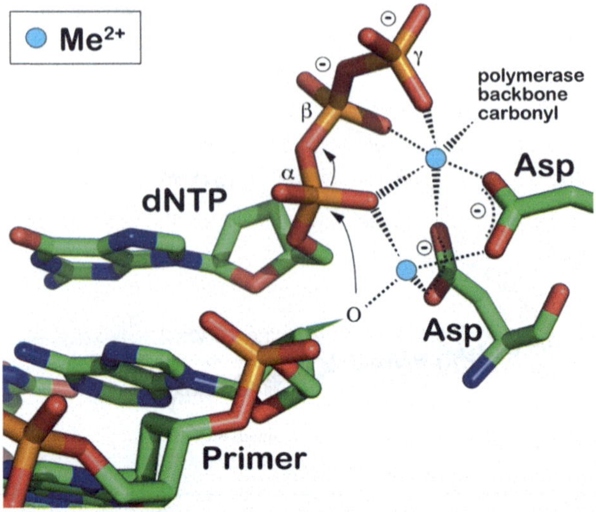

Fig. 8.5 Divalent cations are cofactors for protein enzymes that process nucleic acids. In a generally accepted mechanism for DNA and RNA polymerases (Steitz 1999; Lykke-Andersen and Christiansen 1998; Yin and Steitz 2004; Doublie et al. 1998), two divalent metal cations (Me^{2+}) stabilize the 5′ phosphate(s) of an incoming nucleotide and activate the 3′ hydroxyl of an existing nucleic acid polymer for nucleophilic attack, facilitating phosphodiester bond formation. Mechanism adapted from Steitz (1999), illustrated with PDB 1T7P (Doublie et al. 1998). This structure uses a dideoxy NTP and the position of the primer ribose 3′-OH is approximated. Waters are omitted for clarity

Dafforn 2000; Lee et al. 2000; Ellenberger and Tomkinson 2008; Lykke-Andersen and Christiansen 1998; Yin and Steitz 2004). We have begun to test the hypothesis that prior to the GOE, the dominant divalent cation was Fe^{2+} instead of Mg^{2+} in these protein-based nucleic processing enzymes (Okafor et al. 2017). Under anoxic conditions, Mg^{2+} was removed from three enzymes and replaced with Fe^{2+}. Thermostable Deep Vent (exo-) DNA polymerase, T7 RNA polymerase, and T4 DNA ligase all showed catalytic activity using Fe^{2+} instead of Mg^{2+} as a cofactor.

The ability of a DNA polymerase [Deep vent (exo-)] to amplify a DNA fragment in the presence of various divalent cations was tested by PCR. Product DNA was observed at the same PCR cycle number with Fe^{2+}, Mg^{2+}, or Mn^{2+} cofactors. The ability of T7 RNA polymerase to synthesize RNA was also assayed with Mg^{2+} or Fe^{2+}. A maximum yield was obtained with 0.75 mM Fe^{2+} compared to 6 mM Mg^{2+}, an eightfold difference. The ability of T4 DNA ligase to synthesize RNA was also assayed with Mg^{2+} or Fe^{2+}. The results show that Fe^{2+} can act as a cofactor for this DNA ligase.

In all enzymes investigated, control reactions were performed in the absence of divalent metals. No product was observed in controls, validating the methods used for Mg^{2+} extraction in the study, and confirming that enzyme activity in samples containing Fe^{2+} was due to enzyme utilization of added Fe^{2+} rather than residual Mg^{2+}.

8.7 Theoretical Support for Iron as an Ancient Cofactor of Catalysis

In both ribozymes and protein enzymes, Fe^{2+} appears to be a more potent cofactor than Mg^{2+}. Lower Fe^{2+} concentrations are required to achieve optimal catalysis in RNA polymerases, and increased activity was seen with Fe^{2+} over Mg^{2+} in both L1 ligase and hammerhead ribozymes (Athavale et al. 2012). Quantum mechanics calculations provide explanations for this observation [Table 8.1, (Athavale et al. 2012; Okafor et al. 2017)]. While conformations and geometries are highly similar between Fe^{2+} and Mg^{2+} complexes, differences were observed in metal interactions with water and phosphorus atom chemistry. When interaction energies were decomposed into charge transfer, polarization and exchange, divalent metal (M^{2+}) hexa-aquo complexes showed greater depletion of electrons from Fe^{2+}-coordinated water molecules than Mg^{2+}-coordinated waters. $Fe^{2+}(H_2O)_6$ is a stronger acid than $Mg^{2+}(H_2O)_6$, suggesting a greater frequency of occurrence of the $M^{2+}(H_2O)_5(OH^-)$ species of Fe^{2+} than Mg^{2+}. As most ribozyme mechanisms involve a nucleophilic attack by the deprotonated ribose 2′-OH group (Lilley 2011), the higher acidity of Fe^{2+} in the M^{2+} $(H_2O)_6$ complex may be more favorable for facilitating ribozyme catalysis than Mg^{2+}.

Additionally, the presence of lower lying d orbitals in Fe^{2+} have the effect of increased electron-withdrawing power compared to Mg^{2+}, making the phosphorus of phosphate a better electrophile with Fe^{2+} than with Mg^{2+}. Finally, Fe^{2+} displays increased affinity for oxygen atoms in the first coordination shell. Both of these properties affect rates of nucleophilic attack on phosphorus which modulates the observed rate of catalysis. However, the calculations do not explain why higher Fe^{2+} concentrations inhibit catalysis by a protein enzyme as observed by Okafor et al. (2017).

8.8 Iron-RNA in Extant Biology

Methods for mitigation of iron toxicity are widely distributed in biological systems. The iron-trafficking protein ferritin is ubiquitous in animal cells and found in most aerobic prokaryotes (Barton 2005). In animals, ferritin expression is controlled by iron regulatory proteins (IRPs) that bind the noncoding iron-responsive element (IRE) of ferritin mRNA during iron scarcity and repress translation (Ma et al. 2012). It has been suggested that the ferritin IRE is the ancestor of this type of translational control in more recent key metabolic proteins such as mitochondrial aconitase (mACO; Piccinelli and Samuelsson 2007). Fe^{2+} was shown to affect binding of iron regulatory protein 1 (IRP1) to ferritin and mACO IREs by inducing destabilizing conformational changes in the IREs, which result in decreased binding of IRP1 (Khan et al. 2009). Diminished binding of IRE RNA to IRP1 is also seen with Mg^{2+} but at 100 times the concentration of Fe^{2+} (Khan et al. 2009).

Failure of cellular iron regulatory systems has medical implications. Accumulation of redox-active iron correlates with the development and progression of

Alzheimer's disease (AD) (Smith et al. 2010). Oxidized cytoplasmic RNA is found in neurons vulnerable to AD (Nunomura et al. 1999). RNAs in AD brains are associated with Fe^{2+}; rRNA, and mRNA show twice as much iron binding as tRNA (Honda et al. 2005). Ribosomes purified from AD brains have reduced ability to perform translation, contain Fe^{2+}, and are redox active, potentially serving as "redox centers" for oxidation of cytoplasmic RNA (Honda et al. 2005).

8.9 Iron-RNA Research Methodology

The Fenton reaction, using Fe^{2+} to generate hydroxyl radical, was initially developed by Tullius and coworkers to probe nucleic acid structure (Tullius and Greenbaum 2005; Powers and Noller 1995; Tullius and Dombroski 1985; Berens et al. 1998; Celander and Cech 1990; Latham and Cech 1989; Burkhoff and Tullius 1987). Hydroxyl radical footprinting uses EDTA-chelated iron(II) to decompose H_2O_2, producing hydroxyl radicals, and $^{\bullet}OH$, which cleave nucleic acid backbones (Shcherbakova and Mitra 2009). Changes in the patterns of $^{\bullet}OH$ reactivity can be used to monitor groove width of DNA (Price and Tullius 1993) and compact structures of RNA (Latham and Cech 1989). Fe^{2+} was used to probe divalent binding sites in RNA, replacing Mg^{2+} at ion-binding sites and cleaving RNA in proximity of these sites via Fenton chemistry (Berens et al. 1998). This work showed that Mg^{2+} and Fe^{2+} compete for the same binding sites in the *Tetrahymena* group I intron (Berens et al. 1998). Fe^{2+}-EDTA tethered to a small molecule such as ethidium (Dervan 1986; Vary and Vournakis 1984; Kean et al. 1985) or to deoxyribonucleotides (Moser and Dervan 1987; Chu and Orgel 1985) has also been used to probe nucleic acids.

8.10 Summary

The conditions on early Earth (anoxic, high availability of soluble iron) were conducive for RNA-Fe^{2+} or RNA-Fe^{2+} protein biology and, as illustrated here, could have sustained an early Earth where Fe^{2+} was the primary cationic cofactor for RNA and proteins, later to be replaced by Mg^{2+} as free iron was sequestered.

There is strong evidence that Fe^{2+}, under anoxic conditions, can interact with RNA and proteins in a manner that closely resembles Mg^{2+}. However only a small, recent body of work aims to directly understand the structural and conformational features of RNA-Fe^{2+} complexes and the role of RNA-Fe^{2+}-protein interactions in nucleic acid processing. Fe^{2+} experimental studies are largely complicated by the high oxidation potential of Fe^{2+} to Fe^{3+} (−0.77 V). In the presence of oxygen, Fe^{2+} readily oxidizes to Fe^{3+} and/or promotes RNA degradation. Anoxic laboratory conditions are therefore crucial for experiments addressing the importance of Fe^{2+}.

In comparison with Mg^{2+}, Fe^{2+} is a more potent activator of RNA and at least some nucleic acid processing proteins. The minimum concentration for RNA folding is lower for Fe^{2+} than for Mg^{2+}. At least some ribozymes are more active with Fe^{2+} than with Mg^{2+}. T7 RNA polymerase is far more active at low Fe^{2+} concentrations than in low Mg^{2+}. The minimum concentration for inline cleavage is lower for Fe^{2+} than for Mg^{2+} (Okafor, Hud, and Williams, unpublished). RNA-Fe^{2+} has more diverse chemical functionality, including catalysis of redox chemistry, than RNA-Mg^{2+}.

Recent Fe^{2+} experiments in simulated early Earth conditions have begun to examine the relevance of pH. More RNA-Fe^{2+} studies that vary other environmental factors, such as temperature and buffer components, under early Earth conditions, will be invaluable for increased understanding of Fe^{2+} interactions with RNA and protein pre-GOE.

Acknowledgments This work was supported in part by National Aeronautics and Space Administration grants NNX16AJ28G and NNX16AJ29G.

References

Aguirre JD, Culotta VC (2012) Battles with iron: manganese in oxidative stress protection. J Biol Chem 287:13541–13548
Anbar AD (2008) Oceans. Elements and evolution. Science 322:1481–1483
Anjem A, Varghese S, Imlay JA (2009) Manganese import is a key element of the oxyr response to hydrogen peroxide in *Escherichia coli*. Mol Microbiol 72:844–858
Athavale SS, Petrov AS, Hsiao C, Watkins D, Prickett CD, Gossett JJ, Lie L, Bowman JC, O'Neill E, Bernier CR et al (2012) RNA folding and catalysis mediated by iron (II). PLoS One 7:e38024
Atkins JF, Gesteland RF, Cech TR (eds) (2011) RNA worlds: from life's origins to diversity in gene regulation. Cold Spring Harbor Laboratory Press, Cold Spring Harbor, NY
Barton L (2005) Structural and functional relationships in prokaryotes. Springer, Berlin
Berens C, Streicher B, Schroeder R, Hillen W (1998) Visualizing metal-ion-binding sites in group I introns by iron(II)-mediated fenton reactions. Chem Biol 5:163–175
Bock CW, Markham GD, Katz AK, Glusker JP (2006) The arrangement of first- and second-shell water molecules around metal ions: effects of charge and size. Theor Chem Accounts 115:100–112
Bowman JC, Lenz TK, Hud NV, Williams LD (2012) Cations in charge: magnesium ions in RNA folding and catalysis. Curr Opin Struct Biol 22:262–272
Brion P, Westhof E (1997) Hierarchy and dynamics of RNA folding. Annu Rev Biophys Biomol Struct 26:113–137
Brown ID (1988) What factors determine cation coordination numbers. Acta Crystallogr Sect B 44:545–553
Brown ID (1992) Chemical and steric constraints in inorganic solids. Acta Crystallogr Sect B 48:553–572
Burkhoff AM, Tullius TD (1987) The unusual conformation adopted by the adenine tracts in kinetoplast DNA. Cell 48:935–943
Butcher SE (2011) The spliceosome and its metal ions. Met Ions Life Sci 9:235–251
Cate JH, Gooding AR, Podell E, Zhou K, Golden BL, Kundrot CE, Cech TR, Doudna JA (1996) Crystal structure of a group I ribozyme domain: principles of RNA packing. Science 273:1678–1685
Cate JH, Hanna RL, Doudna JA (1997) A magnesium ion core at the heart of a ribozyme domain. Nat Struct Biol 4:553–558

Celander DW, Cech TR (1990) Iron (II)-ethylenediaminetetraacetic acid catalyzed cleavage of RNA and DNA oligonucleotides: similar reactivity toward single-and double-stranded forms. Biochemistry 29:1355–1361

Chu BC, Orgel LE (1985) Nonenzymatic sequence-specific cleavage of single-stranded DNA. Proc Natl Acad Sci USA 82:963–967

Cole PE, Yang SK, Crothers DM (1972) Conformational changes of transfer ribonucleic acid. Equilibrium phase diagrams. Biochemistry 11:4358–4368

Cotruvo JA, Stubbe J (2011) Class I ribonucleotide reductases: metallocofactor assembly and repair *in vitro* and *in vivo*. Annu Rev Biochem 80:733–767

Deamer D, Weber AL (2010) Bioenergetics and life's origins. Cold Spring Harb Perspect Biol 2: a004929

Derry LA (2015) Causes and consequences of mid-proterozoic anoxia. Geophys Res Lett 42:8538–8546

Dervan P (1986) Design of sequence-specific DNA binding molecules. Science 232:464–471

Doherty AJ, Dafforn TR (2000) Nick recognition by DNA ligases. J Mol Biol 296:43–56

Doublie S, Tabor S, Long AM, Richardson CC, Ellenberger T (1998) Crystal structure of a bacteriophage T7 DNA replication complex at 2.2 Å resolution. Nature 391:251–258

Draper DE (2004) A guide to ions and RNA structure. RNA 10:335–343

Drever JI (1974) Geochemical model for the origin of precambrian banded iron formations. GSA Bull 85:1099–1106

Dupont CL, Yang S, Palenik B, Bourne PE (2006) Modern proteomes contain putative imprints of ancient shifts in trace metal geochemistry. Proc Natl Acad Sci USA 103:17822–17827

Dupont CL, Butcher A, Valas RE, Bourne PE, Caetano-Anolles G (2010) History of biological metal utilization inferred through phylogenomic analysis of protein structures. Proc Natl Acad Sci USA 107:10567–10572

Ellenberger T, Tomkinson AE (2008) Eukaryotic DNA ligases: structural and functional insights. Annu Rev Biochem 77:313

Hanna R, Doudna JA (2000) Metal ions in ribozyme folding and catalysis. Curr Opin Chem Biol 4:166–170

Harel A, Bromberg Y, Falkowski PG, Bhattacharya D (2014) Evolutionary history of redox metal-binding domains across the tree of life. Proc Natl Acad Sci USA 111:7042–7047

Hazen RM, Ferry JM (2010) Mineral evolution: mineralogy in the fourth dimension. Elements 6:9–12

Holland HD (1973) The oceans; a possible source of iron in iron-formations. Econ Geol 68:1169–1172

Holland H (1984) The chemical evolution of the atmosphere and oceans. Princeton University Press, Princeton, NJ

Holland HD (2006) The oxygenation of the atmosphere and oceans. Philos Trans R Soc B Biol Sci 361:903–915

Holland DM, Jacobs SS, Jenkins A (2003) Modelling the ocean circulation beneath the ross ice shelf. Antarct Sci 15:13–23

Honda K, Smith MA, Zhu X, Baus D, Merrick WC, Tartakoff AM, Hattier T, Harris PL, Siedlak SL, Fujioka H et al (2005) Ribosomal RNA in Alzheimer disease is oxidized by bound redox-active iron. J Biol Chem 280:20978–20986

Hsiao C, Tannenbaum M, VanDeusen H, Hershkovitz E, Perng G, Tannenbaum A, Williams LD (2008) In: Hud N (ed) Nucleic acid metal ion interactions. The Royal Society of Chemistry, London, pp 1–35

Hsiao C, Chou I-C, Okafor CD, Bowman JC, O'Neill EB, Athavale SS, Petrov AS, Hud NV, Wartell RM, Harvey SC et al (2013) Iron(II) plus RNA can catalyze electron transfer. Nat Chem 5:525–528

Iyengar V, Woittiez J (1988) Trace elements in human clinical specimens: evaluation of literature data to identify reference values. Clin Chem 34:474–481

Izawa MRM, Nesbitt HW, MacRae ND, Hoffman EL (2010) Composition and evolution of the early oceans: evidence from the tagish lake meteorite. Earth Planet Sci Lett 298:443–449

Johnson CM, Beard BL, Roden EE (2008) The iron isotope fingerprints of redox and biogeochemical cycling in modern and ancient earth. Annu Rev Earth Planet Sci 36:457–493

Johnson-Buck AE, McDowell SE, Walter NG (2011) Metal ions: supporting actors in the playbook of small ribozymes. Met Ions Life Sci 9:175–196

Jones C, Nomosatryo S, Crowe SA, Bjerrum CJ, Canfield DE (2015) Iron oxides, divalent cations, silica, and the early earth phosphorus crisis. Geology 43:135–138

Josephy PD, Eling T, Mason RP (1982) The horseradish peroxidase-catalyzed oxidation of 3,5,3',5'-tetramethylbenzidine. Free radical and charge-transfer complex intermediates. J Biol Chem 257:3669–3675

Kean JM, White SA, Draper DE (1985) Detection of high-affinity intercalator sites in a ribosomal RNA fragment by the affinity cleavage intercalator methidiumpropyl-EDTA-Iron(II). Biochemistry 24:5062–5070

Khan MA, Walden WE, Goss DJ, Theil EC (2009) Direct Fe^{2+} sensing by iron-responsive messenger RNA: repressor complexes weakens binding. J Biol Chem 284:30122–30128

Kozlowski H, Kolkowska P, Watly J, Krzywoszynska K, Potocki S (2014) General aspects of metal toxicity. Curr Med Chem 21:3721–3740

Lanier KA, Petrov AS, Williams LD (2017) The central symbiosis of molecular biology. J Mol Evol 85:8–13

Latham JA, Cech TR (1989) Defining the inside and outside of a catalytic RNA molecule. Science 245:276–282

Leclerc F (2010) Hammerhead ribozymes: true metal or nucleobase catalysis? Where is the catalytic power from? Molecules 15:5389

Lee JY, Chang C, Song HK, Moon J, Yang JK, Kim H-K, Kwon S-T, Suh SW (2000) Crystal structure of Nad(+)-dependent DNA ligase: modular architecture and functional implications. EMBO J 19:1119–1129

Lilley DMJ (2011) Mechanisms of RNA catalysis. Philos Trans R Soc B Biol Sci 366:2910–2917

Lykke-Andersen J, Christiansen J (1998) The C-terminal carboxy group of T7 RNA polymerase ensures efficient magnesium ion-dependent catalysis. Nucleic Acids Res 26:5630–5635

Lyons TW, Reinhard CT, Planavsky NJ (2014) The rise of oxygen in earth's early ocean and atmosphere. Nature 506:307–315

Ma J, Haldar S, Khan MA, Sharma SD, Merrick WC, Theil EC, Goss DJ (2012) Fe^{2+} binds iron responsive element-RNA, selectively changing protein-binding affinities and regulating mRNA repression and activation. Proc Natl Acad Sci USA 109:8417–8422

Maguire ME, Cowan JA (2002) Magnesium chemistry and biochemistry [review]. BioMetals 15:203–210

Martin JE, Imlay JA (2011) The alternative aerobic ribonucleotide reductase of *Escherichia coli*, Nrdef, is a manganese-dependent enzyme that enables cell replication during periods of iron starvation. Mol Microbiol 80:319–334

Merino EJ, Wilkinson KA, Coughlan JL, Weeks KM (2005) RNA structure analysis at single nucleotide resolution by selective 2'-hydroxyl acylation and primer extension (shape). J Am Chem Soc 127:4223–4231

Misra VK, Draper DE (1998) On the role of magnesium ions in RNA stability. Biopolymers 48:113–135

Moser HE, Dervan PB (1987) Sequence-specific cleavage of double helical DNA by triple helix formation. Science 238:645–650

Nunomura A, Perry G, Pappolla MA, Wade R, Hirai K, Chiba S, Smith MA (1999) RNA oxidation is a prominent feature of vulnerable neurons in Alzheimer's disease. J Neurosci 19:1959–1964

Okafor CD, Lanier KA, Petrov AS, Athavale SS, Bowman JC, Hud NV, Williams LD (2017) Iron mediates catalysis of nucleic acid processing enzymes: support for Fe(II) as a cofactor before the great oxidation event. Nucleic Acids Res 45:3634–3642

Penedo JC, Wilson TJ, Jayasena SD, Khvorova A, Lilley DM (2004) Folding of the natural hammerhead ribozyme is enhanced by interaction of auxiliary elements. RNA 10:880–888

Petrov AS, Bowman JC, Harvey SC, Williams LD (2011) Bidentate RNA-magnesium clamps: on the origin of the special role of magnesium in RNA folding. RNA 17:291–297

Petrov A, Bernier C, Hsiao C, Okafor CD, Tannenbaum E, Stern J, Gaucher E, Schneider D, Hud NV, Harvey SC et al (2012) RNA-magnesium-protein interactions in large ribosomal subunit. J Phys Chem B 116:8113–8120

Piccinelli P, Samuelsson T (2007) Evolution of the iron-responsive element. RNA 13:952–966

Popovic M, Fliss PS, Ditzler MA (2015) In vitro evolution of distinct self-cleaving ribozymes in diverse environments. Nucleic Acids Res 43:3789–3801

Powers T, Noller H (1995) Hydroxyl radical footprinting of ribosomal proteins on 16S rRNA. RNA 1:194

Price MA, Tullius TD (1993) How the structure of an adenine tract depends on sequence context: a new model for the structure of T_na_n DNA sequences. Biochemistry 32:127–136

Prousek J (2007) Fenton chemistry in biology and medicine. Pure Appl Chem 79:2325–2338

Pyle AM (1993) Ribozymes: a distinct class of metalloenzymes. Science 261:709–714

Rashin AA, Honig B (1985) Reevaluation of the born model of ion hydration. J Phys Chem 89:5588–5593

Reinhard CT, Planavsky NJ, Gill BC, Ozaki K, Robbins LJ, Lyons TW, Fischer WW, Wang C, Cole DB, Konhauser KO (2017) Evolution of the global phosphorus cycle. Nature 541:386–389

Rittié L, Perbal B (2008) Enzymes used in molecular biology: a useful guide. J Cell Commun Signal 2:25–45

Robertson MP, Scott WG (2007) The structural basis of ribozyme-catalyzed RNA assembly. Science 315:1549–1553

Scott WG (2007) Ribozymes. Curr Opin Struct Biol 17:280–286

Shcherbakova I, Mitra S (2009) Hydroxyl-radical footprinting to probe equilibrium changes in RNA tertiary structure. Methods Enzymol 468:31–46

Smith MA, Zhu X, Tabaton M, Liu G, McKeel DW Jr, Cohen ML, Wang X, Siedlak SL, Dwyer BE, Hayashi T et al (2010) Increased iron and free radical generation in preclinical alzheimer disease and mild cognitive impairment. J Alzheimers Dis 19:363–372

Steitz TA (1999) DNA polymerases: structural diversity and common mechanisms. J Biol Chem 274:17395–17398

Theil EC, Goss DJ (2009) Living with iron (and oxygen): questions and answers about iron homeostasis. Chem Rev 109:4568–4579

Torrents E, Aloy P, Gibert I, Rodriguez-Trelles F (2002) Ribonucleotide reductases: divergent evolution of an ancient enzyme. J Mol Evol 55:138–152

Tullius TD, Dombroski BA (1985) Iron (II) EDTA used to measure the helical twist along any DNA molecule. Science 230:679–681

Tullius TD, Greenbaum JA (2005) Mapping nucleic acid structure by hydroxyl radical cleavage. Curr Opin Chem Biol 9:127–134

Ushizaka S, Kuma K, Suzuki K (2011) Effects of Mn and Fe on growth of a coastal marine diatom *Talassiosira weissflogii* in the presence of precipitated Fe(III) hydroxide and EDTA-Fe(III) complex. Fish Sci 77:411–424

Uudsemaa M, Tamm T (2004) Calculation of hydration enthalpies of aqueous transition metal cations using two coordination shells and central ion substitution. Chem Phys Lett 400:54–58

Vary CP, Vournakis JN (1984) RNA structure analysis using methidiumpropyl-EDTA.Fe(II): a base-pair-specific RNA structure probe. Proc Natl Acad Sci USA 81:6978–6982

Vicens Q, Gooding AR, Laederach A, Cech TR (2007) Local RNA structural changes induced by crystallization are revealed by shape. RNA 13:536–548

Wilkinson KA, Merino EJ, Weeks KM (2005) RNA shape chemistry reveals nonhierarchical interactions dominate equilibrium structural transitions in tRNA(Asp) transcripts. J Am Chem Soc 127:4659–4667

Wilkinson KA, Gorelick RJ, Vasa SM, Guex N, Rein A, Mathews DH, Giddings MC, Weeks KM (2008) High-throughput shape analysis reveals structures in Hiv-1 genomic RNA strongly conserved across distinct biological states. PLoS Biol 6:883–899

Wolfe-Simon F, Starovoytov V, Reinfelder JR, Schofield O, Falkowski PG (2006) Localization and role of manganese superoxide dismutase in a marine diatom. Plant Physiol 142:1701–1709
Wulfsberg G (1991) Principles of descriptive inorganic chemistry. University Science Books, Sausalito, CA
Yang W, Lee JY, Nowotny M (2006) Making and breaking nucleic acids: two-Mg^{2+}-ion catalysis and substrate specificity. Mol Cell 22:5–13
Yin YW, Steitz TA (2004) The structural mechanism of translocation and helicase activity in T7 RNA polymerase. Cell 116:393–404
Zheng H, Shabalin IG, Handing KB, Bujnicki JM, Minor W (2015) Magnesium-binding architectures in RNA crystal structures: validation, binding preferences, classification and motif detection. Nucleic Acid Res 43:3789–3801

Chapter 9
Connections Between Mathematical Models of Prebiotic Evolution and Homochirality

Celia Blanco and Irene A. Chen

Abstract The evolutionary dynamics of prebiotic replicators and the amplification kinetics leading to homochirality share certain features and properties, such as the emergence of a certain type that achieves high abundance. The study of replicator dynamics and the study of the origin of homochirality have both seen numerous advances, both theoretical and experimental, in the last decades. Experimental models formulated in these fields are quite different one from the other, and these fields have traditionally been viewed as separate undertakings. However, despite differences in formalisms, it is remarkable that mathematical descriptions used to explain the behavior of replicating entities can be transformed into the mathematical descriptions of models leading to enantiomeric symmetry breaking. Thus two important phenomena during the origin of life, the selection of replicators and the origin of biological homochirality, share similar dynamics.

9.1 Introduction

In their classic work, Eigen and Schuster (1977) point out, in a single breath, two apparently defining properties of living matter, genetic information and homochirality. The origin of these two properties has spurred many productive lines of research. However, these two subfields have remained fairly distinct from one another, as experimental systems differ substantially. We have recently reviewed elsewhere the recent progress that has been made by adopting a synthetic and mechanistic perspective to these questions, shifting the focus from the specific

C. Blanco (✉)
Department of Chemistry and Biochemistry, University of California, Santa Barbara, CA, USA
e-mail: cblanco@chem.ucsb.edu

I. A. Chen
Department of Chemistry and Biochemistry, University of California, Santa Barbara, CA, USA

Program in Biomolecular Sciences and Engineering, University of California, Santa Barbara, CA, USA

© Springer International Publishing AG, part of Springer Nature 2018
C. Menor-Salván (ed.), *Prebiotic Chemistry and Chemical Evolution of Nucleic Acids*, Nucleic Acids and Molecular Biology 35,
https://doi.org/10.1007/978-3-319-93584-3_9

origin of life on Earth to the mechanisms that could give rise to life more generally (Pressman et al. 2015). This shift is timely given the developing picture of the universe as being quite rich in exoplanets that might have conditions favorable for life. While mathematical models may lack the baroque beauty and realism of experimental science, they can be powerful in revealing general principles and mechanisms that may apply widely. In the spirit of searching for generalizable mechanisms, we review here mathematical models describing the prebiotic evolution of genetic information and the emergence of homochirality. A comparison of these models reveals an underlying commonality that connects the evolutionary dynamics of prebiotic replicators to the amplification dynamics of chiral molecules.

In this discussion, we will assume for convenience that genetic evolution took place in the RNA world in which RNA stored genetic information and catalyzed chemical reactions (Bartel and Szostak 1993; Crick 1968; Orgel 1968; Woese et al. 1966). The RNA world probably included other molecules, such as lipids and simple peptides, but genetic inheritance would have been based on RNA. At some point, the presence of other chemical cycles would have been necessary to generate the compounds needed for RNA replication. However, we will focus on replicator scenarios, taking the simplistic perspective that these chemical cycles are considered only insofar as they may affect RNA fitness. While RNA is used for the sake of discussion, the mathematical models are relatively agnostic regarding the identity of the genetic polymer.

In the absence of any chiral polarization, the synthesis of chiral molecules from achiral starting materials generally results in the formation of mixtures containing equal amounts of both enantiomers (i.e., racemic mixtures). For example, this is the case for the formation of amino acids in the prebiotic synthesis experiments carried out by Miller (1953). In stark contrast, molecules that form the building blocks of life (e.g., amino acids and sugars) essentially exist as a single enantiomer in biology. Empirically, biological systems are homochiral, having a nearly complete chiral imbalance (mirror symmetry breaking), raising the question of how biological homochirality arises (Guijarro and Yus 2009; Blackmond 2010; Cintas 2016).

Several theories have been proposed to explain the origins of biomolecular homochirality (Guijarro and Yus 2009). Proposed mechanisms are either deterministic (i.e., a specific chiral field or influence causes the breaking of mirror symmetry) or random (i.e., the direction of symmetry breaking is random). In deterministic theories, the enantiomer imbalance is created due to an external chiral field or influence. Deterministic mechanisms can also be either local in space or universal, i.e., applying everywhere. Some examples of local deterministic mechanisms are circularly polarized light (CPL), β-Radiolysis, or the magnetochiral effect (Barron 1981, 1986). This mechanism does not have to occur on Earth, but instead may occur elsewhere with the enantiomeric imbalance of organic molecules being delivered to Earth by meteorites and comets (Myrgorodska et al. 2015). On the other hand, the most accepted universal deterministic theory is based on the electroweak interaction. The theory of electroweak interactions predicts a parity violating energy difference (PVED) between the two enantiomers of chiral molecules of 10^{-13}–10^{-21} eV; however, no conclusive energy difference has been reported so far (Bargueno et al. 2011). In comparison, chance or random theories are based on the knowledge

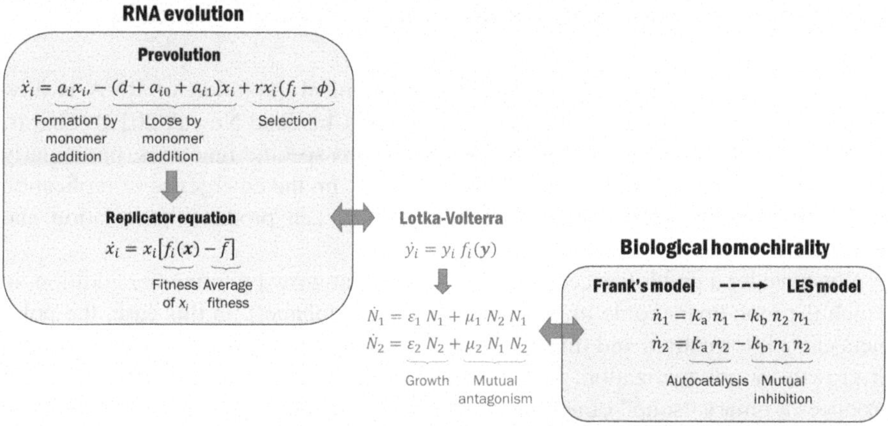

Fig. 9.1 Relationships among models considered

that a perfect racemic mixture is statistically exceedingly improbable for a molecular system of any reasonable size (Mislow 2003). That is, an initial enantiomeric excess is virtually inescapable due to stochastic fluctuations within a racemic system.

Regardless of the type of theory, biological homochirality is believed to arise through the same three sequential stages: (1) the symmetry is broken, either by a chiral field or influence or by random chance on a microscopic scale; (2) once any kind of imbalance has been created, the initial imbalance is amplified; and (3) once a significant enantiomeric enrichment has been produced, the chiral imbalance is transferred through the entire system (e.g., from monomers to polymers); this step is referred to as "transmission."

Although every theory follows the three same steps, the relevance of each step is different for each type of theory. For a deterministic mechanism, the method by which an initial enantiomeric excess is made is important (step 1). In a chance mechanism, by contrast, the source of the initial imbalance is not as important as the need to amplify the initial imbalance through an efficient mechanism; the key process in this case is chiral amplification (step 2). While chance theories obviate the need to invoke chiral physical fields and support the idea that homochirality is a "stereochemical imperative" of molecular evolution (Siegel 1998), both chance and deterministic theories, regardless of the chiral field or force involved, would still require an effective amplification mechanism. Therefore, for mathematical models of the evolution of biological homochirality, the focus of the model is not on the source of the initial enantiomeric imbalance but on the ability of the model to amplify a tiny initial enantiomeric imbalance, such as by autocatalytic reactions.

The chemical reactions involved in the evolution of genetic information and the emergence of homochirality can be translated readily into mathematical models (Fig. 9.1). The purpose of this chapter is to review some models of early evolutionary dynamics and illustrate their relationships with models describing possible amplification mechanisms leading to homochirality in the prebiotic world.

9.2 Polymerization and Replication in "Prelife" Models

A crucial question regarding the origin and chemical evolution of RNA is how chemical kinetics become evolutionary dynamics (Chen and Nowak 2012). That is, how does evolution get started before RNA exhibits specific functions, particularly replication? We review the models by Nowak et al. on the emergence of replication in the simplest possible population dynamics that can produce information and complexity.

We consider a prebiotic scenario of polymers that grow by monomer addition, in which the monomers come in two types (a binary alphabet). In this case, the polymers carry information, and different sequences may or may not differ in their rates of growth by polymerization. For example, we imagine a prebiotic chemistry that produces a binary "soup" of activated monomers, denoted by 0^* and 1^*, which can either become deactivated ($0^* \rightarrow 0$ and $1^* \rightarrow 1$) or form random polymers (binary sequences) of any length via the chemical reactions $i + 0^* \rightarrow i0$ and $i + 1^* \rightarrow i1$. Following this scheme, each binary string, i, has only one possible predecessor, i', and two possible descendants, $i0$ and $i1$. This system has been named "prelife" and the associated dynamics "prevolution" (Nowak and Ohtsuki 2008; Ohtsuki and Nowak 2009). If we denote the abundance of each binary sequence i as x_i, then the chemical kinetics of prelife can be described by the following system of differential equations:

$$\dot{x}_i = a_i x_{i'} - (d + a_{i0} + a_{i1})x_i, \tag{9.1}$$

where a_i is the rate constant of monomer addition (i.e., $i' \rightarrow i$) and includes the concentration of activated monomers, which are assumed to be at steady state, and d is the decay rate. If we consider that every sequence of the same length grows at the same rate ($a_0 = a_1$ and $a_i = a$ for all other i), then all sequences with the same length are found to have the same abundance at equilibrium and the longer sequences are found to be exponentially less abundant. If by contrast, different sequences of the same length grow at different rates (e.g., some reactions occur at a rate, $a_i = 1 + s$, while the other reactions occur at a slower rate $a_i = 1$), then the equilibrium distribution of all sequences depends on the rate difference s. As s increases, the equilibrium abundance of some sequences becomes higher than that of other sequences of the same length. Thus, differences in reaction rate can create downstream asymmetries that are essentially a chemical analog of natural selection.

As Eigen demonstrated earlier for his model of replicators, prelife also exhibits an "error threshold," i.e., a limit to the degree of mutation that can be tolerated while preserving genetic information. In analogy to Eigen's model, we can consider a "master sequence" of length n, which is more fit and abundant than all other sequences of the same length (this situation can be depicted as a "single-peak" fitness landscape). The master sequence is defined by the fact that the reactions leading to it are faster than the other reactions taking place in the system. If every reaction leading to the master sequence makes mistakes by incorporating the wrong monomer with probability w, then the master sequence is selected only if $w < 1/n$.

That is, there is an error threshold for the emergence of the master sequence, namely, that the mutation rate is inversely proportional to the length of the master sequence. In other words, for a certain polymerization accuracy, there is a maximum value for the sequence length over which the master sequence is not preserved anymore and the informational content is destroyed by the mistakes made in polymerization.

To introduce replication to prelife, assuming the parameters are adequate for selection of the master sequence, some sequences are allowed to act as templates for replication (enzymatic or chemical). The system is then described by a set of differential equations based on Eq. (9.1) for prevolutionary dynamics, except with an extra term which represents replication (see Sect. 9.3.1):

$$\dot{x}_i = a_i x_{i'} - (d + a_{i0} + a_{i1})x_i + rx_i(f_i - \phi). \tag{9.2}$$

In this case, f_i is the fitness of sequence i, $\phi = \sum_i f_i x_i / \sum_i x_i$ is the average fitness (this term ensures that the total population in the system remains constant), and the parameter r represents a scale factor between the rates of template-directed replication ("life") and template-independent sequence growth ("prelife"). For small values of r, the dynamics are dominated by prevolution, that is, the abundance of potential replicators is not high enough to affect the equilibrium structure of prelife, dominated by non-templated polymerization. However, there is a critical value for r, over which those sequences that replicate at a faster rate dominate the population, whereas all other sequences that replicate slower are depleted.[1] This critical value, r_c, which evidences a well-defined phase transition between prelife and life, can be obtained by solving for the condition that the net reproductive rate of replicator i, defined as $g_i = r(f_i - \phi) - (d + a_{i0} + a_{i1})$, is positive.

If sequences incorporate mutations (errors) when replicating with probability u, then the net reproductive rate of the master sequence is $g_i = r(f_i q - \phi) - (d + a_{i0} + a_{i1})$, where $q = (1 - u)^n$ represents the replication accuracy (i.e., the probability of error-free replication). The master sequence will be selected only if the replication accuracy exceeds a certain minimum value, $q > (d + a_{i0} + a_{i1})/rf_i$. In other words, life (replicators) is selected over prelife (polymerization) only if the mutation rate, u, is less than a critical value. Therefore, imperfect replication imposes an error threshold for the emergence of life-like growth dynamics that depends on the length of the potential replicators, the relative fitness of the master sequence, and the balance between the rate of polymerization and the rate of replication.

These models clarify that a kind of natural (chemical) selection can precede replicators per se and may be a mechanism for favoring a master sequence. In addition, the mere presence of the mechanism of replication is not enough to favor replicators, as replicating sequences in the system do not always attain much higher abundances than non-replicating sequences of the same length. Interestingly, in prelife scenarios *without* replication, small differences in growth rates result in small differences in abundances, reflecting the expected chemical kinetics. On the

[1]In the limit of large values of r, Eq. (9.2) becomes the standard selection equation of evolutionary dynamics with competitive exclusion (see Sect. 9.3.1).

other hand, in prelife scenarios *with* replication, small differences in replication rates can lead to large differences in abundance, reflecting the dynamics seen in evolution.

Although the prelife models use a binary alphabet for the monomers, similar dynamics are expected in a quaternary system (Sievers and Von Kiedrowski 1994). Prelife models, although abstract in some aspects, are a straightforward approach to modeling that could be applied to non-templated nucleic acid polymerization (Ertem and Ferris 1996; Ferris and Ertem 1992, 1993; Ferris et al. 1996; Rajamani et al. 2008; Monnard and Deamer 2001, 2002) and template-directed polymerization (Sawai and Orgel 1975; Orgel 1992; Manapat et al. 2009; Ohtsuki and Nowak 2009) in different scenarios.

9.3 Models of Replication and Mutation

9.3.1 The Replicator, Replicator-Mutator, and Quasispecies Models

The dynamics of a system composed of self-replicating entities is described by the replicator equation (Hofbauer and Sigmund 1998; Hofbauer et al. 1979; Maynard Smith 1982). An important feature that is captured in the replicator equation is frequency-dependent selection, i.e., that the fitness of a particular sequence depends on the frequency of the other sequences in the population. The replicator equation that describes evolutionary game dynamics of discrete phenotypes reads as follows:

$$\dot{x}_i = x_i[f_i(\mathbf{x}) - \bar{f}], \tag{9.3}$$

where x_i is the frequency of sequence i; f_i is the fitness of x_i and is a function of the distribution of the population, given by the vector $\mathbf{x} = (x_1, \ldots, x_n)$; and $\bar{f}$ represents the average fitness, $\bar{f} = \sum_{j=1}^{n} x_j f_j(\mathbf{x})$. Note that $\sum_i x_i = 1$ by definition. In Eq. (9.3), the fitness function depends on the distribution of the population types, so that the replicator equation can capture the frequency-dependence of fitness. This key feature is the reason why the replicator equation is useful in several different fields [such as population genetics (Hadeler 1981), autocatalytic reaction networks (Stadler and Schuster 1992), game theory (Bomze and Burger 1995), or language evolution (Nowak et al. 2001)]. However, with respect to early evolution, a major caveat is that the replicator equation does not account for the effect of mutations and so does not model the invention of new types. A generalization of the replicator equation which incorporates mutation is given by the replicator-mutator equation:

$$\dot{x}_i = \sum_{j=1}^{n} x_j f_j(\mathbf{x}) q_{ji} - x_i \bar{f}, \tag{9.4}$$

where q_{ij} is the probability that replication of sequence i gives rise to sequence j. The replicator equation (Eq. 9.3) is clearly a particular case of the replicator-mutator equation (Eq. 9.4), in which the fitness of each x_i is a function of the distribution of the population $\boldsymbol{x}$, but there is no mutation. Presumably the different replicators eventually established cooperation (e.g., increased fitness of i with increased abundance of j) during early evolution.

While the replicator-mutator equation describes a system in which fitness depends on the frequency of other types, the quasispecies model describes a system in which mutations are allowed but the fitness of any sequence does not depend on the fitness of other sequences. The quasispecies can be visualized as a family of closely related sequences (or genotypes) that exist in a scenario in which there is replication and mutations (that is, mistakes can be made when replicating sequences). Consider a master sequence, x_m, whose fitness is much higher than all competing sequences, such that its abundance persists at a level higher than that of all other sequences. With every round of replication, x_m will generate a population of mutants closely related to itself, making it impossible for x_m to completely eliminate its competitors (members of its own quasispecies). Since the replicator equation does not include mutation, the quasispecies model may be a more adequate formalism to study early evolution. For a detailed discussion on the mathematical equivalences between these replicator models, or for other uses of the models, see Page and Nowak (2002).

9.3.2 *Lotka-Volterra Equations of Interacting Species*

The Lotka-Volterra equations (Lotka 1920; Volterra 1926) (also commonly known as the predator-prey equations) are frequently used in ecology to describe the interactions among n different species. The model can be described by the following differential equation for each species:

$$\dot{y}_i = y_i f_i(\boldsymbol{y}), \tag{9.5}$$

where y_i is the abundance of species i and f_i is the fitness (or reproductive rate) of each species, which is a function of the distribution of the population abundance, given by the vector $\boldsymbol{y} = (y_1, \ldots, y_n)$. One may see that these equations describe frequency-dependent selection, like the replicator models discussed above. Interestingly, as Hofbauer et al. pointed out almost two decades ago (Hofbauer and Sigmund 1998), using the barycentric transformation $x_i = y_i/(1 + y)$ for $i = 1, \ldots, n - 1$, and $x_n = 1/(1 + y)$, where y is defined as $y = \sum_{i=1}^{n-1} y_i$, it is readily shown that the replicator equation for n phenotypes (Eq. 9.3) is equivalent to the Lotka-Volterra equations for $n - 1$ species (Eq. 9.5).

One of the simplest forms of the Lotka-Volterra model considers only reproduction and mutual antagonism effects, and two different species, N_1 and N_2, which have net rates of increase (birth minus death) ε_1 and ε_2. The change of N_1 and N_2

population numbers over time can be described by the following pair of differential equations (Lotka 1925, 1932):

$$\dot{N}_1 = \varepsilon_1 N_1 + \mu_1 N_2 N_1, \tag{9.6}$$
$$\dot{N}_2 = \varepsilon_2 N_2 + \mu_2 N_1 N_2. \tag{9.7}$$

The signs of μ_1 and μ_2 describe the interaction between the two species. If μ_1 or μ_2 is negative, the interaction is unfavorable and antagonistic for that species; if μ_1 or μ_2 is positive, the interaction is favorable for that species; and if μ_1 and μ_2 are 0, the interaction is neutral for that species. When the interaction between species is mutually unfavorable ($\mu_1 < 0$ and $\mu_2 < 0$), and assuming that the rates of reproduction are both positive ($\varepsilon_1 > 0$ and $\varepsilon_2 > 0$), the system is found to have a saddle point about which the slightest variation leads to the complete extinction of one of the two species. Thus, if initial conditions are such that $N_1 > N_2$ (or $N_2 > N_1$), the system evolves toward an asymptotic solution, in which only one of the two species survives (competitive exclusion). This feature is reminiscent of chiral amplification, in which an initially small imbalance in numbers is amplified to total enantiomeric excess. We examine models of homochirality to address this similarity in detail.

9.4 Models of Absolute Asymmetric Synthesis

9.4.1 Frank Model of Homochirality

In a footnote, Volterra specified that he did not consider the degenerate case (equal rates and parameters for both species; in this case $\varepsilon_1 = \varepsilon_2$ and $\mu_1 = \mu_2$) because such a situation is of "infinitesimally small probability" (Volterra 1926; English Translation in Chapman 1931). However, as it was recently noted by Ribo and Hochberg (2015), this special case of the Lotka-Volterra two-species competitive exclusion model for two distinguishable but degenerate species is identical to the degenerate case of enantiomerism that was later considered by Frank in a model for spontaneous asymmetric synthesis in chemical systems (Frank 1953).

In an attempt to explain the origin of biological homochirality, Frank proposed a simple model in which two chemical substances, n_1 and n_2, which are enantiomers of each other, act as autocatalysts for their own production (with rate $k_a > 0$) and as inhibitors for the production of their optical enantiomer (with rate $k_b > 0$). The system is described by the following pair of differential equations (where concentration brackets are omitted in the notation for simplicity):

$$\dot{n}_1 = k_a n_1 - k_b n_2 n_1, \tag{9.8}$$
$$\dot{n}_2 = k_a n_2 - k_b n_1 n_2. \tag{9.9}$$

As in the Lotka-Volterra model, the analytical solutions for Eqs. (9.8) and (9.9) show that every starting condition different from $[n_1]_0 = [n_2]_0$ will lead to one of the asymptotes $[n_1] = 0$ or $[n_2] = 0$. Thus, the equality of $[n_1]_0$ and $[n_2]_0$ represents a

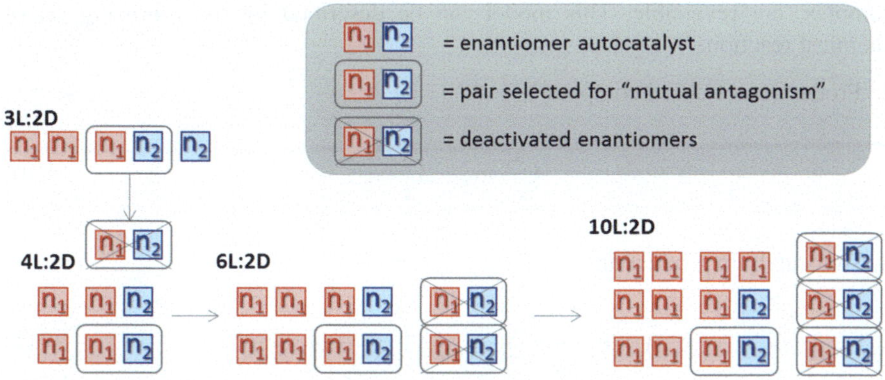

Fig. 9.2 Homochirality emerging according to the Frank model, based on autocatalytic replication and mutual inhibition of enantiomers. Adapted from Blackmond (2010)

condition of unstable equilibrium. With this in mind, the ecological competitive exclusion principle, originally derived from the Lotka-Volterra two-species model, can be regarded as a consequence of sufficiently antagonistic interactions between two (biological) species that are competing for common, finite resources. In the Frank model—and generally in the literature on absolute asymmetric synthesis—this antagonistic interaction between (chemical) species is usually referred to as mutual inhibition.

Figure 9.2 shows how the autocatalytic production of enantiomers n_1 and n_2, coupled with a step of mutual inhibition, in which n_1 and n_2 interact with each other in such a way that results in the removal of both from the system, can propagate and amplify an initial imbalance of one of the enantiomers present in the system [Eqs. (9.8)–(9.9)]. In this sense, this antagonistic interaction (represented by the mutual inhibition between enantiomers) essentially decreases the racemic content in the system, making the enantiomeric imbalance more evident. For low concentrations of monomers, the mutual inhibition between enantiomers will remove the monomers from the system before a significant enrichment can be observed. However, if the concentration of enantiomers is large enough, the amplification process can be sustained and the enantioselective autocatalysis of one of the enantiomers (n_1 in the example of Fig. 9.2) will eventually dominate the system.

9.4.2 *A Realistic Model of Homochirality*

The original Frank model did not account for reversibility, as it should do for realistic chemical scenarios. A more realistic version of the model was later proposed (Kondepudi and Nelson 1983), in which the autocatalytic species are derived from an achiral precursor, and all steps, including the formation of the mutual inhibition complex, are reversible. This model can be described by the following set of chemical reactions[2]:

1. Production of chiral compound:

$$A \underset{k_{-1}}{\overset{k_1}{\rightleftharpoons}} n_1, \quad A \underset{k_{-1}}{\overset{k_1}{\rightleftharpoons}} n_2. \tag{9.10}$$

2. Autocatalytic production:

$$A + n_1 \underset{k_{-2}}{\overset{k_2}{\rightleftharpoons}} n_1 + n_1, \quad A + n_2 \underset{k_{-2}}{\overset{k_2}{\rightleftharpoons}} n_2 + n_2. \tag{9.11}$$

3. Hetero-dimerization:

$$n_1 + n_2 \underset{k_{-3}}{\overset{k_3}{\rightleftharpoons}} n_1 n_2. \tag{9.12}$$

Equation (9.10) represents the formation of chiral product from achiral precursors. Equation (9.11) corresponds to the (first order) autocatalytic reaction described by the first term in Eqs. (9.8) and (9.9). Finally, Eq. (9.12) corresponds to the mutual inhibition of the two species described by the second term in Eq. (9.8–9.9). This mutual inhibition can be expressed as a reaction in which the interaction of the two chiral species produces an achiral product that is usually removed from the system or converted back to chiral monomers through reversible reactions (Eq. 9.12). The elimination of the heterodimer can be neglected if the mutual inhibition is irreversible. However, under realistic chemical conditions, including reversibility in the system, some thermodynamic constraints must be considered and fulfilled (Blackmond and Matar 2008). The equilibrium constants for the direct production and the autocatalytic reactions in Eqs. (9.10)–(9.12) are given by:

$$K_1 = \frac{k_1}{k_{-1}} = \frac{[n_1]}{[A]} = \frac{[n_2]}{[A]}, \quad K_2 = \frac{[k_2]}{[k_{-2}]} = \frac{[n_1]^2}{[A][n_1]} = \frac{[n_2]^2}{[A][n_2]},$$

and therefore the system must satisfy the interesting constraint that $k_1/k_{-1} = k_2/k_{-2}$.

The Lotka-Volterra and Frank models are both described by the same general differential equations. However, distinctions can be seen in the more realistic version of the Frank model. In contrast to biological transformations, chemical reactions are reversible, and the constraints on the reaction rate constants are required to fulfill the principle of micro-reversibility. In addition, several versions of Frank's original model have been proposed during the last decades: considering only one achiral

[2]Although in this chapter, for the purpose of illustrating the connection to other models, we will use the notation n_1 and n_2 to refer to the enantiomeric species, the notation of L and D is customarily used in the literature on absolute asymmetric synthesis.

precursor (A), as in Eq. (9.10) (Plasson et al. 2007; Saito and Hyuga 2005), or two achiral precursors (A and B) (Kondepudi and Nelson 1983), neglecting the reaction in Eq. (9.10) compared with the autocatalytic reaction in Eq. (9.11) (Frank 1953; Saito and Hyuga 2004; Iwamoto 2003), or using the direct continuous elimination of both chiral compounds L and D from the system to model the mutual inhibition in Equation (9.12) and the removal of the achiral heterodimer from the system (Iwamoto 2003). These models differ in detail but all exhibit the basic property of mutual inhibition (or antagonism) which leads to all-or-nothing selection of one chemical (or biological) species.

For a system of reversible reactions under conditions that allow a chemical thermodynamic equilibrium to be reached (i.e., a closed system with uniform matter, temperature, and energy distributions), the racemic state is the state of maximum entropy. Homochirality therefore appears as the result of a temporary asymmetric amplification [i.e., a chiral excursion (Blanco et al. (2011)], which may be kinetically trapped. In this case, the system described by Eqs. (9.10)–(9.12) is capable of amplifying an initially tiny statistical enantiomeric excess, from ee $\sim 10^{-8}\%$ to practically 100%, leading to a long duration chiral excursion at nearly 100% ee, before someday approaching the lowest energy racemic state at thermodynamic equilibrium. On the other hand, in a system with a nonuniform energy distribution (e.g., energy absorption by only some of the species of the system, or open to matter exchange with the surroundings), depending on the conditions, the final stable stationary state may be chiral. To describe this, a key parameter is used, defined as $g = k_{-2}/k_3$; and the chiral state is found to be stable if $g < g_c \leq 1$, where $g_c = \left(\sqrt{1+16h} - 1\right)/8h$ and $h = (k_1\ k_3[\mathrm{A}]/(k_2[\mathrm{A}] - k_{-1})^2$. Thus, a necessary but not sufficient condition to achieve a final stable chiral state is $k_3 > k_{-2}$. That is, the chiral state is stable if and only if the heterochiral complex forms more quickly than the homochiral reversion to the achiral precursor (Crusats et al. 2009), or in different words, if the racemic content of the system decreases faster than the decay of the homochiral state.

Frank stated in his 1953 paper that "A laboratory demonstration may not be impossible" and he was right. More than 40 years after Frank proposed his original model, the first experimental demonstration of absolute asymmetric synthesis was made when Soai and coworkers reported spontaneous generation of enantiomeric excess in the autocatalytic addition of diisopropylzinc to prochiral pyrimidine carbaldehydes (Soai et al. 1995). Furthermore, and as Frank predicted, this reaction was shown to yield the autocatalytic product in very *high* enantiomeric excess (over 90%) even if starting from a very *low* enantiomeric excess (2%) in the original catalyst. Shortly after the initial discovery, Soai's group reported enantiomeric excesses as high as 85% for a reaction initiated with an enantiomeric excess of 0.1% (Shibata et al. 1998). This intriguing behavior stands out not only as a paradigm of absolute asymmetric synthesis (Avalos et al. 1998; Feringa and Van Delden 1999) but also as an experimental proof of concept for the abiotic emergence of biological homochirality (Weissbuch et al. 2005).

The kinetic schemes derived from the Frank model reproduce the mirror-symmetry-breaking behavior of the Soai reaction. More importantly, from the kinetic viewpoint, the enantioselective autocatalysis at the monomer level seems to be consistent with all reported experimental results of the Soai reaction (Islas et al. 2005). More recently, Mauksch, Tsogoeva, and coworkers have found experimental evidence for both asymmetric autocatalysis (Mauksch et al. 2007a; Amedjkouh and Brandberg 2008) and for spontaneous mirror symmetry breaking (SMSB) (Mauksch et al. 2007b) in the organocatalytic Mannich reaction, a process that takes place in conditions much closer to equilibrium than those of the mostly irreversible Soai dialkylzinc addition.

9.4.3 Limited Enantioselectivity (LES)

In a Frank-like chemical reaction network, a necessary (but not sufficient) condition to achieve a final stable chiral state is that the heterochiral interaction between products and catalysts is favored compared to the homochiral interaction [$k_3 > k_{-2}$ in Eqs. (9.10)–(9.12)]. This seems to be the case in the majority of chiral organic compounds (following the high number of chiral compounds that crystallize as racemic crystals, compared to those yielding a racemic mixture of enantiopure crystals or racemic conglomerates) (Collet et al. 1981). However, this is not the case for some significant compounds in prebiotic chemistry, as, for example, several amino acids. To explain the emergence of chirality in enantioselective autocatalysis for compounds which do not follow Frank-like schemes, the limited enantioselectivity (LES) model (Avetisov and Goldanskii 1996) was proposed as a mechanism for SMSB. The basic model is composed of coupled enantioselective and non-enantioselective autocatalysis and is described by the following chemical transformations:

1. Production of chiral compound:

$$A \underset{k_{-1}}{\overset{k_1}{\rightleftharpoons}} n_1, \quad A \underset{k_{-1}}{\overset{k_1}{\rightleftharpoons}} n_2. \tag{9.13}$$

2. Autocatalytic production:

$$A + n_1 \underset{k_{-2}}{\overset{k_2}{\rightleftharpoons}} n_1 + n_1, \quad A + n_2 \underset{k_{-2}}{\overset{k_2}{\rightleftharpoons}} n_2 + n_2. \tag{9.14}$$

3. Limited enantioselectivity:

$$A + n_1 \underset{k_{-3}}{\overset{k_3}{\rightleftharpoons}} n_1 + n_2, \quad A + n_2 \underset{k_{-3}}{\overset{k_3}{\rightleftharpoons}} n_1 + n_2. \tag{9.15}$$

Equations (9.10) and (9.11) in the Frank model are identical to Eqs. (9.13) and (9.14) in the LES model. However, the catalytic production of the opposite enantiomeric product (cross-catalysis rather than autocatalysis) in LES (in Eq. 9.15) substitutes for the mutual inhibition reaction in the Frank model (in Eq. 9.12). In both models (Frank and LES), the enantioselective autocatalysis (Eqs. 9.11 and 9.14) is coupled to a reaction that leads to the decrease of the racemic composition of chiral catalysts. In Frank, this decrease is achieved through the "mutual inhibition" between enantiomers in Eq. (9.12) (with rate k_3) and in LES through the inverse reaction of the non-enantioselective autocatalysis in Eq. (9.15) (with rate k_{-3}). Both reactions lead to the same mathematical terms.

The equilibrium constants for the direct production and the autocatalytic reactions in this case are given by:

$$K_1 = \frac{k_1}{k_{-1}} = \frac{[n_1]}{[A]} = \frac{[n_2]}{[A]}, \quad K_2 = \frac{[k_2]}{[k_{-2}]} = \frac{[n_1]^2}{[A][n_1]} = \frac{[n_2]^2}{[A][n_2]}, \quad K_3 = \frac{[k_3]}{[k_{-3}]} = \frac{[n_1][n_2]}{[A][n_1]} = \frac{[n_2][n_1]}{[A][n_2]},$$

so the system is under the thermodynamic constraint $k_1/k_{-1} = k_2/k_{-2} = k_3/k_{-3}$.

A linear stability study reveals that in this case, the key parameter can be approximated to $g = k_{-2}/k_{-3}$; and the chiral state is found to be stable if $g < g_c \leq 1$, where $g_c \approx (1 - k_3/k_2)/(1 + 3k_3/k_2)$ (Ribo and Hochberg 2008). However, this cannot be achieved when the thermodynamic constraint is fulfilled. Previous reports had claimed SMSB in this model; however, as Blackmond et al. already pointed out in the past (Blackmond and Matar 2008; Blackmond 2009), contradictory reports concerning this were the consequence of the use of a set of reaction rate constants which do not fulfill the thermodynamic constraint mentioned above.

For a system to be maintained in a chiral stationary state, final conditions must be those of constant energy exchange with the surroundings and maximum entropy. Thus, although the thermodynamic constraint forbids any SMSB from occurring in this model (in either closed or open systems), it has been suggested that this can occur in the presence of additional reagents (Blanco et al. 2013a) or in temperature gradients if the autocatalysis and limited enantioselective catalysis are compartmentalized within regions of low and high temperature, respectively (Blanco et al. 2013b)—in such a way that the thermodynamic constraints become compatible with conditions for a final stable chiral state.

We note that the inverse reaction of the non-enantioselective autocatalysis ($n_1 + n_2 \overset{k_{-3}}{\rightarrow} A + n_1$ and $n_1 + n_2 \overset{k_{-3}}{\rightarrow} A + n_2$), can be regarded as a predator-prey interaction, in which, interestingly, both species play the role of prey and predator at the same time. This dual role reflects the fact that n_1 and n_2 correspond to the two enantiomeric species of the same molecule, so this case must correspond to a completely symmetrical situation. Thus, although LES cannot explain SMSB, it may be interesting to study its properties to understand mutual predator-prey interactions.

9.5 Conclusion

Some new scenarios for the emergence of SMSB in compounds that do not follow a Frank-like scheme have arisen during the last years. These scenarios correspond to the deracemization of racemic mixtures of crystals (Noorduin et al. 2008, 2009; Viedma 2005) and the crystallization from boiling solutions (Viedma and Cintas 2011; El-Hachemi et al. 2011). Furthermore, a recent example on sublimations suggests that the same principle is probably also applicable to other phase transitions (Viedma et al. 2011). All these experiments, as well as both the Frank model and the LES model, correspond to cases of SMSB in bifurcation scenarios, where the racemic state is metastable and the more stable final state corresponds to a stationary chiral state. In all cases, small statistical fluctuations about the ideal racemic composition are amplified, taking the system out of the metastable racemic state and driving it into one of the two degenerate stationary chiral states. In the absence of any external polarization dictating the sign of the outcome, the sign of the chiral final state follows a stochastic distribution, as is the case in all the abovementioned experiments.

Although the replicator equation, the LV equations, and the Frank model are used for different purposes and may appear to have little resemblance to each other, it is interesting to note that the three models have equivalent mathematical descriptions. In particular, winner-take-all outcomes prevail in certain parameter regimes in these scenarios. As with the prelife models, these sharp transitions seem to characterize living systems, whether they consist of biological or chemical species.

References

Amedjkouh M, Brandberg M (2008) Asymmetric autocatalytic Mannich reaction in the presence of water and its implication in prebiotic chemistry. Chem Commun 44:3043–3045

Avalos M, Babiano R, Cintas P, Jimenez JL, Palacios JC, Barron LD (1998) Absolute asymmetric synthesis under physical fields: facts and fictions. Chem Rev 98:2391–2404

Avetisov V, Goldanskii V (1996) Mirror symmetry breaking at the molecular level. Proc Natl Acad Sci U S A 93:11435–11442

Bargueno P, De Tudela RP, Miret-Artes S, Gonzalo I (2011) An alternative route to detect parity violating energy differences through Bose-Einstein condensation of chiral molecules. Phys Chem Chem Phys 13:806–810

Barron LD (1981) Optical-activity and time-reversal. Mol Phys 43:1395–1406

Barron LD (1986) True and false chirality and absolute asymmetric-synthesis. J Am Chem Soc 108:5539–5542

Bartel DP, Szostak JW (1993) Isolation of new ribozymes from a large pool of random sequences. Science 261:1411–1418

Blackmond DG (2009) "If pigs could fly" chemistry: a tutorial on the principle of microscopic reversibility. Angew Chem Int Ed 48:2648–2654

Blackmond DG (2010) The origin of biological homochirality. Cold Spring Harb Perspect Biol 2: a002147

Blackmond DG, Matar OK (2008) Re-examination of reversibility in reaction models for the spontaneous emergence of homochirality. J Phys Chem B 112:5098–5104
Blanco C, Stich M, Hochberg D (2011) Temporary mirror symmetry breaking and chiral excursions in open and closed systems. Chem Phys Lett 505:140–147
Blanco C, Crusats J, El-Hachemi Z, Moyano A, Hochberg D, Ribo JM (2013a) Spontaneous emergence of chirality in the limited enantioselectivity model: autocatalytic cycle driven by an external reagent. Chemphyschem 14:2432–2440
Blanco C, Ribo JM, Crusats J, El-Hachemi Z, Moyano A, Hochberg D (2013b) Mirror symmetry breaking with limited enantioselective autocatalysis and temperature gradients: a stability survey. Phys Chem Chem Phys 15:1546–1556
Bomze IM, Burger R (1995) Stability by mutation in evolutionary games. Games Econ Behav 11:146–172
Chapman RN (1931) Animal ecology, with special reference to insects. McGraw-Hill, New York
Chen IA, Nowak MA (2012) From prelife to life: how chemical kinetics become evolutionary dynamics. Acc Chem Res 45:2088–2096
Cintas P (2016) Homochirogenesis and the emergence of lifelike structures. Chirality in supramolecular assemblies. Wiley, New York
Collet A, Gabard J, Jacques J, Cesario M, Guilhem J, Pascard C (1981) Synthesis and absolute-configuration of chiral (C-3) cyclotriveratrylene derivatives – crystal-structure of (M)-(-)-2,7,12-triethoxy-3,8,13-tris-[(R)-1-methoxycarbonylethoxy]-10,15-dihydro-5h-tribenzo[a,D,G]-cyclononene. J Chem Soc Perkin Trans 1:1630–1638
Crick FHC (1968) Origin of genetic code. J Mol Biol 38:367
Crusats J, Hochberg D, Moyano A, Ribo JM (2009) Frank model and spontaneous emergence of chirality in closed systems. ChemPhysChem 10:2123–2131
Eigen M, Schuster P (1977) Hypercycle – principle of natural self-organization. A. Emergence of hypercycle. Naturwissenschaften 64:541–565
El-Hachemi Z, Crusats J, Ribo JM, McBride JM, Veintemillas-Verdaguer S (2011) Metastability in supersaturated solution and transition towards chirality in the crystallization of $NaClO_3$. Angew Chem Int Ed 50:2359–2363
Ertem G, Ferris JP (1996) Synthesis of RNA oligomers on heterogeneous templates. Nature 379:238–240
Feringa BL, Van Delden RA (1999) Absolute asymmetric synthesis: the origin, control, and amplification of chirality. Angew Chem Int Ed 38:3419–3438
Ferris JP, Ertem G (1992) Oligomerization of ribonucleotides on montmorillonite – reaction of the 5′-phosphorimidazolide of adenosine. Science 257:1387–1389
Ferris JP, Ertem G (1993) Montmorillonite catalysis of RNA oligomer formation in aqueous-solution – a model for the prebiotic formation of RNA. J Am Chem Soc 115:12270–12275
Ferris JP, Hill AR, Liu RH, Orgel LE (1996) Synthesis of long prebiotic oligomers on mineral surfaces. Nature 381:59–61
Frank FC (1953) On spontaneous asymmetric synthesis. Biochim Biophys Acta 11:459–463
Guijarro A, Yus M (2009) The origin of chirality in the molecules of life: a revision from awareness to the current theories and perspectives of this unsolved problem. Royal Society of Chemistry, Cambridge
Hadeler KP (1981) Stable polymorphisms in a selection model with mutation. SIAM J Appl Math 41:1–7
Hofbauer J, Sigmund K (1998) Evolutionary games and population dynamics. Cambridge University Press, Cambridge, New York, NY
Hofbauer J, Schuster P, Sigmund K (1979) A note on evolutionary stable strategies and game dynamics. J Theor Biol 81:609–612
Islas JR, Lavabre D, Grevy JM, Lamoneda RH, Cabrera HR, Micheau JC, Buhse T (2005) Mirror-symmetry breaking in the Soai reaction: a kinetic understanding. Proc Natl Acad Sci U S A 102:13743–13748

Iwamoto K (2003) Spontaneous appearance of chirally asymmetric steady states in a reaction model including Michaelis-Menten type catalytic reactions. Phys Chem Chem Phys 5:3616–3621
Kondepudi DK, Nelson GW (1983) Chiral symmetry-breaking in non-equilibrium systems. Phys Rev Lett 50:1023–1026
Lotka AJ (1920) Undamped oscillations derived from the law of mass action. J Am Chem Soc 42:1595–1599
Lotka AJ (1925) Elements of physical biology. Williams and Wilkins, Baltimore
Lotka AJ (1932) Contribution to the mathematical theory of capture: I. Conditions for capture. Proc Natl Acad Sci U S A 18:172–178
Manapat M, Ohtsuki H, Burger R, Nowak MA (2009) Originator dynamics. J Theor Biol 256:586–595
Mauksch M, Tsogoeva SB, Martynova IM, Wei SW (2007a) Evidence of asymmetric autocatalysis in organocatalytic reactions. Angew Chem Int Ed 46:393–396
Mauksch M, Tsogoeva SB, Wei SW, Martynova IM (2007b) Demonstration of spontaneous chiral symmetry breaking in in asymmetric Mannich and Aldol reactions. Chirality 19:816–825
Maynard Smith J (1982) Evolution and the theory of games. Cambridge University Press, Cambridge, New York
Miller SL (1953) A production of amino acids under possible primitive earth conditions. Science 117:528–529
Mislow K (2003) Absolute asymmetric synthesis: a commentary. Collect Czechoslov Chem Commun 68:849–864
Monnard PA, Deamer DW (2001) Nutrient uptake by protocells: a liposome model system. Orig Life Evol Biosph 31:147–155
Monnard PA, Deamer DW (2002) Membrane self-assembly processes: steps toward the first cellular life. Anat Rec 268:196–207
Myrgorodska I, Meinert C, Martins Z, D'hendecourt LL, Meierhenrich UJ (2015) Molecular chirality in meteorites and interstellar ices, and the chirality experiment on board the ESA cometary rosetta mission. Angew Chem Int Ed 54:1402–1412
Noorduin WL, Izumi T, Millemaggi A, Leeman M, Meekes H, Van Enckevort WJP, Kellogg RM, Kaptein B, Vlieg E, Blackmond DG (2008) Emergence of a single solid chiral state from a nearly racemic amino acid derivative. J Am Chem Soc 130:1158
Noorduin WL, Vlieg E, Kellogg RM, Kaptein B (2009) From Ostwald ripening to single chirality. Angew Chem Int Ed 48:9600–9606
Nowak MA, Ohtsuki H (2008) Prevolutionary dynamics and the origin of evolution. Proc Natl Acad Sci U S A 105:14924–14927
Nowak MA, Komarova NL, Niyogi P (2001) Evolution of universal grammar. Science 291:114–118
Ohtsuki H, Nowak MA (2009) Prelife catalysts and replicators. Proc R Soc B Biol Sci 276:3783–3790
Orgel LE (1968) Evolution of genetic apparatus. J Mol Biol 38:381
Orgel LE (1992) Molecular replication. Nature 358:203–209
Page KM, Nowak MA (2002) Unifying evolutionary dynamics. J Theor Biol 219:93–98
Plasson R, Kondepudi DK, Bersini H, Commeyras A, Asakura K (2007) Emergence of homochirality in far-from-equilibrium systems: mechanisms and role in prebiotic chemistry. Chirality 19:589–600
Pressman A, Blanco C, Chen IA (2015) The RNA world as a model system to study the origin of life. Curr Biol 25:R953–R963
Rajamani S, Vlassov A, Benner S, Coombs A, Olasagasti F, Deamer D (2008) Lipid-assisted synthesis of RNA-like polymers from mononucleotides. Orig Life Evol Biosph 38:57–74
Ribo JM, Hochberg D (2008) Stability of racemic and chiral steady states in open and closed chemical systems. Phys Lett A 373:111–122
Ribo JM, Hochberg D (2015) Competitive exclusion principle in ecology and absolute asymmetric synthesis in chemistry. Chirality 27:722–727

Saito Y, Hyuga H (2004) Complete homochirality induced by nonlinear autocatalysis and recycling. J Phys Soc Jpn 73:33–35
Saito Y, Hyuga H (2005) Chirality selection in open flow systems and in polymerization. J Phys Soc Jpn 74:1629–1635
Sawai H, Orgel LE (1975) Oligonucleotide synthesis catalyzed by Zn2+ ion. J Am Chem Soc 97:3532–3533
Shibata T, Yamamoto J, Matsumoto N, Yonekubo S, Osanai S, Soai K (1998) Amplification of a slight enantiomeric imbalance in molecules based on asymmetric autocatalysis: the first correlation between high enantiomeric enrichment in a chiral molecule and circularly polarized light. J Am Chem Soc 120:12157–12158
Siegel JS (1998) Homochiral imperative of molecular evolution. Chirality 10:24–27
Sievers D, Von Kiedrowski G (1994) Self-replication of complementary nucleotide-based oligomers. Nature 369:221–224
Soai K, Shibata T, Morioka H, Choji K (1995) Asymmetric autocatalysis and amplification of enantiomeric excess of a chiral molecule. Nature 378:767–768
Stadler PF, Schuster P (1992) Mutation in autocatalytic reaction networks – an analysis based on perturbation-theory. J Math Biol 30:597–632
Viedma C (2005) Chiral symmetry breaking during crystallization: complete chiral purity induced by nonlinear autocatalysis and recycling. Phys Rev Lett 94:065504
Viedma C, Cintas P (2011) Homochirality beyond grinding: deracemizing chiral crystals by temperature gradient under boiling. Chem Commun 47:12786–12788
Viedma C, Noorduin WL, Ortiz JE, De Torres T, Cintas P (2011) Asymmetric amplification in amino acid sublimation involving racemic compound to conglomerate conversion. Chem Commun 47:671–673
Volterra V (1926) Variazioni e fluttuazioni del numero d'individui in specie animali conviventi. Mem Accad Naz Lincei 2:31–113
Weissbuch I, Leiserowitz L, Lahav M (2005) Stochastic "mirror symmetry breaking" via self-assembly, reactivity and amplification of chirality: relevance to abiotic conditions. In: Walde P (ed) Prebiotic chemistry: from simple amphiphiles to protocell models, vol 259. Springer, Berlin, pp 123–165
Woese CR, Dugre DH, Dugre SA, Kondo M, Saxinger WC (1966) On fundamental nature and evolution of genetic code. Cold Spring Harb Symp Quant Biol 31:723

Chapter 10
Network Theory in Prebiotic Evolution

Sara Imari Walker and Cole Mathis

Abstract One of the most challenging aspect of origins of life research is that we do not know precisely what life is. In recent years, the use of network theory has revolutionized our understanding of living systems by permitting a mathematical framework for understanding life as an emergent, collective property of many interacting entities. So far, complex systems science has seen little direct application to the origins of life, particularly in laboratory science. Yet, networks are important mathematical descriptors in cases where the structure of interactions matters more than counting individual component parts—precisely what we envision happens as matter transitions to life. Here, we review a few notable examples of the use of network theory in prebiotic evolution, and discuss the promise of systems approaches to origins of life. The end goal is to develop a statistical mechanics useful to origins of life—that is, one that deals with interactions of system components (rather than merely counting them) and is therefore equipped to model life as an emergent phenomena.

S. I. Walker (✉)
Beyond Center for Fundamental Concepts in Science, Arizona State University, Tempe, AZ, USA

School of Earth and Space Exploration, Arizona State University, Tempe, AZ, USA

ASU-SFI Center for Biosocial Complex Systems, Arizona State University, Tempe, AZ, USA

Blue Marble Space Institute for Science, Seattle, WA, USA
e-mail: sara.i.walker@asu.edu

C. Mathis
Beyond Center for Fundamental Concepts in Science, Arizona State University, Tempe, AZ, USA

Department of Physics, Arizona State University, Tempe, AZ, USA
e-mail: cole.mathis@asu.edu

© Springer International Publishing AG, part of Springer Nature 2018

C. Menor-Salván (ed.), *Prebiotic Chemistry and Chemical Evolution of Nucleic Acids*, Nucleic Acids and Molecular Biology 35,
https://doi.org/10.1007/978-3-319-93584-3_10

10.1 Introduction

One of the most challenging aspects of origin of life research is identifying those properties of life likely to be characteristic not only of life as it exists today, after >3.5 billion years of evolutionary refinement, but also at its origin. In reality, the problem is harder than even this, as we must identify properties of life that could have preceded its origin *and* could also be responsible for driving the transition to the living state. So far, the community of origin of life researchers has seen tremendous progress in synthesizing different molecular components of life, including lipids and amino and nucleic acids. This approach assumes the properties of life preceding it should include some of current life's basic molecular components (although it is still debated which ones).[1] However, to become life, these molecular components would necessarily have to interact to generate chemical systems exhibiting the emergence of increasingly "lifelike" properties. But, what are the emergent "lifelike" properties prebiotic chemists should focus on?

In biology, it is often the case we deal with highly complex interacting systems, with hundreds or thousands of components (see, e.g., Fig. 10.1). Understanding the fundamental processes driving the large-scale organization of living systems is therefore no small challenge (and it is more challenging still to distill and import relevant ideas to prebiotic chemistry). Network theory has become an indispensable tool for making sense of the mess of biology by reducing the study of complex interacting systems to the study of the statistical properties of their graphical representation. A graph (or network) is a set of nodes and edges, sometimes with additional attributes and structure. A simple example is shown in Fig. 10.2. In chemistry nodes could be molecules, where two molecules are connected by an edge if they participate in the same reaction. As a mathematical abstraction, networks have found utility in describing the large-scale statistical properties of living systems from the functioning of cells to the organization of cities. For example, studies of biochemical networks led to the discovery of the "scale-free" structure of metabolism (see Sect. 10.2), which describes a heterogeneity in the global organization of chemical reactions associated with bioenergetics, common to all three domains of life (Jeong et al. 2000). In addition to revealing organizational properties, once the structure of the graph is known, its generative and evolutionary mechanisms can be identified (Barabasi and Albert 1999) and its robustness and stability properties characterized (Larhlimi et al. 2011). Network theory therefore provides a set of mathematical tools, which could be utilized to understand not only the organization of living networks but also how this organization emerges in the first

[1]It is important to point out it is an assumption of our theories for the origin of life that the process started with molecules we would identify as biological. Alternative hypotheses, such as Cairns-Smith's "clay world" (Cairns-Smith 1986), make different assumptions. It is a reasonable assumption to make, but in the field of origins where we remain largely in the dark about exactly what happened, it is important to be aware of the starting points we adopt to make traction on the problem.

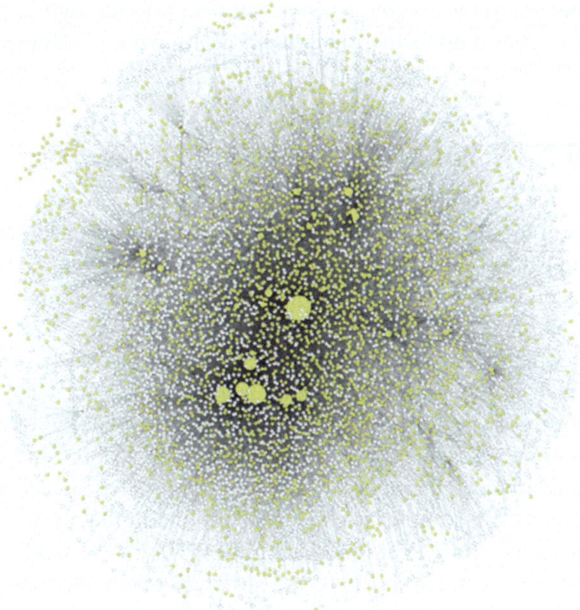

Fig. 10.1 A network representation of global biochemistry, containing thousands of compounds cataloged across organism on Earth. Highlighted in yellow are compounds (represented by nodes in the network) common to all three domains of life. Figure adopted from Kim et al. (2018)

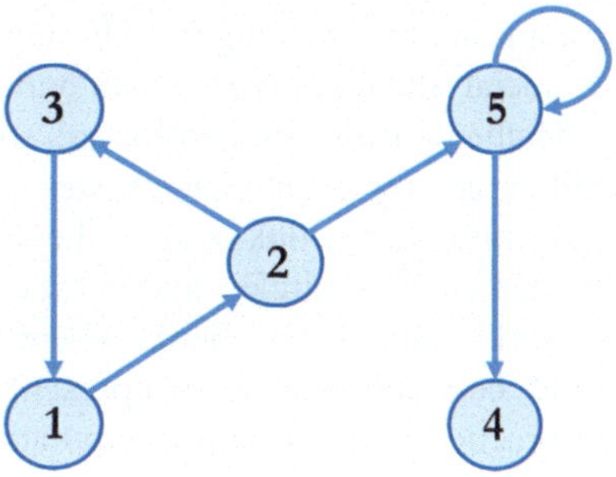

Fig. 10.2 A network with nodes (circles) and edges connecting them (arrows). Here, additional attributes and structure include labeling of nodes and arrows on the edges, respectively

place. Combining novel techniques from systems chemistry with the mathematical formalism of network science will enable prebiotic chemists to develop new concepts with testable consequences (Cronin and Walker 2016).

In this chapter, our goal is to introduce the concepts of network theory to prebiotic chemists as a mathematical formalism for making sense of prebiotic systems and as a tool to identify the processes driving the emergence of life. We first review the basics of network theory, discussing some of the successes in its application to chemical and biological systems. Our focus is on properties of biochemical networks of use to prebiotic chemists interested in studying the emergence of more "lifelike" chemical systems in the lab. We also discuss future directions for both how the study of biochemical networks might inform origin and how the study of chemical networks relevant to the origin of life might also provide tighter constraints on what network properties could truly distinguish living from nonliving organization.

10.2 What Is a Network?

The phrase "World Wide Web" vividly captures the complex web of interactions connecting computers across the globe. Many other biological and technological systems share similar weblike structure, being comprised of many heterogeneous interconnected components (see, e.g., Fig. 10.1). Over the past several decades, it was realized a new statistical mechanics was necessary to describe such systems, which goes beyond the nineteenth-century statistical mechanics of idealized non-interacting particles to include the topology of interactions among system components and their resultant dynamics (Albert and Barabási 2002) The natural mathematical framework for developing such a theory is network theory, which projects the complex web of interactions in real systems onto an abstract representation as a graphical object (Barabási 2016). Networks are important mathematical descriptors in cases where the structure of interactions matters more than counting individual component parts—precisely what we envision happens as nonliving matter transitions to life. Due to its utility in concisely describing complex, interacting systems, network theory has been applied to an increasing number of systems in fields ranging from biology (Barabási and Oltvai 2004), to engineering, to the social sciences (Wasserman and Faust 1994).

Mathematically, networks are studied using the tools of graph theory, where entities are represented by *nodes* (also called vertices) and their interactions by *edges* (also called links), as in the simple network shown in Fig. 10.2 where nodes are depicted as circles and edges as arrows. Familiar examples include social networks, such as Facebook, where the nearly two billion individuals on Facebook could be mathematically represented by nodes and their friendships by edges (in practice it is difficult computationally to construct and analyze networks this large, but many networks of interest are smaller than Facebook or subnetworks can be studied). In a graph-theoretic representation of Facebook, an individual would be connected to every individual they "like," and network dynamics might include studying how the structure of interactions changes as individuals "like" and "unlike" one another or as new individuals are added to the network and others lost from it. Likewise, chemical species reacting with one another form networks within cells where nodes represent molecular species or reactions and edges represent connections of molecular species to reactions they participate in (see Fig. 10.3).

There are in fact many different ways to represent a network. For example, the Facebook network mentioned above could be represented as a *directed* network, meaning that the connections between nodes are not symmetric but instead reflect the directionality of "like" relations. Cole "liking" Sara's Facebook page does not imply Sara also "likes" Cole's page. The network in Fig. 10.2 is an example of a directed network, where edges are represented by arrows delineating the

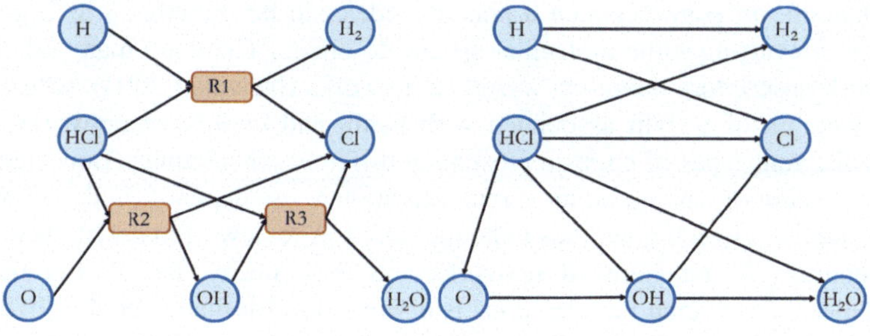

Fig. 10.3 Two different graphical representations of the same chemical reaction network. On the left is a bipartite substrate-reaction network, and on the right is shown the same system represented as a unipartite substrate-substrate network (see text for descriptions)

directionality of the relationship (in the Facebook example, a "like" relationship might point from the node labeled "Cole" to the node labeled "Sara"; if Sara also "likes" Cole's page, there would also be an arrow connecting the same two nodes but pointing in the opposite direction). Networks may also be *undirected*, where edges do not encode the directionality of the relationship. The choice of network is often motivated by the problem of interest and the measures one is interested in calculating (e.g., there is a richer literature of network measures for undirected networks as they are simpler, but the trade-off is they do not capture as much information as directed ones).

There are many different ways to graphically represent chemical systems, each permitting quantitative analysis of different aspects of global organization and in turn identification of the role of specific molecules in the robustness and function of living systems. Two examples of the most commonly implemented graphical representations for chemical systems are shown in Fig. 10.3, which both represent the following sequence of reactions:

$$\mathrm{H} + \mathrm{HCl} \rightarrow \mathrm{H_2} + \mathrm{Cl}$$
$$\mathrm{HCl} + \mathrm{O} \rightarrow \mathrm{Cl} + \mathrm{OH}$$
$$\mathrm{HCl} + \mathrm{OH} \rightarrow \mathrm{Cl} + \mathrm{H_2O}$$

The left panel of Fig. 10.3 shows a *bipartite* network, called a reaction-substrate graph, where substrates (reactants and products) and their reactions are both nodes and edges connect substrates to their relevant reactions. Bipartite networks are so-called because there exist two distinct types of nodes in the network: here, molecular species represent one type of node (circles), while chemical reactions represent the other (squares). The representation of the same network in the right panel of Fig. 10.3 is an example of a *unipartite*, substrate-substrate graph, where reactions are abstracted away and reactants are directly connected to products by an edge (if they are from the same reaction). A further refinement in the unipartite graph is representing edges as undirected (loosing directionality in the relationship of substrates and products as discussed above). In the substrate-substrate representation, an edge between a pair of nodes can be thought of as a group of processes converting some molecular species to others. There are many other types of network descriptors including *weighted* networks (Newman 2004), where edges have a strength of weight associated with them, and *multilayer* networks, which contain different types of edges representing different connections (Boccaletti et al. 2014). Selection of which graphical representation to use depends on the question of interest and the relevant quantities to be measured. A review of many of the different representations of biochemical networks and their utility and shortcomings as applied to different scientific questions is discussed in Montañez et al. (2010).

10.2.1 Measuring Statistical Properties of Networks

In a seminal paper published in 2000, Jeong et al. reported metabolic networks of 43 distinct organisms—representing all three domains of life—are *scale-free* (Jeong et al. 2000), meaning their degree distributions roughly follow a power-law $P(k) \sim k^{-\alpha}$. An example of a scale-free network is shown in Fig. 10.4a and the corresponding power-law degree distribution in Fig. 10.4c. Here, $P(k)$ is the probability a given molecular species participates in k reactions. In graph theory k is called the *degree* of a node, corresponding to the number of edges connected to it. The *degree distribution,* or degree sequence, is the probability distribution of node degree taken over an entire network.[2] In the simple example network of Fig. 10.1, the degrees are 2, 2, 3, 1, and 3 for nodes 1, 2, 3, 4, and 5, respectively, yielding a degree distribution of $P(k) = 1/5$, 2/5, and 2/5 for $k = 1$, 2, and 3, respectively. This distribution has a *mean degree* $\langle k \rangle = 2.2$, and there are no outlier nodes with a significantly higher degree than the others. In this respect, the network is fairly homogeneous (the network in Fig. 10.2 is of course is too small to make statistically meaningful statements, but it serves for illustrative purposes). For a longtime it was thought most networks were homogeneous, but in the late 1990s and early 2000s, it was discovered most real-world biological and technological networks are in fact very *heterogeneous*, with *heavy-tailed* degree distributions consistent with power-law or lognormal fits [see, e.g., Barabasi (2009) for perspective]. In many real-world networks, most nodes have very few connections, but a few nodes called *hubs* have many connections and link less connected nodes together. The discovery of the power-law scaling in metabolic networks by Jeong et al. was part of this watershed moment in our understanding of the organization of biological and technological systems, but the significance of this property and its evolutionary origins still remain poorly understood.

In our Facebook example, a very small minority of Facebook's >2 billion users are hubs, such as Mark Zuckerburg with 98,885,179 "likes" (as of writing).

[2]The degree distribution is calculated by determining the frequency of the degree for each node, and is often normalized by dividing by the total number of edges in the graph, which can be interpreted as a probability of connection and the resulting distribution interpreted as probability distribution.

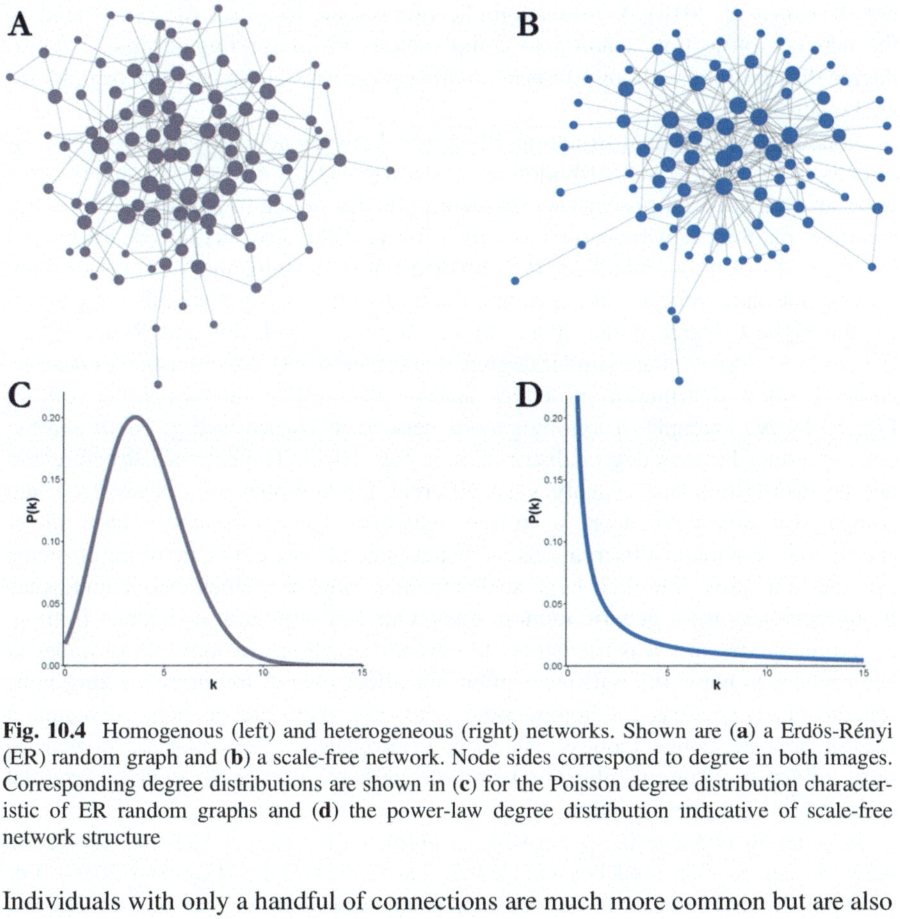

Fig. 10.4 Homogenous (left) and heterogeneous (right) networks. Shown are (**a**) a Erdös-Rényi (ER) random graph and (**b**) a scale-free network. Node sides correspond to degree in both images. Corresponding degree distributions are shown in (**c**) for the Poisson degree distribution characteristic of ER random graphs and (**d**) the power-law degree distribution indicative of scale-free network structure

Individuals with only a handful of connections are much more common but are also much less connected (e.g., by comparison the authors each have only a few hundred "likes"). As with social networks, metabolic networks also contain hubs, which include highly utilized molecules in biochemistry such as H_2O and ATP (Andreas Wagner 1998). These molecules participate in hundreds of reactions, with a comparably high node degree, whereas the mean degree of metabolic networks globally is in the range of just 2–5 connections (Jeong et al. 2000; Kim et al. 2018). Projecting metabolism onto a substrate-substrate network representation yields fits for the degree sequence that in general follow a power-law fit, indicative of scale-free structure. However, rigorously confirming a power-law fit for a given degree distribution is a challenging technical problem and an active area of research in the statistical inference community. In recent years, new tools have been developed for reliably determining cases of scaling consistent with a true power-law behavior, as opposed to other heavy-tailed degree distributions, such as lognormal (see, e.g., Clauset et al. 2009). Our recent analysis applying these tools to a dataset of >28,000 biochemical networks extracted from genomic and metagenomic data revealed a majority of biochemical networks can plausibly be fit to true power-law scaling, but not all (Kim et al. 2018). A further complication is scale-free structure can depend on the network projection, leading to complications in interpreting results of fits to degree distributions without reference to other properties of the network or randomized controls.

Nonetheless, important structural differences between networks can often be seen directly from the degree distribution and other topological measures: in many cases these are indicative of properties that seem to be distinctive to living networks. For example, random networks, such as Erdös-Rényi (ER) networks, are characterized by degree distributions which are Poisson distributed, meaning that most nodes share roughly the same number of edges and the probability is exponentially suppressed for the highest degree nodes (the $P(k){\sim}e^{-k}$ for $k \gg 1$) (Erdös and Rényi 1959). Because most nodes share similar degree, random networks are described as *homogenous* in their distribution of edges among nodes (like our small network in Fig. 10.1). An example of a homogenous network is shown in Fig. 10.4b and the corresponding Poisson degree distribution in Fig. 10.4d. The network structure and degree distribution are visually very different for homogeneous networks when compared to heavy-tailed or "scale-free" networks. Due to these structural differences, the systematic observations of heterogeneous networks in living systems provides a window into their large-scale organizational properties that distinguishes living networks from generic random ones. One key structural difference from an evolutionary standpoint is robustness to random mutation: random loss of nodes in heterogeneous networks will most often not affect overall topology so long hubs remain intact, whereas for homogenous networks, there are no hubs to maintain overall network connectedness. As such, heterogeneous networks are in general more robust to random failure or mutation, perhaps motivating their preferential selection in living organization.

In order to characterize a network, a number of different statistics about the network can be measured beyond degree distribution alone (Barabási 2016). For example, the mean degree of the network, as mentioned above, can be calculated as the mean value of the degree distribution and provides information about how connected each node in the network is on average. For directed networks this can be broken down into two contributing terms: the mean in-degree (number of edges pointing into a node) and mean out-degree (number of edges emanating from a node), which represent sinks and sources in chemical transformation space. Another important network statistic, the *average shortest path length,* measures the average number of steps it would take to get from one node to any other node in the network by taking steps along its edges. That is, it quantifies the minimal number of chemical transformations it takes, on average, to convert one molecular species to another.

Networks with a low average shortest path length are sometimes said to have a *small-world* property because it is relatively easy to get from one node to any other: one need only traverse a few steps. Readers may be familiar with the term "six degrees of separation" to describe this small-world property in human social networks. For a while it was thought metabolism too had the small-world property, meaning it only should take a handful of chemical reactions to transform any molecule to any other in a biochemical network (Wagner and Fell 2001). However, subsequently, it was discovered metabolism is not small in a study performing detailed analysis of the network structure of *E. coli* (Arita 2004). In the context of prebiotic chemistry, we need an accurate picture of the structure of biochemical networks in order to identify how they can be generated in the absence of life—only then will we be able to map this to the appropriate chemical and physical properties driving prebiotic network evolution (not a small task!).

There are other network measures too that could aid in this effort. Another important statistic is *betweenness centrality*, which measures how often a particular node is on the shortest path between all other nodes in the network. Nodes with high betweenness centrality can sometimes be low degree but nonetheless essential to dynamics and function since they play a key structural role by connecting many otherwise disconnected or distant nodes. In many cases however, high betweenness centrality is correlated with high degree (hubs). For example, in a network representation of social media interactions, one might expect Barack Obama to have high betweenness centrality as he is a highly connected node (hub) through which many other individuals are connected. In biochemical networks, molecules like H_2O and ATP tend to have both very high degrees and high centrality, due to their fundamental roles in aqueous organic chemistry and metabolism, respectively. Hubs with high betweenness are therefore often among the most vulnerable nodes in a network for directed attack (rather than random loss): targeting removal of such nodes can lead to a network breaking apart into smaller isolated graphs. This is a chief vulnerability of the Internet (Cohen et al. 2001) and is often more technically discussed in terms of breaking apart the *largest connected component* of a network. A *connected component* is a subgraph of a network (e.g., a subset of nodes) where there exists a path between any two nodes in the subgraph. For understandable reasons, metabolic networks are dominated by a single large connected component [see, e.g., supplement of Kim et al. (2018) for size of largest connected component in biochemical networks]. The idea of connected components becomes important in discussions of graph-theoretic models of the origins of life, such as autocatalytic sets, which also form as connected components (discussed in Sect. 10.3). The early emergence of molecules with high betweenness centrality may have therefore been critical to rapid formation of connected components in prebiotic evolution. Identification of molecules fulfilling this role prebiotically could therefore provide insights the emergence of many key structural properties of living networks.

Many of the measures discussed so far track statistical properties of individual nodes, or paths between two nodes, but there are also many measures for higher-order properties of networks. For example, *clustering coefficient* tracks how many tightly knit communities exist within a given network, typically measured by counting the number of complete triangles connecting three nodes. Networks with high clustering coefficients have many clusters of nodes with above-average connections between them (relative the rest of the network). Complete triangles represent one example of a *network motif*. Network motifs are subgraphs which have specific connection patterns and which are overrepresented in biological systems with respect to randomized networks (Milo 2002). They were first uncovered in networks as diverse as those from biochemistry, ecology, neurobiology, and engineering and have been proposed as a means to uncover the building blocks of functional networks (Alon 2003). From this perspective, they are an important concept for prebiotic evolution—identifying the network motifs which readily form under abiotic conditions and could combine to form more complex, lifelike systems would advance our understanding of key structural properties needed for assembly of living networks. As just one example, we recently constructed all possible three-member networks of cooperating RNA using reaction rate data from a real RNA system based on the *Azoarcus* ribozyme (Mathis et al. 2017b). The goal was to determine the types of cooperation possible when building prebiotic networks from their component parts. Here cooperation was defined in terms of the structure of the subgraph (see Mathis et al. 2017b). Our results demonstrate the triplet network interactions among genotypes (nodes) in the real *Azoarcus* ribozyme system were intrinsically biased to favor cooperation due to the particular distribution of catalytic rate constants in the real system, as compared to other possible distributions for the rate constants. This example demonstrates how coupling properties of chemistry with network structure can provide new insights into the emergent properties of prebiotic networks, such as whether we should expect them to be cooperative.

10.2.2 Generative and Evolutionary Models

Knowledge of topological properties, such as the small-world property or scale-free structure, can provide insights into how networks with those properties can arise in the first place. To get at the interesting properties, we must first identify what features are expected to arise randomly. There are many different models to generate random networks for comparison. These models are known as *random graphs*. We introduced above the first class of random graphs to be formalized, the ER graph, which was developed by Erdös and Rényi in 1959 (called Erdös-Rényi or ER random graphs) (Erdös and Rényi 1959). ER random graphs are defined by the number of nodes (n) and number of edges (v) they contain. A single instance of an ER random graph is generated by starting with n unconnected nodes and randomly assigning edges between nodes with *equal and independent* probability p, until v edges exist. For a very long time, it was assumed that ER random graphs represented ideal null models for network organization. However, as more empirical examples of networks were accrued through the 1970s and early 1980s, it became apparent that ER random graphs failed to produce statistical features common in real-world networks, such as high clustering coefficients, small-world topology, and heterogeneous degree distributions discussed above.

The degree distribution of ER random networks is always homogeneous, described by a Poisson distribution. These provided a stark contrast with real-world networks which are observed to have small-world properties and heterogeneous degree distributions. To address this, in 1998 Watts and Strogatz published a random graph model combining some properties of ER random graphs with regular graphs (which are similar to lattice structures) in order to generate networks with high clustering coefficients and small-world properties, much like real-world systems (Watts and Strogatz 1998). However, these failed to produce the heterogeneous, heavy-tailed degree distributions characteristic of many real-world systems. In 1999 Barabási and Albert introduced a model using preferential attachment to generate networks with the desired scaling properties (Barabasi and Albert 1999). Preferential attachment models start with a small network of nodes and add nodes one at a time to the network by preferentially attaching new nodes to nodes with high degree. In the context of life's chemical networks, such a growth model would imply metabolic networks grow by adding new metabolites, with the most highly connected nodes also being the most likely candidates for being the oldest. Among these ancient, highly connected nodes in metabolism are intermediates of glycolysis and the tricarboxylic acid cycle (TCA); consistent with the hypothesis of Morowitz and later Smith and Morowitz, the evolution of biochemistry is recapitulated in intermediary metabolism (with TCA being among the most ancient components) (Smith and Morowitz 2016). However, modifications to the Barabási-Albert preferential attachment model are necessary to explain the network evolution of biochemistry: the model always produces exactly scale-free networks, whereas observed biochemical networks have heavy tails but are not precisely scale-free (see Kim et al. 2018; Clauset et al. 2009). A number of models have been developed to address this gap (e.g., Bianconi and Barabási 2000). Identifying prebiotically relevant random graph models will be an important step toward understanding the transition from nonliving to living matter.

10.3 Prebiotic Chemical Networks: Prospects and Promise

An important feature of the Erdös-Rényi (ER) model described in the previous section is the existence of a phase transition as the probability of two nodes being connected by an edge increases. At a critical connection probability, *pc*, corresponding to a critical mean degree, ER graphs transition from having many disconnected components to being dominated by one large connected component. Although this transition occurs within an abstract mathematical object, it has implications for the origins of life. Kauffman was the first to recognize this link (Kauffman 1993). In a stroke of insight, he realized a similar process should exist in chemistry: if enough reactions are possible in a given chemical system, one should end up with a large connected set of reactions. Chemical reaction networks should therefore exhibit a phase transition much like the ER transition, where increasing the number of possible reactions among a set of molecules induces a transition from many disconnected networks to a large connected one. To model this process, he considered abstract proteins represented as binary sequences of "0"s and "1"s, often referred to as *binary polymer models* in the artificial chemistry literature. An example autocatalytic network of binary polymers is represented in Fig. 10.5. Kauffman showed that if there is a small, independent, and identical probability

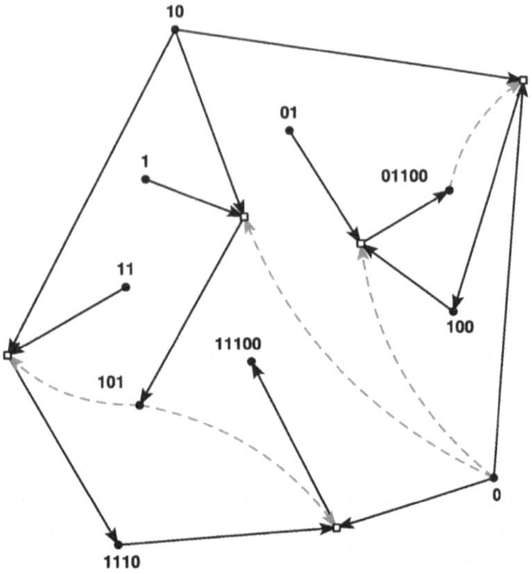

Fig. 10.5 Example of an autocatalytic network within the binary polymer model, where polymers consisted of two monomer species "0" and "1." In this multilayer graph, black nodes represent molecules, white nodes represent reactions, black arrows connect molecules to reactions they participate in, and gray-dashed arrows point from catalysts to reactions they catalyze. Figure adopted from Hordijk et al. (2012)

p that any protein up to a length L catalyzes any given reaction, then the probability P of finding a connected set, such as the one in Fig. 10.5, increases as the length L of the longest sequences increases. In fact, in the mathematical model, the probability approaches 100% as the max length L of the proteins grows, even if the probability of any given protein being catalytic active, quantified by the p, is made arbitrarily small. Placed within the broader context of network theory discussed in Sect. 10.2, Kauffman was not only looking for connected components but a specific motif. His interest was in collectively reproducing sets of molecules: these are network motifs composed of closed cycles of reactions. Analyzing chemical reaction graphs for the existence of these motifs forms the foundation of autocatalytic set theory, the first systematic application of network science to prebiotic chemistry.

Other elements of network science have been suggested in prebiotic chemistry over the years, although they are often not studied with the formalism of graph theory. As one example, in 1978, Eigen and Shuster introduced the hypercycle as a proposed solution to the error threshold problem in prebiotic evolution (Eigen and Schuster 1978). The error threshold sets a fundamental bound on the minimal amount of information necessary to transmit between successive generations for heredity to be possible (Eigen 1971). For a prebiotic replicator, such as a self-copying RNA, this bound is approximately a mutation rate $\mu = 1/L$,where μ is the mutation rate per monomer and L is the length of the sequence (in reality the critical mutation rate depends on the shape of the fitness landscape). In prebiotic systems, before the evolution of error-correction mechanisms of modern cells, error rates were high. The intrinsically high error rates of nonenzymatic templated replication place a strict limit on the amount of information an individual sequence can faithfully copy before error-correcting enzymes evolved. In the words of Szathmary and Maynard Smith, this encompasses a catch-22 for prebiotic evolution: "no enzymes before genes, no genes before enzymes" (Smith and Szathmáry 1995). The hypercycle, as

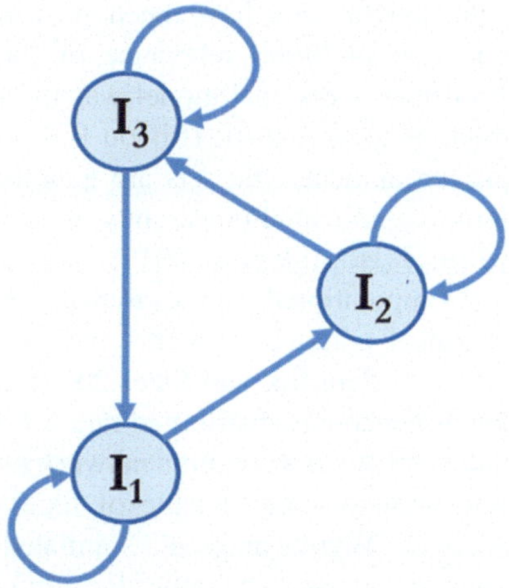

Fig. 10.6 A graphical representation of a three-node hypercycle composed of three replicators I_1, I_2, and I_3. Self-loops indicate self-replication and directed arrows between nodes indicate direction of catalysis

first imagined by Eigen, was proposed as a resolution to this paradox. His idea was to couple two or more replicating species, where each was capable of promoting the replicative efficiency of the other forming a cyclic graph with self-loops; see Fig. 10.6. This constitutes a simple network, where every chemical species can be represented as a node and catalytic connections by edges. Thinking graphically, a solution to the error threshold problem is to distribute information over a network of interacting molecules, rather than storing all of it within a single molecule. For a number of decades, Eigen's idea remained hypothetical, but recently hypercycle networks have been demonstrated in real chemical systems of interacting RNA molecules, providing an important empirical window into understanding potential early stages of evolution and cooperation in molecular systems (Vaidya et al. 2012).

10.3.1 Autocatalytic Set Theory

Chemically, autocatalytic sets are collections of molecules, where every molecule in the set is produced by a reaction catalyzed by another molecule in the set. Graphically, autocatalytic sets represent a class of subgraphs, or network motifs, with directed paths forming closed cycles. It is in this respect the hypercycle can be considered an example of an autocatalytic set. Since the early work of Kauffman, Eigen, and others exploring how closed cycles might lead to collectively reproducing systems, there have been a number of efforts to both develop better theoretical and experimental approaches to understanding these systems. Of note, Kauffman's original idea has been formalized within the context of RAF theory. RAF is short for *reflexively autocatalytic and food-generated.* RAF sets are graphical structures forming closed cycles with inputs for food to the cycle (Hordijk and Steel 2004; Hordijk et al. 2012). Within this formalism a variety of properties of autocatalytic sets have been proven over the past few years, strengthening the potential prebiotic relevance of the theory. In particular, a major criticism of Kauffman's original model was the required level of catalysis, which was deemed too high to be realistic (Lifson 1997). Within the RAF formalism, Hordjik et al. have proven autocatalytic sets are guaranteed for realistic levels of catalysis and when more complicated constraints of real-world chemistries are imposed (such as base pairing) (Hordijk et al. 2011).

Computational and analytical results show autocatalytic sets are common in chemical reaction networks with random independent and identically distributed catalysts (Hordijk and Steel 2017) and also occur in more realistic scenarios with heterogeneously distributed rates of catalysis (Hordijk et al. 2014). However, while autocatalysis is a common network motif, multiple studies have shown that relatively few networks are capable of fixating dynamically when kinetics are taken into account. Wynveen et al. simulated the dynamics of a binary polymer system constructed using the same algorithm as Kauffman. They demonstrated that relatively few networks were able to depart from an expected maximum entropy state, meaning that networks composed of random and identically distributed catalysis rarely display "lifelike" dynamics (Wynveen et al. 2014). Similarly, Filisetti et al. found only a small fraction of autocatalytic sets which were able to increase the abundance of their constituents above a background level expected nonenzymatically (Filisetti et al. 2012).

As theory improves and makes closer contact with experiment, the challenge ahead will be to understand better the properties of real biochemical networks and how those arise prebiotically. It has already been confirmed RAFs exist in real biochemical networks, such as the metabolic network of *E. coli* (Sousa et al. 2015). Additionally, some work has been done to connect RAF theory to the structure of real biochemical networks. When Kauffman originally devised autocatalytic set theory, it was thought most real-world networks were homogenous, like the ER model. However, as discussed above, it was subsequently discovered heterogeneous networks are more common in real-world systems. Recent work has also shown RAFs are common in catalytic networks with power-law distributed catalysis (Hordijk et al. 2014), which more closely resembles the distribution of catalysis in metabolic networks. Future work should further the connections between these abstract models and the properties of real biochemical networks.

10.3.1.1 Evolvability of Autocatalytic Sets

A key transition in the origin of life on Earth was the emergence of Darwinian evolution via natural selection (Nowak and Ohtsuki 2008). Natural selection requires mechanisms to generate variation among individuals, which can then be selected. For single molecule replicators or single cells, these requirements are easy to satisfy as the "unit" of evolutionary selection is readily identifiable (a replicating sequence or cell, respectively). However, for collectively reproducing systems without a well-defined boundary of "self" and "other," the concepts of individuality and heredity are poorly defined. It is not yet even clear well-defined units for selection exist in such systems. As such, the evolvability of catalytic networks has been a subject of intense debate in origin of life research. At stake is whether catalytic networks are indeed a viable alternative to genetic polymers as the first hereditary system capable of Darwinian evolution.

Among models proposed for catalytic network evolution is the “lipid world” scenario proposed by Segré, Lancet, and collaborators (Segré et al. 2001). The model system includes simulated random networks with lognormal distributed catalytic efficiencies, meant to capture aspects of the asymmetry of catalytic efficiency in real systems. The lognormal distribution of catalysis can be modeled by a strongly connected, weighted network. From this model, Segré et al. have shown in some situations these networks are capable of evolution by natural selection (Segre et al. 2000). However, using the same model, Vasas et al. have shown that, in general, these networks cannot be evolved to generate arbitrary steady states (Vasas et al. 2010). The problem arises because random networks generated using the lognormal catalytic distributions contain subtle motifs which prevent the maintenance of variation between competing networks. This led Vasas et al. to claim that autocatalytic networks, in general, are not evolvable. In subsequent work, Vasas et al. investigated the evolvability of autocatalytic sets similar to those first suggested by Kauffman (Vasas et al. 2012). These more recent results suggest autocatalytic sets can indeed evolve in a limited sense, as long as they contain multiple *viable cores*. Viable cores are a specific network motif composed of completely connected catalytic subgraphs. Taken together, these results indicate it may be possible for catalytic networks to evolve in the absence of genes, but the details are sensitive to network topology in ways genetic propagation of information isn’t (perhaps one selective factor in the transition to genetic heredity during early evolution).

The jury is still out on whether general, evolvable models of catalytic networks are possible and what the key properties of such networks might be. In an attempt to summarize the current state of the field, and to project what network properties might emerge as those most essential to defining evolvability, Nghe et al. recently identified six key network parameters to focus research efforts (Nghe et al. 2015). Among these were the concepts of viable cores. Other parameters include familiar concepts in prebiotic chemistry, such as resource availability, and compartmentalization. Resource availability is essential to maintaining collective reproduction (e.g., the “food” in RAF sets), and compartmentalization is essential for forming selectable units (this could occur via localization on a surface and need not be physical compartmentalization). Other parameters may be less familiar and are more intrinsic to network organization, including its connectivity, controllability, and scalability. Connectivity, combined with the availability of resources determines how effectively molecular species outside of viable cores can be produced. Very sparse, poorly connected networks will have limited evolvability, as there are no paths for transitions between graphs. In the other extreme, networks that are too densely connected will generate non-specific tars. Controllability can be implemented in chemical system through dynamic feedback. These feedbacks stabilize network dynamics against random perturbations, enhancing the robustness of chemical networks in fluctuating environments and can play a critical role in inheritance. As one example, Kaneko and collaborators have worked out a model catalytic network, where reproduction is controlled by a "minority" population of molecules regulating the behavior of the rest of the network. Their proposal is that these minority molecules played the role of primitive genes (Kaneko and Yomo 2002). The scalability of chemical networks refers to an ability to grow in size while maintaining functional modules. In order to scale efficiently, prebiotic chemical networks must be sparse, meaning most nodes have few connections, reducing the likelihood new functional modules will interfere with the rest of the network. This introduces a tension between scalability and evolvability as sparsity favors one but not the other. In order to understand the evolution of primitive chemical networks, future studies must constrain these six key parameters in real networks, and theory must be developed to better understand how each impacts evolutionary outcomes.

10.3.2 Autocatalytic Sets in the Lab

Identifying and exploring dynamic, complex chemical networks represents a major analytical challenge for organic chemists. The most interesting aspects of complex (bio)chemical networks are due to interactions between tens to thousands of dynamically coupled reactions, meaning any given network cannot be understood as the sum of many isolated reactions. Many of the standard techniques for characterizing reaction products depend on identifying single molecule products which can be compared to lab standards. Understanding the properties of chemical networks depends on first understanding how the topology of the networks is related to their dynamics. Luckily, there are many questions in this area that can be addressed using different chemical models, and the earliest investigations of chemical networks have come from polypeptide systems as well as RNA systems.

In 2003 Ashkenasy et al. predicted and constructed a complex network of peptide fragments (Ashkenasy et al. 2004). The authors had previously demonstrated a reaction between electrophilic (E) and nucleophilic (N) peptide fragments could be promoted by a template peptide (T). The peptides form a quaternary complex, and in isolated reactions they had shown that the efficiency of the reactions could be predicted by the stability of the complex, which could be in turn estimated from the template structure. Using nine different templates, Ashkenasy et al. constructed a weighted network with the templates as the nodes and while the edges represented the predicted catalytic pathways, a schematic representation of this network is shown in Fig. 10.7. When they implemented this network in the lab, they found that some of the predicted edges were not realized due to competition for shared substrates. This network represents one of the earliest physical instantiations of Kauffman's autocatalytic set theory. This work demonstrates real chemical networks are not simply the sum of all possible reaction pathways but include emergent properties arising due to the complex interplay between topology (catalytic efficiency) and dynamics

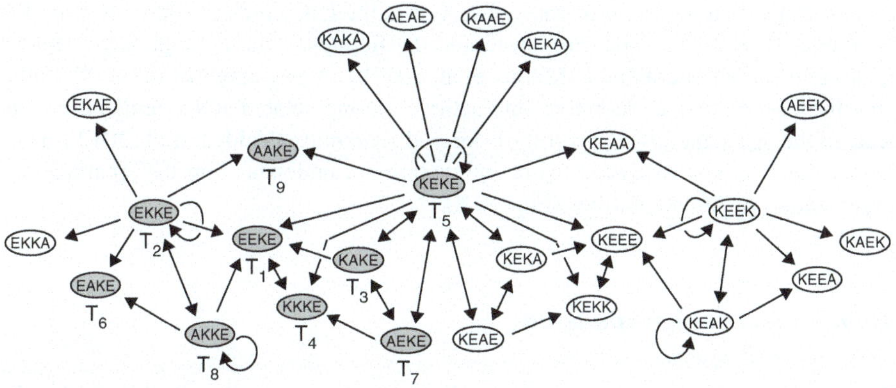

Fig. 10.7 Illustration of a self-organized peptide network composed of 25 nodes joined by 53 edges. Nodes are different peptide templates, while edges represent catalytic activity. Adapted from Ashkenasy et al. (2004)

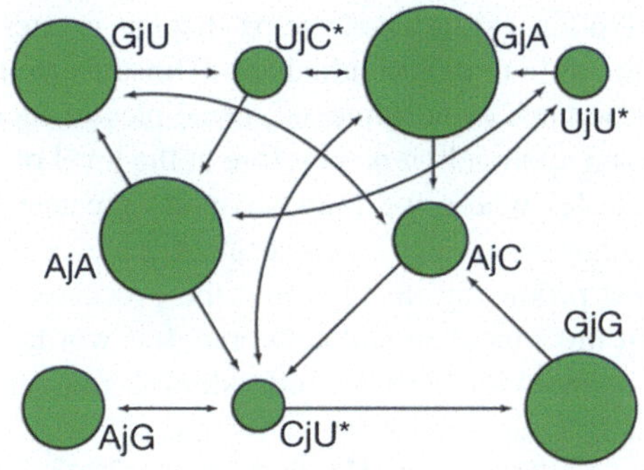

Fig. 10.8 Network structure of cooperative RNA hypercycle. Node labels correspond to different genotypes, while edges represent catalytic activity. Node sizes correspond to steady-state abundances. Adapted from Vaidya et al. (2012)

(resource availability) [see, e.g., also Vaidya et al. (2013) for a combined theory-experiment model of the role of limited resources in RNA systems] (Fig. 10.7).

In 2012 Vaidya et al. demonstrated hypercycle networks could form spontaneously in RNA networks, using the *Azoarcus* ribozyme system, one of which is shown in Fig. 10.8 (Vaidya et al. 2012). The *Azoarcus* ribozyme is a ~200 nt RNA sequence capable of self-assembly. By varying four different bases in the ribozyme, 48 different genotypes can be made. Each genotype can catalyze its own assembly as well as the assembly of other genotypes with different efficiencies. Lehman et al. demonstrated strongly cooperative triplet motifs could form, within the larger 48 node network. Subsequent studies have shown the dynamics of those motifs can be described using the tools of evolutionary game theory, suggesting the networks are evolvable. Our work on the degree of cooperation within triplet motifs demonstrated that the catalytic rates observed in the lab, which are derived from the energetics of RNA base pairing, promoted the frequency these cooperative triplets relative to selfish alternatives (Mathis et al. 2017b). A key result of the 2012 study was the observation cooperative interactions among related RNA replicators can lead to the spontaneous formation of ordered dynamics (Vaidya et al. 2012). This highlighted the key role of network interactions in understanding the spontaneous organization of lifelike entities (Fig. 10.8).

10.3.3 Network Expansion

Autocatalytic sets were envisioned as self-generating networks, which collectively can act as selectable evolutionary units. A different approach to the application of network-theoretic ideas to understanding the early evolution of biochemistry is to instead consider the properties of ecosystem or biosphere-level models (without specific knowledge of individual evolutionary units). One motivation for this approach is the observation extant ecosystems display greater regularity in terms of their stability and function than individuals do (Dinsdale et al. 2008). If we are to uncover general, and perhaps even universal, principles of biological organization through a network-based approach, it is therefore at the level of ecosystems (or even the biosphere as a whole) where we may have the greatest success (Smith and Morowitz 2016). Under this view, compartment-free, ecosystem-level models could provide the most promising insights into the processes governing the emergence and evolution of life's biochemical networks. It is worth noting we also do not know the level of complexity where "individuals" first emerged in prebiotic evolution.

One approach, first developed by Handorf et al. (2005), implements network expansion algorithms to explore the temporal order of incorporation of metabolic pathways in global biochemistry (Handorf et al. 2005). Network expansion leverages the availability of databases, such as the Kyoto Encyclopedia of Genes and Genomes (KEGG) (Ogata et al. 1999), which provide publicly accessible catalogs containing a large majority of all known biochemical reactions (the network in Fig. 10.1 was generated from all enzymatic reactions cataloged in KEGG). The algorithm proceeds by recursively determining the set of all possible molecules (the "scope"), which can be produced from an initial "seed set" of molecules. Each time the algorithm is iterated, the newly produced products of reactions are added to the graph, expanding the network. Prebiotically relevant seed sets might include simple molecular species proposed in different origin of life scenarios, such as H_2CO, CH_3SH, NH_3, and $P_2O_7^{4-}$, for example (see, e.g., Handorf et al. 2005). Starting from a given seed set, and iteratively expanding the network along all possible enzymatically catalyzed reactions, permits asking questions about the ordering and interdependency of the expansion of biochemical networks at a global scale. For example, Raymond and Segre utilized network expansion on a biosphere-level network representing global biochemistry to uncover a critical role for O_2 in permitting the emergence of metabolic pathways associated with complex life (Raymond and Segre 2006). Network expansion has also been implemented to study the coevolution of enzymes and metabolic pathways, revealing enzymatic novelty emerging in punctuated clusters corresponding to enzyme classes (Schütte et al. 2010).

More recently, network expansion has been applied directly to a problem of relevance to the origin of life—could a primitive core metabolism exist in the absence of phosphate? The answer, according to Goldford et al., is "yes" (Goldford et al. 2017). Starting from a set of prebiotically plausible seed molecules, exclusive of phosphate, they identified a phosphate-independent core metabolism, which could in principle support synthesis of a broad array of bioessential compounds. This model lends support to the concept of a "thioester world," preceding the use of ATP as the major energy currency of life. Other origin of life theories could similarly be tested with network expansion algorithms to identify possible ancestral networks, which could then be leveraged to generate new hypotheses about different origin of life scenarios.

10.3.4 Graph Grammars and Generative Models

One challenge of the approach provided by network expansion is it is not predictive but can only retrodict the potential pathways by which metabolism could have expanded through evolutionary history. Ideally, we should be able to predict all possible chemical pathways and networks and then identify the possible paths traversed in transitioning from nonlife to life and in the subsequent evolution of life. With this knowledge in hand, it would be easier to ask questions about why life arose and what its characteristic properties are. To do this, in addition to knowing the biochemistry of life, we must have some knowledge of the chemical networks *not* selected by life. Predictive theory in chemistry is an area of intensive research, with much progress made but much further to go. One promising area is the development of graph grammars as applied to predicting transformations on chemical structure. In this approach, a molecule itself is mathematically represented as a graph, where atoms correspond to nodes (vertices) and bonds to edges in the graphical representation of a molecule. Reactions are then modeled as rewiring transformations that transform the graph into another graph (representative of a different molecule) (Andersen et al. 2017). As an example of the application of graph grammars to prebiotic chemistry, this formalism has been applied to HCN polymerization, demonstrating the combination of graph grammars with experimental data can lead to guide exploration of different chemical pathways and roots to open-ended evolution (Andersen et al. 2013).

As systems biologists, network theorists, physicists, biochemists, and others collaboratively illuminate the structure of biochemical networks and the generative mechanisms which produce them, prebiotic chemists are charged with the task of explaining the origins of that structure. In order to effectively explain the topological properties of living networks, prebiotic network scientists will need to choose random network models to compare their networks against. The ubiquity of heterogeneous degree distributions suggest the Barabasi-Albert model might be a good place to start; however, the modularity described by Jeong et al. implies random hierarchical graphs might be better suited (Ravasz et al. 2002). Each random graph model contains within it implicit assumptions about the generative mechanisms involved in networks. For a given network and a set of questions about it, the appropriate random models will be different. For example, we recently compared the network structure of biochemical networks at the scale of individuals, ecosystems, and the biosphere as a whole (Kim et al. 2018) (see discussion below). For this work, the appropriate random graph for individual organismal biochemical networks involved constructing networks by randomly sampling biochemical reactions from the KEGG database, while the appropriate random graph model for ecosystems instead constructed networks by randomly sampling whole genome networks and merging them.

Prebiotic chemists, who straddle abiotic organic chemistry and biochemical networks, will need to develop appropriate random models to understand the transition from nonliving to living networks. Graph grammars provide a useful framework for understanding the generative mechanisms underlying organic chemistry, while thermodynamic calculations can generate networks of plausible geochemical reactions. An important next step in understanding the large-scale structure of biochemical reaction networks will involve deploying machine learning techniques to infer the generative mechanisms underlying those networks. Once those generative mechanisms are identified, comparing them to expected abiotic mechanisms will allow prebiotic chemists to separate the roles of chance and necessity in the evolution of biochemical networks.

10.4 Future Directions

So far, we have discussed general background in network theory and provided some applications where it has been successfully applied to modeling origin of life processes. Up to now, most research investigating the network structure of biochemical and chemical networks has focused purely on their graphical properties. To understand the physical and chemical principles underlying the origins of life, closer contact must be made with understanding—in terms of physics and chemistry—why particular network architectures are selected by life (Walker 2017). One approach is to identify universal structural properties of living networks. The "scale-free" property is one such candidate property. But, as we have discussed, fitting degree distributions is only one way to gain insights into the structure of a network and is challenging to interpret because of ambiguities in identifying the correct fit for a given distribution. Nonetheless, Jeong et al.'s work demonstrating a universal network structure for metabolism across all three domains of life does hint there exist organizational properties of biochemistry common to all life on Earth. Just as astrobiologists discuss the "universal" nature of biochemical components—e.g., all known life is composed of DNA, RNA, proteins, etc.—in informing models for origins of life, we must also consider the "universal" nature of biochemical organization, e.g., the network structure of biochemical reactions.

10.4.1 Universal Properties of Biochemical Networks

One major hurdle for making claims about universality is the common ancestry of all life on Earth. When we talk about universal properties of life, we mostly mean universal properties of life on Earth, and not necessarily properties truly universal to life, characteristic of any life in our universe. Such principles, if they exist, would form the foundation of a new research field in universal biology (Sterelny 2015; Goldenfeld et al. 2017). Discovery of alien life would obviously enable us to identify such properties, if they exist. But, in the absence of discovering alien life, is there any way we might confidently make claims of universality? This is a question the origins of life field is primed to address. We must understand life and its universal properties in order to definitively say how such systems can arise in the first place. One advantage of network thinking is we need not think of life as a level-specific phenomenon in the same way we must if we focus on chemistry alone as the defining feature of life. We are accustomed to thinking of life as a chemical phenomenon, i.e., defined by the "right" chemical building blocks. But, shifting our thinking to organizational properties of networks, and their informational properties, allows studying recurring properties of life across different scales of organization. If common patterns are found in how living matter organizes across scales within the biosphere, it increases our confidence those patterns of derivative of universal laws, rather than shared common ancestry.

We recently analyzed the structure of biochemical networks across multiple levels of organization in the biosphere ranging from the chemical reaction networks within cells, to ecosystems, to the biosphere as a whole (Kim et al. 2018). Biochemical reaction networks were constructed using annotated genomic data from 21,637 bacteria taxa, 845 archaea taxa, 77 eukaryotic taxa, and 5587 metagenomes, using methods developed by Jeong et al. (2000). A biosphere-level network was constructed from all enzymatically catalyzed reactions cataloged in the Kyoto Encyclopedia of Genes and Genomes (KEGG) database, which is the network shown in Fig. 10.1. Analyzing the topological structure of these networks as a function of network size (number of compounds) and level or organization reveals universal structural properties across all biochemical networks on Earth. These are described by universal scaling laws; scaling behavior is shown in Fig. 10.9. Scaling laws are often cited as a candidate for universal biology as they unify trends across different biological organisms and scales of organization (Gisiger 2001; West 1999a). Familiar examples of scaling behavior from physics include critical phenomena near phase transitions, where physical properties such as heat capacity, correlation length, and susceptibility all follow power-law behavior. The scaling of

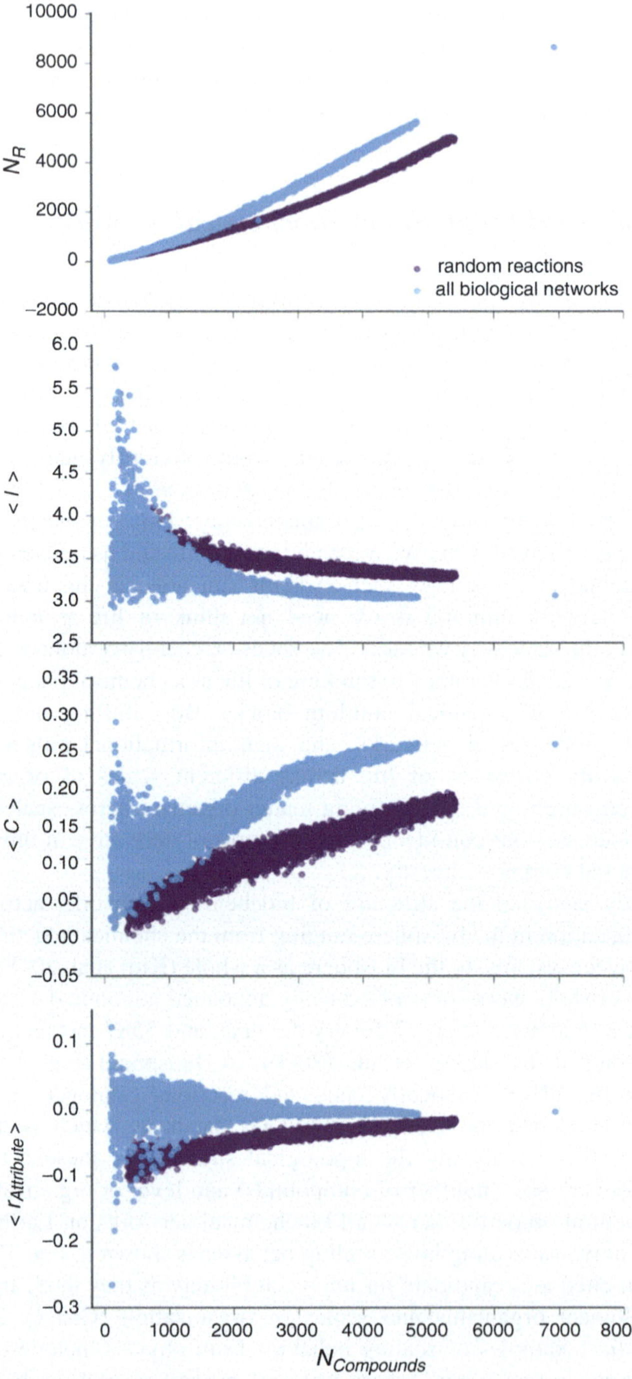

Fig. 10.9 Scaling of network attributes with network size. Biological networks (blue) scale differently from random collections of biochemical reactions (purple). Figure adapted from Kim et al. (2018)

network structure across levels of organization is different than the power-law relationship for degree distribution of scale-free networks described in the previous section because it applies *across* networks and not just *within* networks of individual organisms.

Randomly sampling reactions from known biochemistry to construct networks of similar size to organismal and ecosystem-level biochemical networks does not reproduce the scaling observed for living networks (Kim et al. 2018). This suggests it is the particular manner in which reactions are organized in living matter, and not the compounds or set of reactions alone, which yield the distinctive properties of living systems. Network growth models, such as preferential attachment, can reproduce some aspects of the architecture of life, but are not physically or chemically motivated and do not explain the constraints given rise to an observed network architecture. Scaling relations, due to their ability to "predict" the values of system parameters based on other measured quantities, represent one of the closest approaches so far to a predictive theoretical biology, akin to theoretical physics. Using the observation that cells and organisms are constrained in their growth by resource distribution networks, predictive models can be generated that accurately provide values for the scaling exponents observed in a number of diverse biological systems (West 1999b). Similar predictive models should be generated for biochemical network scaling (a work in progress) and would provide insights into universal constraints on biochemical architecture, which likely played a role in shaping the earliest networks in the transition from nonlife to life.

10.4.2 Information and Controllability

One of the most widely discussed, distinctive characteristics of life is its "informational" properties (Walker and Davies 2013; Yockey 2005; Kuppers 1990). The study of biology is replete with informational analogies, such as coding, signaling, sensing, interpretation, etc. As such, information theory is increasingly being utilized to characterize living systems across all scales (Davies and Walker 2016). An open question is how useful the concept of information is for origin of life research. A first step is to identify in what sense information could distinguish living networks from nonliving ones. The static network picture discussed throughout much of this chapter is not readily amenable to analysis from an informational perspective as many measures from information theory rely on knowledge of the dynamic properties of a system. One class of models for biological networks where information theory is readily applied is so-called random Boolean network (RBN) models (Wang et al. 2012), which are most commonly used to model gene regulatory networks where genes can be represented in one of two states "1" (activated) or "0" (inhibited). While this may seem an abstract representation, such models have been successful in systems biology and are widely applied. They may also have some utility in prebiotic evolution as peptide networks have been shown to capable to execute simple Boolean logic (Ashkenasy and Ghadiri 2004). In a recent study by one of us, it was shown Boolean models for biological gene regulatory networks (GRN) do in fact display different patterns in how information is processed when compared to random networks with similar topological properties, e.g., random networks with the same degree distribution (Kim et al. 2015; Walker et al. 2016). This suggests at least some of life's biochemical networks might be optimized for information processing, and these optimization properties may go above and beyond the topological structure of the network alone for networks encoding function.

A relevant question for origins is why biology is optimized in this manner. In the case of the GRN models, the distinctive informational properties are associated with their controllability. The models under study were the GRNs describing the state of genes responsible for regulating the cell cycle in the fission yeast *Schizosaccharomyces pombe* and the budding yeast *Saccharomyces cerevisiae*. The Boolean models correctly reproduce the sequence of gene expression patterns observed in dividing fission and budding yeast, respectively. Regulating a small subset of nodes, called the control kernel, drives both networks toward their resting phenotype. That is, by regulating just a few nodes, one can control the function of the entire network. Intriguingly, these nodes also dominate the distinctive informational properties of these networks, suggesting a relationship between information processing and controllability in the biological networks, which is not generally present in random graphs.

In Nghe et al., information control was recognized as among the six key network parameters we must understand better in order to build an evolutionary theory of catalytic networks (Nghe et al. 2015). Most origins of life research so far has focused on the role of genes in information storage and propagation and not their role in information processing or as regulators of biological function in a dynamic system. However, control is essential to biological function, for example, in maintaining homeostasis. As with the toy model of the cell cycle networks, there likely is a connection between the function of early genes as control elements in early cells and their role in information processing and storage. Hints of this are apparent in models exploring these concepts. For example, Kaneko and colleagues discovered a key role for "minority molecules" in regulating reproduction of catalytic networks in a protocell model (Kamimura and Kaneko 2010). The minority molecules are kinetically slower components in an autocatalytic network and were suggested to play the role of primitive genes, regulating reproduction of the entire system. More models and more empirical work studying how networks might evolve control nodes are necessary to understand how very primitive biochemical networks first evolved regulatory feedback and may provide insights into the early evolution of genetic function.

10.4.3 A Network Theory of Planetary Biospheres

In the previous sections, we have talked about two distinct layers of biochemical networks—metabolic networks describing all of the catalyzed (programmed) chemical reactions transforming molecular compounds within cells and gene regulatory networks which regulate cellular function (e.g., do the programming). While we often study these systems separately, in reality, they are tightly coupled. The biochemical network organization of the biosphere emerges due to the structure of reactions which are *enzymatically catalyzed*—that is, the subset of the Earth's chemistry life controls. That control is itself implemented through gene regulatory networks, which represent a "higher level" in life's hierarchical organization. As we go up in the hierarchy of structure and function in biological systems, we see similar motifs of interacting networks, where some biological networks regulate the function of others. The phenomena of life itself may be thought of as a hierarchy of interacting networks. One critical question for origin of life research is to uncover how such a hierarchy emerges in the first place.

In order to answer this question, we must consider the coupling of Earth's biological networks to their geochemical and atmospheric context. At some level, terrestrial biochemistry should be continuous with terrestrial geochemistry, implying geochemistry should represent the bottom level of life's hierarchy (Shock and Boyd 2015). In a similar vein, the biosphere's coupling to atmospheric chemistry has driven the most dramatic planetary scale changes in Earth's history (Sessions et al. 2009). If this strong coupling between life and its planetary environment is considered fundamental to living processes, the emergence of feedbacks between "life" and environment must be an important process even prior to life's emergence, perhaps even driving it (Mathis et al. 2017a). To develop quantitative frameworks for understanding the emergence of life as a planetary process, a network theory of biogeochemistry is necessary. One possible mathematical framework for formalizing such a theory is multiplex networks. In a multiplex (or multilayer) network, nodes are connected by different types of edges (to be contrasted with bipartite networks where different types of nodes are connected by edges) (Boccaletti et al. 2014). Multiplex networks are often visualized as several networks layered on top of one another. A multiplex network of planetary chemistry would involve several scales of organization, connecting a network of geochemical reactions (at the "bottom") to metabolic processes, to atmospheric chemistry (at the "top"). While biochemical networks are well characterized, geochemical networks and atmospheric networks remain relatively unexplored. Some work on the network structure of planetary atmospheres has revealed topological differences between Earth's atmospheric reaction network and that of other planetary bodies (with atmospheres) in our solar system, such as Mars, Venus, and Titan (Sole and Munteanu 2004; Gleiss et al. 2001); suggestive network theory can distinguish properties of living worlds from those of nonliving worlds. Developing a network theory of planetary (biogeo)chemistry would also allow astrobiologists to incorporate information about exoplanets, such as atmospheric spectra and planetary composition, into a unified framework that will be essential for characterizing alien biosignatures.

10.5 Conclusions

In the last century, prebiotic chemists have focused on identifying the molecular aspects of biochemistry which may have played prominent roles in the origin of life on Earth. However, while there is much debate about whether or not all life will share a common chemistry, it is less debated that life will display some form of organization (Schrodinger 1944). The organizational principles of life likely transcend levels of organization within the biosphere: we see evidence of the role of information in organizing living matter within cells, in intracellular signaling in living tissues, in ecosystems, and in societies. Adopting a view whereby it is the structure of interactions and transformations which defines the living state, and how that is mediated by information, life becomes a property not just of the chemistry within our cells but of the organization of that chemistry. Network science provides a natural quantitative framework to study this. By shifting to a system-level perspective, and embracing the tools of provided by network science, prebiotic chemistry will be able to not only understand the synthesis of molecules relevant to life on the primitive Earth but, perhaps more importantly, how those molecules collectively drove the emergence of the first living systems.

References

Albert R, Barabási A-L (2002) Statistical mechanics of complex networks. Rev Mod Phys 74 (1):47–97. https://doi.org/10.1103/RevModPhys.74.47

Alon U (2003) Biological networks: the tinkerer as an engineer. Science 301(5641):1866–1867. https://doi.org/10.1126/science.1089072

Andersen JL, Andersen T, Flamm C, Hanczyc MM, Merkle D, Stadler PF (2013) Navigating the chemical space of HCN polymerization and hydrolysis: guiding graph grammars by mass spectrometry data. Entropy 15(10):4066–4083. https://doi.org/10.3390/e15104066

Andersen JL, Flamm C, Merkle D, Stadler PF (2017) An intermediate level of abstraction for computational systems chemistry. Philos Trans R Soc A Math Phys Eng Sci 375 (2109):20160354. https://doi.org/10.1098/rsta.2016.0354

Arita M (2004) The metabolic world of *Escherichia coli* is not small. Proc Natl Acad Sci U S A 101 (6):1543–1547. https://doi.org/10.1073/pnas.0306458101

Ashkenasy G, Ghadiri MR (2004) Boolean logic functions of a synthetic peptide network. J Am Chem Soc 126(36):11140–11141. https://doi.org/10.1021/ja046745c

Ashkenasy G, Jagasia R, Yadav M, Ghadiri MR (2004) Design of a directed molecular network. Proc Natl Acad Sci U S A 101(30):10872–10877. https://doi.org/10.1073/pnas.0402674101

Barabasi A-L (2009) Scale-free networks: a decade and beyond. Science 325(5939):412–413. https://doi.org/10.1126/science.1173299

Barabási A-L (2016) Network science. Cambridge University Press, Cambridge

Barabasi L, Albert R (1999) Emergence of scaling in random networks. Science 286:509–513. https://doi.org/10.1126/science.286.5439.509

Barabási A-L, Oltvai ZN (2004) Network biology: understanding the cell's functional organization. Nat Rev Genet 5(2):101–113. https://doi.org/10.1038/nrg1272

Bianconi G, Barabási A-L (2000) Competition and multiscaling in evolving networks. Europhys Lett 54(May):436–442. https://doi.org/10.1209/epl/i2001-00260-6

Boccaletti S, Bianconi G, Criado R, del Genio CI, Gómez-Gardeñes J, Romance M, Sendiña-Nadal I, Wang Z, Zanin M (2014) The structure and dynamics of multilayer networks. Phys Rep 544(1):1–122. https://doi.org/10.1016/j.physrep.2014.07.001

Cairns-Smith AG (1986) Clay minerals and the origin of life. Cambridge University Press, New York

Clauset A, Shalizi CR, Newman MEJ (2009) Power-law distributions in empirical data. SIAM Rev 51(4):661–703. https://doi.org/10.1137/070710111

Cohen R, Erez K, Ben-Avraham D, Havlin S (2001) Breakdown of the Internet under intentional attack. Phys Rev Lett 86(16):3682–3685. https://doi.org/10.1103/PhysRevLett.86.3682

Cronin L, Walker SI (2016) Beyond prebiotic chemistry: what dynamic network properties allow the emergence of life? Science 352(6290):1174–1175. https://doi.org/10.1126/science.aaf6310

Davies PCW, Walker SI (2016) The hidden simplicity of biology. Rep Prog Phys 79(10):102601. https://doi.org/10.1088/0034-4885/79/10/102601

Dinsdale EA, Edwards RA, Hall D, Angly F, Breitbart M, Brulc JM, Furlan M et al (2008) Functional metagenomic profiling of nine biomes. Nature 452(7187):629–632. https://doi.org/10.1038/nature06810

Eigen M (1971) Self-organisation of matter and the evolution of biological macromolecules. Naturwissenschaften 58:465–523

Eigen M, Schuster P (1978) The hypercycle. a principle of natural self-organisation. Part C: the realistic hypercycle. Naturwissenschaften 65(2):341–369. https://doi.org/10.1007/BF00420631

Erdös P, Rényi A (1959) On random graphs. Publ Math 6:290–297. https://doi.org/10.2307/1999405

Filisetti A, Graudenzi A, Serra R, Villani M, Füchslin RM, Packard N, Kauffman SA, Poli I (2012) A stochastic model of autocatalytic reaction networks. Theory Biosci 131(2):85–93. https://doi.org/10.1007/s12064-011-0136-x

Gisiger T (2001) Scale invariance in biology: coincidence or footprint of a universal mechanism? Biol Rev Camb Philos Soc 76(2). Arizona State University Libraries:S1464793101005607. doi: https://doi.org/10.1017/S1464793101005607

Gleiss PM, Stadler PF, Wagner A, Fell D a (2001) Relevant cycles in chemical reaction networks. Adv Complex Syst 4(02n03):207–226. https://doi.org/10.1142/S0219525901000140

Goldenfeld N, Biancalani T, Jafarpour F (2017) Universal biology and the statistical mechanics of early life. Philos Trans R Soc A 375(2109):20160341. https://doi.org/10.1098/RSTA.2016.0341

Goldford JE, Hartman H, Smith TF, Segrè D (2017) Remnants of an ancient metabolism without phosphate. Cell 168(6):1126–1134.e9. https://doi.org/10.1016/j.cell.2017.02.001

Handorf T, Ebenhöh O, Heinrich R (2005) Expanding metabolic networks: scopes of compounds, robustness, and evolution. J Mol Evol 61(4):498–512. https://doi.org/10.1007/s00239-005-0027-1

Hordijk W, Steel M (2004) Detecting autocatalytic, self-sustaining sets in chemical reaction systems. J Theor Biol 227(4):451–461. https://doi.org/10.1016/j.jtbi.2003.11.020

Hordijk W, Steel M (2017) Chasing the tail: the emergence of autocatalytic networks. BioSystems 152:1–10. https://doi.org/10.1016/j.biosystems.2016.12.002

Hordijk W, Kauffman SA, Steel M (2011) Required levels of catalysis for emergence of autocatalytic sets in models of chemical reaction systems. Int J Mol Sci 12(5):3085–3101. https://doi.org/10.3390/ijms12053085

Hordijk W, Steel M, Kauffman S (2012) The structure of autocatalytic sets: evolvability, enablement, and emergence. Acta Biotheor 60(4):379–392. https://doi.org/10.1007/s10441-012-9165-1

Hordijk W, Hasenclever L, Gao J, Mincheva D, Hein J (2014) An investigation into irreducible autocatalytic sets and power law distributed catalysis. Nat Comput 13(3):287–296. https://doi.org/10.1007/s11047-014-9429-6

Jeong H, Albert R, Ottval ZN, Barabási AL (2000) The large scale organization of metabolic networks. Nature 407(6804):651–654

Kamimura A, Kaneko K (2010) Reproduction of a protocell by replication of a minority molecule in a catalytic reaction network. Phys Rev Lett 105(26):1–4. https://doi.org/10.1103/PhysRevLett.105.268103
Kaneko K, Yomo T (2002) On a kinetic origin of heredity: minority control in a replicating system with mutually catalytic molecules. J Theor Biol 214(4):563–576. https://doi.org/10.1006/jtbi.2001.2481
Kauffman S (1993) The origins of order: self-organization and selection in evolution. Oxford University Press, New York
Kim H, Davies P, Walker SI (2015) New scaling relation for information transfer in biological networks. J R Soc Interface 12(113):20150944. https://doi.org/10.1098/rsif.2015.0944
Kim H, Smith H, Mathis C, Raymond J, Walker SI (2018) Universal scaling across biochemical networks on earth. bioRxiv. https://doi.org/10.1101/212118
Kuppers B-O (1990) Information and the origin of life. MIT Press, Cambridge, MA
Larhlimi A, Blachon S, Selbig J, Nikoloski Z (2011) Robustness of metabolic networks: a review of existing definitions. BioSystems 106(1):1–8. https://doi.org/10.1016/j.biosystems.2011.06.002
Lifson S (1997) On the crucial stages in the origin of animate matter. J Mol Evol 44(1):1–8. https://doi.org/10.1007/PL00006115
Mathis C, Bhattacharya T, Walker SI (2017a) The emergence of life as a first-order phase transition. Astrobiology 17(3):266–276. https://doi.org/10.1089/ast.2016.1481
Mathis C, Ramprasad S, Walker S, Lehman N (2017b) Prebiotic RNA network formation: a taxonomy of molecular cooperation. Life 7(4):38. https://doi.org/10.3390/life7040038
Maynard Smith J, Szathmáry E (1995) The major transitions in evolution. Oxford University Press, Oxford
Milo R (2002) Network motifs: simple building blocks of complex networks. Science 298 (5594):824–827. https://doi.org/10.1126/science.298.5594.824
Montañez R, Medina MA, Solé RV, Rodríguez-Caso C (2010) When metabolism meets topology: reconciling metabolite and reaction networks. BioEssays 32(3):246–256. https://doi.org/10.1002/bies.200900145
Newman MEJ (2004) Analysis of weighted networks. Phys Rev E Stat Nonlinear Soft Matter Phys 70(5 2):1–9. https://doi.org/10.1103/PhysRevE.70.056131
Nghe P, Hordijk W, Kauffman SA, Walker SI, Schmidt FJ, Kemble H, Yeates JAM, Lehman N (2015) Prebiotic network evolution: six key parameters. Mol BioSyst 11(12):3206–3217. https://doi.org/10.1039/c5mb00593k
Nowak MA, Ohtsuki H (2008) Prevolutionary dynamics and the origin of evolution. Proc Natl Acad Sci U S A 105(39):14924–14927. https://doi.org/10.1073/pnas.0806714105
Ogata H, Goto S, Sato K, Fujibuchi W, Bono H, Kanehisa M (1999) KEGG: Kyoto encyclopedia of genes and genomes. Nucleic Acids Res 27(1):29–34. https://doi.org/10.1093/nar/27.1.29
Ravasz E, Somera AL, Mongru DA, Oltvai ZN, Barabasi A-L (2002) Hierarchical organization of modularity in metabolic networks. Science 297(5586):1551–1555. https://doi.org/10.1126/science.1073374
Raymond J, Segre D (2006) The effect of oxygen on biochemical networks and the evolution of complex life. Science 311(5768):1764–1767. https://doi.org/10.1126/science.1118439
Schrodinger E (1944) What is life? Cambridge University Press
Schütte M, Skupin A, Segrè D, Ebenhöh O (2010) Modeling the complex dynamics of enzyme-pathway coevolution. Chaos 20(4):1–12. https://doi.org/10.1063/1.3530440
Segre D, Ben-Eli D, Lancet D (2000) Compositional genomes: prebiotic information transfer in mutually catalytic noncovalent assemblies. Proc Natl Acad Sci U S A 97(8):4112–4117. https://doi.org/10.1073/pnas.97.8.4112
Segré D, Ben-Eli D, Deamer DW, Lancet D (2001) The lipid world. Orig Life Evol Biosph 31 (1–2):119–145. https://doi.org/10.1023/A:1006746807104
Sessions AL, Doughty DM, Welander PV, Summons RE, Newman DK (2009) The continuing puzzle of the great oxidation event. Curr Biol 19(14):R567–R574. https://doi.org/10.1016/j.cub.2009.05.054

Shock EL, Boyd ES (2015) Principles of geobiochemistry. Elements 11(6):395–401
Smith E, Morowitz H (2016) The origin and nature of life on earth: the emergence of the fourth geosphere. Cambridge University Press, Cambridge
Sole RV, Munteanu A (2004) The large-scale organization of chemical reaction networks in astrophysics. EPL (Europhys Lett) 68(2):170
Sousa FL, Hordijk W, Steel M, Martin WF (2015) Autocatalytic sets in *E. coli* metabolism. J Syst Chem 6(1):4. https://doi.org/10.1186/s13322-015-0009-7
Sterelny K (2015) Universal biology. Br Soc Philos Sci 48(4):587–601
Vaidya N, Manapat ML, Chen IA, Xulvi-Brunet R, Hayden EJ, Lehman N (2012) Spontaneous network formation among cooperative RNA replicators. Nature 491(7422):72–77. https://doi.org/10.1038/nature11549
Vaidya N, Walker SI, Lehman N (2013) Recycling of informational units leads to selection of replicators in a prebiotic soup. Chem Biol 20(2):241–252. https://doi.org/10.1016/j.chembiol.2013.01.007
Vasas V, Szathmary E, Santos M (2010) Lack of evolvability in self-sustaining autocatalytic networks constraints metabolism-first scenarios for the origin of life. Proc Natl Acad Sci U S A 107(4):1470–1475. https://doi.org/10.1073/pnas.0912628107
Vasas V, Fernando C, Santos M, Kauffman S, Szathmáry E (2012) Evolution before genes. Biol Direct 7(1):1. https://doi.org/10.1186/1745-6150-7-1
Wagner A (1998) The large-scale structure of metabolic networks: a glimpse of life's origin? Wiley Periodicals 8(1): 15–19
Wagner A, Fell DA (2001) The small world inside large metabolic networks. Proc R Soc B Biol Sci 268(1478):1803–1810. https://doi.org/10.1098/rspb.2001.1711
Walker SI (2017) Origins of life: a problem for physics, a key issues review. Rep Prog Phys 80 (9):092601. https://doi.org/10.1088/1361-6633/aa7804
Walker SI, Davies PCW (2013) The algorithmic origins of life. J R Soc Interface 10(79):20120869. https://doi.org/10.1098/rsif.2012.0869
Walker SI, Kim H, Davies PCW (2016) The informational architecture of the cell. Philos Trans R Soc A Math Phys Eng Sci 374(2063). https://doi.org/10.1098/rsta.2015.0057
Wang RS, Saadatpour A, Albert R (2012) Boolean modeling in systems biology: an overview of methodology and applications. Phys Biol 9(5):055001. https://doi.org/10.1088/1478-3975/9/5/055001
Wasserman S, Faust K (1994) Social network analysis: methods and applications. Cambridge University Press, Cambridge
Watts DJ, Strogatz SH (1998) Collective dynamics of 'small-world' networks. Nature 393:440–442
West GB (1999a) The fourth dimension of life: fractal geometry and allometric scaling of organisms. Science 284(5420):1677–1679. https://doi.org/10.1126/science.284.5420.1677
West GB (1999b) The origin of universal scaling laws in biology. Phys A Stat Mech Appl 263 (1–4):104–113. https://doi.org/10.1016/S0378-4371(98)00639-6
Wynveen A, Fedorov I, Halley JW (2014) Nonequilibrium steady states in a model for prebiotic evolution. Phys Rev E 89(2):22725. https://doi.org/10.1103/PhysRevE.89.022725
Yockey H (2005) Information theory, evolution and the origin of life. Cambridge University Press, Cambridge

The manufacturer's authorised representative in the EU is Springer Nature Customer Service Centre GmbH, Europaplatz 3, 69115 Heidelberg, Germany. If you have any concerns regarding our products, please contact ProductSafety@springernature.com

Printed and bound by CPI Group (UK) Ltd, Croydon, CR0 4YY
05/10/2026
02228144-0006